WINNING A FUTURE WAR

War Gaming and Victory in the Pacific War

Norman Friedman

Naval History and Heritage Command, Department of the Navy
Washington, DC

In memory of Captain Hugh Nott, USN

Contents

Acknowledgments ix

Introduction 1

1. Naval Transformation 6
   Exercises: Full-Scale Fleet Problems and Games at Newport 8
   Naval Aviation as a Driver Toward Transformation 11
   The Inter-War Navy and Its World 16
   The Strategic Problem 20
   Naval Arms Control 23
   Ships 27

2. The Naval War College and Gaming 31

3. War Gaming and War Planning 43
   The "Applicatory System" 43
   War Gaming 51
   War Gaming at the Inter-War War College 54
   Simulation 55
   Some Limits of Gamed Reality 62
   Using War Gaming 63
   War Gaming and War Planning 65

4. War Gaming and Carrier Aviation 73
    Guessing What Aircraft Could Do. 75
    Gaming and Early Carriers. 78
    Reeves and Operating Practices. 93
    Putting It Together—the *Yorktown* Class 97
    Aftermath 103

5. The War College and Cruisers 109
    Evaluating Alternatives. 115
    Cruisers at War: Three Years of Red-Blue Warfare. 129
    Postscript: The Fate of the Flight-Deck Cruiser. 139

6. Downfall. 142

7. Conclusion: Games Versus Reality in the Pacific 160

Appendixes. 181
    A: Playing the Games. 181
    B: War Game Rules—Aircraft 183
        The Airplanes. 186
        Carrier Air Operation 186
        Bombing. 188
        Bombs Versus Carriers 192
        Torpedo Bombing 193
        Air-to-Air Combat. 194
        Anti-Aircraft Firepower. 196
        Aircraft Navigation and Reliability 198

Notes 201

Bibliography. 253

Index 256

Acknowledgments

This book began with the observation by Captain Jerry Hendrix, USN (Ret.), who was then director of the Naval History and Heritage Command (NHHC), that there was no history of inter-war gaming at the Naval War College in Newport comparable to the detailed histories of the full-scale Fleet Problems (fleet exercises). I am grateful to Captain Hendrix both for initiating this project and for impelling me to collect material covering the full range of games played at the college; I would never have done so without the sponsorship of the Naval History and Heritage Command. In turn, the gaming material became the basis for this book. I should emphasize that it was the full range of material that led me to the conclusions in this book, not any one particular game. I am grateful to Dr. Evelyn Cherpak, now retired, who, as archivist at the Naval War College, made it possible for me to review the gaming material and to copy a great deal of it. I had been visiting Dr. Cherpak and her archive for some years, but had always shied away from the sheer mass of gaming material in the past. The advent of the digital camera, and the impetus provided by Captain Hendrix, changed that. After Dr. Cherpak retired, I benefitted from the assistance of interim archivist Scott Reilly and his assistant (later archivist) Dara Baker. Since important material, including gaming material, was held elsewhere, I visited several other archives. Thus, I am also grateful to the archivists of the National Archives, both in downtown Washington, DC, and at College Park, Maryland; to the Admiralty Librarian Jenny Wraight; and to the archivists at the British National Archives at Kew. For their very enlightening insights and comments on the ongoing project, I would like to thank my friends Chris Carlson, Dr. Thomas Hone, Trent Hone, David Isby, and Christopher C. Wright. At the Naval History and Heritage Command, Dr. Ryan Peeks provided a wide variety of assistance as well as very helpful comments. Earlier, I benefitted from the efforts of Curtis Utz, who was my contract officer at NHHC. I would also like to thank Kristina Gianotta, who, as head of the Histories Branch, managed this project. Far from last, I would like to thank my wife, Rhea, for her sustained loving and patient encouragement and support.

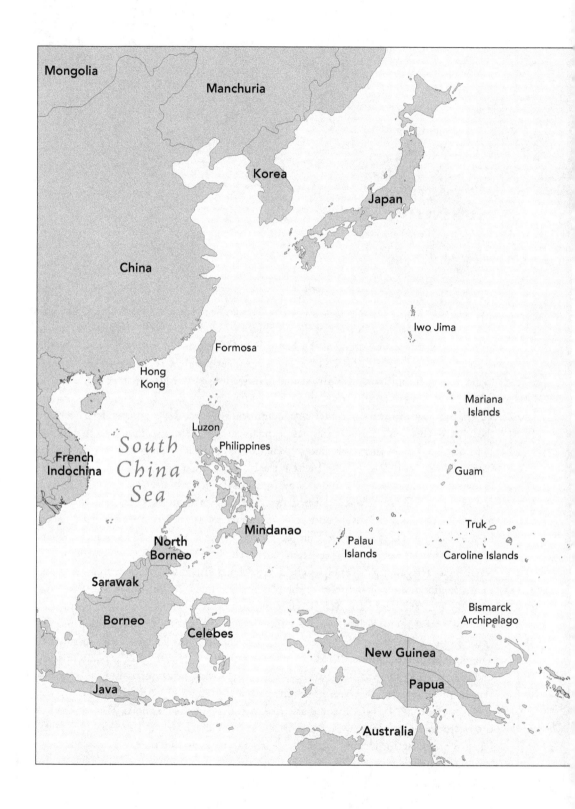

PACIFIC OCEAN

Midway Islands

Hawaiian Islands

Marshall Islands

Solomon Islands

Makin

Gilbert Islands

Solomon Islands

Guadalcanal

Ellice Islands

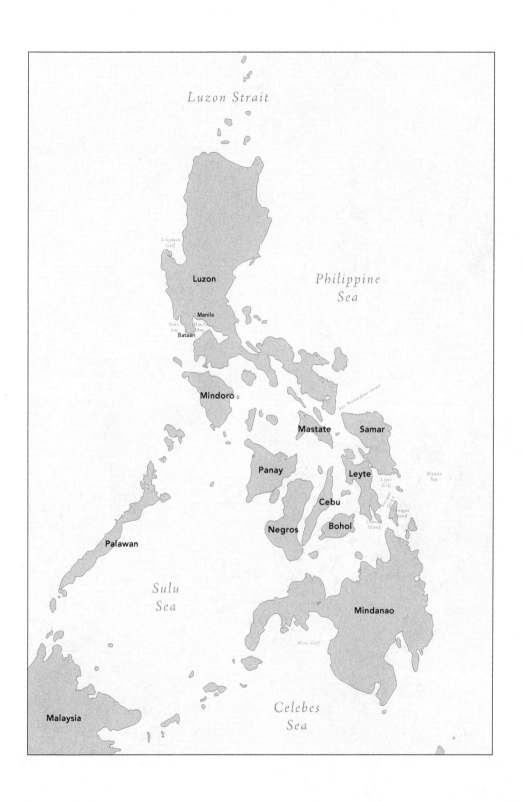

WINNING A FUTURE WAR

Introduction

To win the Pacific War, the U.S. Navy had to transform itself technically, tactically, and strategically. It had to create a fleet capable of the unprecedented feat of fighting and winning far from home, without existing bases, in the face of an enemy with numerous bases fighting in his own waters. Much of the credit for the transformation should go to the war gaming conducted at the U.S. Naval War College. Conversely, as we face further demands for transformation, the inter-war experience at the War College offers valuable guidance as to what works, and why, and how.

The fruits of this transformation are so commonplace now that we may easily forget how radical it was. The U.S. Navy emerged from World War I as a battleship fleet similar to other navies. The British had demonstrated that naval aircraft could be a vital auxiliary to the battleships, but anything more was a distant prospect. The war had demonstrated that an amphibious operation could be mounted in the face of resistance, but not that it would be particularly effective. In 1943–45, carriers were the accepted core of the U.S. fleet, and amphibious operations against enemy shore defenses were routinely conducted. Indeed, without them it would have been impossible to fight World War II.

If it seems obvious that any naval officer aware of the march of technology would have developed the massed carriers and the amphibious fleet, the reader might reflect that the two other major navies failed to do so. The Japanese did create a powerful carrier striking force, but they made no real effort to back it up with sufficient reserves to keep it fighting. They developed very little amphibious capability useful in the face of shore defenses: They could not, for example, have assaulted their own fortified islands, let alone Normandy or southern France. The British built carriers, but accepted very small carrier air groups because, until well into World War II, they saw their carriers mainly as support for their battle fleet. Like the Japanese, they did not develop an amphibious capability effective against serious defense. Each of the three navies was staffed by excellent officers, often with the widest possible experience. What set the U.S. Navy apart?

War gaming at the U.S. Naval War College at Newport, Rhode Island, seems to have been a large part of the answer.

The games played by students there were a vital form of training, but at least as importantly, the games served as a laboratory for the U.S. Navy. It seems to have been significant that, until 1934, the Naval War College was part of the Office of the Chief of Naval Operations (OPNAV) rather than part of the naval school system. Gaming experience fed back into full-scale exercises (Fleet Problems), and full-scale experience fed back into the detailed rules of the games, which were conceived as a way of simulating reality as closely as possible. Game data were also fed to the U.S. Navy's war planners, all of whom had graduated from the War College and thus had considerable game experience. The successes and failures of simulation give some guidance into what is needed in current and future games.

The two most important new developments of the inter-war period were aviation and those involved in amphibious assault. In the case of aviation, the technology involved changed rapidly, so its potential could be grasped in only a limited way from full-scale experience. Gaming revealed some important problems (one of which Naval War College graduate Captain, later Admiral, Joseph M. Reeves solved in a way that proved decisive at Midway, where three U.S. carriers operated as many aircraft as four Japanese). Gaming caused another Naval War College student, then-Commander Roscoe C. MacFall, to devise the circular formations that proved very successful during the Pacific War. Yet another Naval War College student, Commander (later Fleet Admiral) Chester B. Nimitz, who had seen the formations in action, convinced the fleet to adopt them. No full-scale fleet battle experiment could have revealed not only the problems, but tested solutions that took years to implement. The War College taught all its students, not just those interested in aviation, how to fight an air-sea war. Its success was reflected in the capability of officers not trained in carrier operations to direct them. Admiral Raymond F. Spruance is the great, but not unique, case in point.

The games showed that naval aircraft would suffer tremendous wastage in action. Before war broke out, the U.S. Navy created the necessary programs to produce sufficient numbers of pilots. The Japanese did not; they did not look beyond a single battle to the longer campaign in which it would be set. Once their experienced pilots were dead, so was their naval airpower (except for kamikazes) despite their attempts to make up carrier losses.

Games also showed that carrier flight decks would often be knocked out. When the gaming began, it was assumed that such damage would have to be repaired in a shipyard; a carrier once knocked out was out of action for an entire campaign. The U.S. Navy uniquely sought quickly repairable flight decks, which made it possible for U.S. carriers to keep fighting after suffering battle damage. Without this innovation, the United States would have had no operable carriers in the Pacific during vital periods in 1942–43. Japanese carriers lacked this feature, which is why the two best ones, *Shōkaku* and *Zuikaku*, damaged at the Battle of the Coral Sea in May 1942, were not available to fight at Midway in June.

The games highlighted problems of scouting in the vastness of the Pacific, and thus encouraged the U.S. Navy to supplement carriers with large numbers of shore-based naval aircraft, initially seaplanes but later land plane bombers under naval control.

Amphibious assault was a different matter. The technology changed between wars, but the key issue was how it might fit into the larger U.S. strategy of trans-Pacific assault. The key contribution of the War College gamers was to show that a strategy that did *not* include seizing bases en route to the Far East was doomed. This insight motivated the Marines' shift toward seizing and defending advanced bases across the Pacific, and it led to development of the specialized ships and craft and techniques that made amphibious attack a commonplace during World War II. Without one crucial game played in 1933, it is not at all clear that any of this work would have been done.

To anyone looking back at the U.S. Navy of the inter-war years, the revolution is almost invisible. Throughout the period, the U.S. Navy continued to consider the battleship the core of its fleet, just as other major navies did. Aircraft were important, but they were an auxiliary. Gaming explains what was really happening behind this appearance. Gaming forced participants to consider not just the decisive battle that the fleet hoped to fight—a classic battleship-on-battleship engagement—but also the campaign leading up to it. To get at the Japanese fleet, the U.S. Navy had to cross the Pacific, passing through chains of islands the Japanese controlled.

Gaming presented not just what the U.S. Navy wanted to do, but also what the Japanese would choose to do to stop its advance. It became clear in games that the U.S. Navy would have to fight a series of battles against Japanese aircraft and submarines as it pressed westward. These secondary battles were a fair representation of what actually happened in the Pacific in World War II, as the Japanese reserved their own battle fleet for the decisive battle they expected to fight. Incidentally, many of the tactics later attributed to Japanese national character, such as risky night torpedo attacks, emerged clearly in the games as the inevitable results of the Japanese strategic and tactical situation, rather than reflections of their military culture.

Why did gaming work? It was hardly the only way a navy could look toward the future. There were plenty of highly intelligent, experienced officers with excellent military judgment. They often found the results of gaming rather artificial, even rigged in favor of impractical technology. When they confronted straightforward developments of ships and weapons they had used for years, their judgment was probably much better than what games might suggest. However, when they confronted essentially new technologies, particularly rapidly changing aircraft, they were not nearly as well prepared to judge. They also confronted air enthusiasts who made their own inflated claims that traditional (meaning well-developed) naval forces were obsolete. Because it was better balanced, gaming offered a much more acceptable (and better) way to sense what a future air-sea war would be like.

Gaming reached a high point of influence in 1930–34. For most of that period, the Chief of Naval Operations (CNO), Admiral William V. Pratt, was a former president of the Naval War College, who was familiar with the virtues of gaming. Contemporary records show that there was widespread skepticism in the fleet. The fleet saw, for example, the obvious limitations of existing aircraft, some of which were dramatically reduced in game rules (to sense what could be done within a few years). Not surprisingly, the gamers at the Naval War College lost most of their influence after the game-friendly CNO retired and the Naval War College was moved from a position as OPNAV's laboratory into the Navy educational system. The building program advocated in 1939–40, which was battleship- and not carrier-heavy, dramatizes the reaction to what the War College games had taught. As it happened, the approach of war and also pressure by an air-minded President Franklin D. Roosevelt changed the situation. However, the Naval War College gamers had already won a larger victory. When the United States entered World War II, all but one flag officer were Naval War College graduates—former war gamers who had seen the sort of war they would now be fighting. The new techniques were already in place and the strategy demonstrated in that 1933 game had been accepted.

Enough had been done to make it reasonable to see war gaming at Newport as a key element of victory in World War II. That is most obvious in the Pacific War, but in Europe amphibious landings were decisive. The games motivated development of the necessary techniques, which were tested in full-scale Fleet Landing Exercises (FLEXes). Thus, the games can also be thanked for victory in Europe.

This book differs radically from other accounts of the inter-war Navy. It emphasizes the role of the Naval War College as the U.S. Navy's primary think tank rather than as its senior educational institution. The think tank role is evident in statements by the College's first post-1918 president, Admiral William S. Sims, in its pre-1914 role, in much of the gaming material, and in the implicit relationship between the Naval War College and the OPNAV War Plans Division, whose officers were its graduates. Gaming was the core of the think tank role, because gaming was the Navy's laboratory, in a way that full-scale exercises could not be. It was extremely important that attendance at the Naval War College was in effect a prerequisite for service in the OPNAV War Plans Division—the games told the officers who created war plans what a future war would be like. As the fleet's strategic and tactical laboratory alongside the full-scale Fleet Problems, the Naval War College had enormous impact on the way the U.S. Navy prepared for a future war.

This book also differs from previous accounts of inter-war gaming at Newport in several ways, of which three seem particularly important. First, it examines the games primarily as the U.S. Navy's laboratory rather than primarily as a means of training. Second, it is based on detailed examination of *all* the surviving senior course games. Typically, only a few games have been examined. Although the key game that changed Pacific strategy has generally figured prominently in histories, other games have not fared nearly so well. Third, it examines the influence of the games on specific areas of naval development, particularly how the U.S. Navy learned to use naval airpower. Only an examination of the totality of the games makes it possible to show how they influenced the inter-war Navy and its conduct during World War II. It seems clear in retrospect that gaming had its greatest tactical impact in showing how naval airpower would work (not merely promoting it in general) and strategically in showing the vital need for advanced bases en route to the Far East.

The educational aspect of this book emphasizes how the Naval War College sought to shape the way officers thought about the problems they faced. It seems clear that its approach differed radically from that of the senior educational institutions created by the closest equivalent to the U.S. Navy, the Royal Navy.

The games, and the U.S. Navy as a whole, were products of a very specific time. Between the world wars, the United States avoided alliances of any kind, although, in the late 1930s, some games and other documents show an interest in them. It was not imagined that a European war would suddenly leave the major colonies in the Far East almost defenseless. Thus, the Far East wars the war gamers envisaged were fought in the presence of powerful neutrals, particularly the British Empire, which could be expected to affect the prosecution of any war. This mind-set is probably the best explanation for the one gap in inter-war gaming, namely the absence of any large-scale submarine war against Japanese trade (although some games do include small-scale submarine warfare of this type).

1. Naval Transformation

Between 1919 and 1941, the U.S. Navy transformed itself from a powerful if unsophisticated force into the fleet that won a two-ocean war. The great puzzle of U.S. naval history is how that was accomplished. This book argues that war gaming at the U.S. Naval War College made an enormous, and perhaps decisive, contribution. For much of the inter-war period, the Naval War College was the Navy's primary think tank. War gaming was the means the college used to test alternative strategies, tactics, and warship types, in a way that full-scale exercises (the Fleet Problems) could not. There is considerable evidence that the lessons the Naval War College drew from gaming were applied by the fleet. For example, crucial choices in U.S. aircraft carrier design can be traced to experience gained in war games at the college.

The think tank perspective is a new way of looking at the inter-war Naval War College and the war games that formed the core of its curriculum. Historians have generally treated war gaming (if they have treated it at all) as an educational tool, the question being how effective it was.[1] It seems clear that the inter-war Navy valued the college. Attendance was, in effect, a prerequisite for selection to the OPNAV War Plans Division.[2] Of the flag officers commanding combat formations in the 1941 fleet, only one had not attended the War College. After the war, Fleet Admiral Chester W. Nimitz, who had commanded in the central Pacific, pointed out that at one time or another, war games had replicated nearly everything that happened during the Pacific War. Officers who had fought simulated battles were well prepared to fight real ones in the same places, often in similar situations. Surviving records at the Naval War College clearly show the changing relationship between the game scenarios and the outside world, and the increasing sophistication of the game rules intended to simulate reality. They do not show, as naval records from other official sources do, how deeply game experience influenced overall U.S. naval development between wars. This influence is somewhat obscured by frequent protestations by the Naval War College (typically by its president) that it was in no way a war-planning center. That was true in a formal sense, but what the college did mattered enormously.

Although the influence of both the Naval War College's gaming and of the college itself declined after 1933, most of the key decisions shaping the wartime U.S. Navy had already been taken. The two most important ones were on the role of naval aviation and the form of the U.S. war plan against Japan. Furthermore, war gaming helped instruct non-aviators such as Admiral Raymond Spruance about the proper role of carrier aviation. They successfully applied these lessons during the coming war. Similarly, the War College bears some credit for the shift from the "through ticket to Manila" to the step-by-step strategy the United States employed during World War II. This shift was based on gaming experience, particularly the disastrous outcome of the already noted 1933 game.

The inter-war U.S. Navy had to solve two linked problems. One was that it had virtually no experience of naval warfare on which to draw. Its last conventional naval war had been fought against Spain in 1898, before nearly all the weapons and ships of 1919 had even been conceived. Its World War I experience was limited almost entirely to anti-submarine warfare, in many cases in a subordinate position to the Royal Navy, the foremost navy of the time. A second problem was that the recently ended world war had produced further new weapons, particularly aircraft and long-range submarines, whose potential was difficult to estimate.

The problems of the inter-war era may seem dated, but they have very real current equivalents. No one has fought a major war using modern weapons. No one really knows how, for example, cyber weapons will affect such a war. We can and certainly do guess, but it takes a simulated war against a sophisticated opponent to tease out what is likely to happen and to stimulate us to take the appropriate measures. Only a simulated opponent will seek to take full advantage of all available weapons in ways we may not expect.

The story of inter-war simulation—gaming—is worth following because it shows how a particular approach to this problem worked. We can see how well simulation served the U.S. Navy because we know its outcome: the successes and the limitations of the simulation as applied to a real war. What we might expect from successful simulation would be insights, which could not have been gained without it. In the case at hand, they ranged from the realization that the fleet should and could seize intermediate bases in the Pacific—the basis of much of what was done in 1943–44—to important insights into the way to operate aircraft carriers (not to mention the sort of carriers worth buying) and into what other sorts of ships the fleet needed. It is also possible to contrast the success of the U.S. Navy with the problems suffered by the Royal Navy, a comparable sea service that did not use gaming as a primary planning tool.

* * *

Exercises: Full-Scale Fleet Problems and Games at Newport

The U.S. Navy found two linked solutions to the problem of preparing for future war. One was the conventional one: full-scale exercises, which were called Fleet Problems. The other was war gaming at Newport. Each had complementary strengths and weaknesses, both for prediction and for creating a modern officer corps.

It took full-scale exercises to see how ships and aircraft and some of their weapons—and above all the personnel manning the fleet—would perform in reality. Nothing but a full-scale exercise could model how well a cruiser or a division of cruisers would perform at sea. Gamers at Newport used reports of such full-scale behavior to write the detailed rules under which they operated. In some cases it was simply impossible to model full-scale warfare. For example, despite considerable input, the war gamers at Newport were never able to model air-to-air engagements in any realistic way. A full-scale exercise could get closer to reality, although even it could not include the full effects of air-to-air combat. Neither full-scale exercises nor games could quite solve the problem. Officers commanding during an exercise could get a fair idea of what full-scale combat might be like, but they could not use their weapons freely and they could not risk damage to their own ships. Thus, for example, they were not permitted to try dangerous night attacks. Moreover, in the relatively impoverished inter-war Navy, it was impossible to run what might now be called end-to-end tests of important weapons. The best-known case was torpedoes. The inter-war Navy developed what it thought was a radically upgraded torpedo using a magnetic exploder. It was tested only once, and that test turned out to have been unintentionally deceptive.[3]

Against all their virtues, full-scale exercises could not simulate extended campaigns; their duration was necessarily limited. They had to be scripted to the extent that, because they were expensive training opportunities, the opposing forces had to be brought into contact. Thus, they could not even hint at the scouting problems that proved so difficult and so important during the Pacific War. Because the whole fleet had to participate, full-scale exercises generally simulated fleet-on-fleet battle rather than any lesser engagements the fleet might have to fight to get to that point. On a subtler level, it was difficult for most participants to gain a sense of how the exercise as a whole proceeded, how the opposing fleets used their elements in the required complementary way. What they saw, they nearly always saw from a necessarily narrow perspective. Afterward, they might read the official summary of the Fleet Problem, but that was hardly the same thing as experiencing it from a broad perspective.

To a considerable extent, too, a full-scale exercise was a public event. However well the Navy might try to conceal what happened, a great deal inevitably leaked out. At the very least, the progress of the exercise could be tracked by intercepting radio messages used during it.[4] The Japanese could and did monitor U.S. Fleet Problems and at times managed to deduce important aspects of U.S. war plans.[5] Moreover, during the inter-war period, there was considerable political sensitivity to anything suggesting an offensive

The great question of interwar U.S. naval history is how the U.S. Navy transformed itself from a battleship-oriented navy to the victorious carrier navy of 1943–45. When Admiral Sims reopened the War College in 1919, the battleship was indisputably the core of the fleet. Here U.S. battleships perform Battle Practice, 1921. (NHHC NH 124139)

outlook on the part of the United States, yet the main naval scenario, a Pacific war, could be won only by offensive action.[6] That did not imply that war planners or gamers were hotheads seeking war, only that if a war broke out, they knew they had to carry the fight to the enemy. The terms have changed, but this type of political sensitivity, as opposed to what might happen if war actually broke out, is certainly still relevant.

A war game provided the opposite, broad, perspective. It was the only way to simulate a whole naval campaign, or even a lengthy part of a protracted war. The shared experience of protracted campaign games provided many within the inter-war Navy a sense of what a future war would be like. During the inter-war period, the Navy's General Board was largely responsible for the characteristics (broad design requirements) of new ships. It held hearings to gain the views of the fleets and of the technical experts. Often, its senior member asked the same question: How will this ship function in the Orange (Japanese) war? The U.S. Navy's war planners produced detailed plans for particular approaches to the war, but they changed frequently as senior officers replaced each other and offered different perspectives. It was the shared experience of war gaming at Newport that made it possible for all involved in the hearings, and in other Navy activities, to understand the character of an Orange war in a meaningful sense.

The transformed fleet: "Murderer's Row" of fleet carriers at Ulithi, 8 December 1944. From the front, they are: *Wasp* (CV-18), *Yorktown* (CV-10), *Hornet* (CV-12), *Hancock* (CV-19), and *Ticonderoga* (CV-14), of which the first three were named after carriers sunk in 1942. (National Archives 80-G-294131)

Gaming offered another advantage. Those involved received practical experience in the way in which the different elements of a fleet—its ships and its aircraft and its shore supporting structure—worked together. This was a very different perspective from that offered by a full-scale exercise toits participants. It was particularly important in a navy whose officers tended to specialize for much of their careers. It took games to give a cruiser officer like Captain (later Admiral) Raymond F. Spruance a sense of what carriers and their aircraft could and could not do. It took a game to show just how much (and what kind) of scouting a war spread over much of the Pacific would demand (as well as the consequences of failed scouting plans). The same games offered a valuable sense of how other kinds of ships, such as destroyers and submarines, might operate both in support of, and in opposition to, the U.S. fleet. In many games, one of the objectives was to familiarize students with the full range of ships, aircraft, and naval weapons, both U.S. and enemy. Using them, even on a game floor, offered far more insight than reading about them or listening to lectures about them.

Perhaps most important, a game was a secure way to investigate tactics and strategy, far from the eyes of potential enemies—and also far from civilians at home who might be alarmed by the drastic tactics and decisions a major war demanded. Conclusions that might be drawn from games did not have to be publicized and therefore could be drawn out much more freely than those implied by large-scale exercises.

The other great influence on interwar U.S. naval transformation was the shift from an Atlantic to a Pacific perspective. The pre-1919 U.S. Navy had had a Pacific presence, but the primary consideration in development was the threat of an Atlantic war. In 1919, the fleet moved to the Pacific, and so did the focus of U.S. naval strategic thinking. Here, the battleship *New Mexico* (BB-40) enters Miraflores Lake in the Panama Canal, 26 July 1919. (NHHC NH 50296)

Naval Aviation as a Driver Toward Transformation

For the inter-war Navy, aircraft were not too different from such radical current technologies as cyber in the way they affect the modern Navy. Aircraft had proven their value during World War I, and at its end they promised a good deal more. No one could predict how they might evolve over the next decades, yet fleets had to be adapted to their needs and potential. We look back and see a battleship navy barely acknowledging the value of aircraft, which seem very primitive compared to those that fought World War II. However, during the inter-war period, the U.S. Navy was widely seen as the most air-minded in the world, with a very active aircraft research and development program.[7] For example, it was the U.S. Navy that developed and emphasized the new technique of dive bombing, a form of precision attack that proved very effective during World War II. War-gaming experience had other vital impacts on the way in which the U.S. Navy used its aircraft and also on the way it designed its carriers, with important consequences for key World War II battles such as Midway and the Philippine Sea.

Naval aviation grew much more rapidly than might have been expected because the Washington Naval Treaty left the United States with two huge carriers converted from cancelled battle cruisers: *Lexington* (CV-2) and *Saratoga* (CV-3). *Lexington* is shown leaving San Diego on 14 October 1941 with Brewster Buffalo fighters forward, Dauntless dive bombers amidships, and Devastator torpedo bombers aft on her flight deck. (National Archives 80-G-416362)

As the inter-war era opened, the U.S. Navy was interested in aircraft carriers, which the British had invented during World War I, but it had little idea of just what it wanted. The Washington Naval Treaty allowed it to convert the two huge battlecruisers *Lexington* and *Saratoga* into carriers, far larger than anything envisaged beforehand.[8] What could or should they be designed to do? How would comparable enemy carriers affect the U.S. fleet? Much could be learned by using simulated carriers. In retrospect, it seems that gaming experience prior to the completion of the two carriers motivated two of the most important U.S. tactical developments of this period: The method of operating aircraft that gave U.S. carriers a commanding lead in capacity, and the circular formations that made carrier operation effective during the Pacific War.

After World War II, the Japanese wrote that through the inter-war period they had followed the U.S. Navy step by step.[9] Their belief that naval aviation might be effective had been inspired by the U.S. Navy. That might have seemed laughable as U.S. and Allied forces suffered badly at the hands of Japanese naval aviators through much of 1942, but it makes more sense given the triumphant development of U.S. naval aviation during the war.

The two big carriers demonstrated what large numbers of naval aircraft could do, such as when the new *Saratoga* raided the Panama Canal Zone in 1928 during that year's Fleet Problem. Without these ships, the potential of naval aviation would have been far more theoretical. The Fleet Problems and the games at Newport were complementary. Newport showed what might be done with current and planned forces; the Fleet Problems helped show what was practical. *Saratoga* is shown landing aircraft, 6 June 1935. The War College had nothing to do with the terms of the Washington Conference, hence with her origin. However, the U.S. practice of moving aircraft forward as they landed (as here) rather than striking them below was very much a War College–inspired innovation and it was crucial to U.S. carrier success in World War II. No other carrier navy tried this technique before World War II. Note the black stripe on the ship's funnel, which distinguished her from her sister *Lexington*. (National Archives 80-G-651292)

The great question for the three inter-war navies that operated substantial carrier forces was how to integrate them into their fleets. The U.S. Navy envisaged extended carrier-versus-carrier and carrier-versus-land-based aircraft battles not because its leadership was more visionary than those of other navies, but because the kind of war it was likely to face would potentially include such battles. It learned this from the campaign-level war games fought out at Newport. No other form of visualization could include not only what the U.S. Navy wanted to do, but what a likely enemy would do to frustrate it—no other form of prediction included the enemy's vote.

Differences between what the Japanese and U.S. navies did in this area suggest that the U.S. approach, which involved large-scale gaming, was decisively better. Both navies sought to develop strong naval air arms. The inter-war U.S. Navy created a corps of se-

The crucial U.S. carrier innovation, which required arresting gear and a wire barrier, was invented on board the small experimental prototype carrier *Langley* (CV-1) at the behest of the Captain (later Admiral) Joseph M. Reeves, Commander, Battle Fleet Aircraft. *Langley* is shown at Cristobal, Canal Zone, on 1 March 1930 with 24 aircraft on deck, far more than she originally carried, and a credit to Reeves' ingenuity. Reeves, in turn, learned how important it was to speed up aircraft landings when he commanded the Blue air force in a war game at Newport. (National Archives 80-G-185915)

nior aviators by requiring that aviation activities (including the Bureau of Aeronautics and carriers) be commanded by aviators: Earning wings became a valuable career choice. The Japanese did not go nearly so far, but they certainly encouraged air training. In both navies, aviators argued that conventional surface ships, particularly battleships, were obsolete. We look back and see "battleship admirals" opposing the air-minded future. In the Japanese navy, aviators went so far as to advocate elimination of battleships. They clashed with the "fleet faction," which poured money into the ultra-large battleships *Yamato* and *Musashi* that Japan completed at the beginning of the Pacific War. In retrospect, the Japanese naval aviators were over-optimistic as to what their aircraft could do. No such faction formed in the U.S. Navy. That may be because U.S. aviators realized that the all-or-nothing argument was unrealistic. Carriers and surface warships were complementary. They could see that demonstrated by what seemed to be simulated realistic experience at Newport.

It is striking that some U.S. officers who did not have flight training, such as Raymond Spruance and Frank Jack Fletcher, nevertheless proved to be able commanders of fleets built around carrier forces. Spruance went from being a cruiser commander to commanding one of the two carrier task forces at Midway, and then to commanding the Third Fleet—whose main firepower lay in its carriers—in the invasion of Saipan and the Battle of the Philippine Sea. Somehow he and other surface officers managed to understand the ongoing revolution in naval affairs. This book argues that the key difference was extensive gaming at Newport, which provided all participants with an understanding of all of the tools of naval warfare, including the new air tool.

One argument often raised against the inter-war Naval War College is that it missed the changing balance between carriers and battleships. On the eve of war in the Pacific, the U.S. Navy had only seven fleet carriers. The Japanese had ten. It is forgotten that the U.S. figure had been dictated largely by the dead hand of the Washington Naval Treaty, signed in 1922. When the United States began to rearm in the 1930s, new programs were described as a means of building up to a modern "treaty" navy, meaning a fleet of under-age (as defined by the treaty) ships matching the tonnages set in 1922. At that time, carriers and naval aircraft were only barely (if at all) out of the experimental stage. The rather generous tonnage set in 1922 for carriers could be attributed to the Royal Navy, which considered them essential based on its World War I experience. The U.S. Navy would have been satisfied with a much lower figure—but the record shows that the U.S. Navy had almost no input into the treaty process. The War College certainly supported attempts to devise air-capable ships not limited so severely, particularly the abortive flight-deck cruiser and projected (but then abandoned) conversions of merchant ships. These ships were employed in war games and demonstrated their value.

By the 1930s, the question within the U.S. Navy, as stated when the General Board tried to draw up long-term building programs, was not whether carriers and naval aircraft would supersede battleships, but when. The performance and potential of carrier aircraft was increasing rapidly, but in 1939 it was by no means obvious that they could sink modern battleships in the open sea. It took large numbers of aircraft from multiple carriers to sink the huge Japanese battleships *Musashi* and *Yamato* as late as 1944 and 1945. Carriers were much softer targets. For example, it was accepted that they could be disabled or sunk by dive bombers, but the weapons from these same aircraft would be unable to penetrate a battleship's protective deck (torpedoes were a different matter). The Pacific campaign was a carrier war both because the Japanese eliminated the U.S. battle line at Pearl Harbor and also because the Japanese refused to risk their battle force until 1944. That was by no means inevitable. That the Naval War College and most naval officers continued to emphasize battleships in 1941 should not be a surprise, nor should it be considered proof of some failure by the college. On the contrary, war games from the early 1920s on generally involved large numbers of carriers, their numbers often boosted by assumed fast liner conversions.

Two important U.S. naval aviation innovators, both connected with the Naval War College, pose together in front of aerial photography equipment at Naval Air Station North Beach, 27 December 1928: Admiral William V. Pratt (left) and Rear Admiral Joseph M. Reeves (right). Pratt served as president of the Naval War College; he was the only president to become Chief of Naval Operations. War games convinced him that the Navy, bound by treaty, could never have enough carriers. He helped invent the abortive flight-deck cruiser as a way to surmount that obstacle (another was emergency wartime conversion of existing merchant ships). Reeves invented the crucial carrier operating technique of landing aircraft into a barrier (to place them in a deck park forward). As a consequence, U.S. carriers operated more aircraft than their foreign equivalents, to the point that the three U.S. carriers at Midway had much the same number of aircraft as four Japanese ships. (NHHC NH 75876)

The Inter-War Navy and Its World

The inter-war fleet that transformed itself was different enough from the modern U.S. Navy to require some explanation.[10] Between about 1905 and 1914, the U.S. Navy moved up and down between second and third place among the world's navies, in each case far behind the Royal Navy. With the collapse of the Imperial German Navy at the end of World War I, the U.S. Navy was second only to the British, and the expected postwar resumption of Congress' gargantuan 1916 building program promised to bring the American fleet close to parity.

Until 1914, U.S. naval attention was concentrated on the possibility of German aggression in the new world. With Germany gone after 1918, but Japan rapidly building up a new navy, attention shifted to the Pacific, where it remained for most of the inter-war period.

To a modern reader, the most striking difference between the present and the inter-war period is probably that the inter-war United States had no allies. Indeed, it strongly rejected the idea of peacetime alliances. That made the role of powerful neutrals, particularly the British Empire, an important strategic issue in a way utterly unfamiliar in the present day. U.S. strategists seem to have been entirely unaware that after 1921 the British considered the Japanese their likeliest future enemies, based on demonstrated Japanese interest in dismembering the British Empire in Asia. Although the inter-war U.S. Navy did not know it, the British took much the same view as the U.S. Navy of an end game in the war they expected to fight against Japan. The Navy (and the U.S. government) was also unaware, at least until the late 1930s, that the British saw the United States as their only worthwhile prospective ally. When U.S. naval officers were finally permitted to engage in talks with the British in 1938, those talks had to be kept secret from the U.S. public, much of which still imagined that the British had drawn the United States into World War I for their own purposes.

The inter-war United States was, moreover, hardly the world's dominant superpower. The U.S. Navy had to seize control of the sea as a first priority. The modern U.S. Navy enjoys effective sea control over most of the world's oceans; its problem is how to exploit that control to effect events on land. During the Cold War, the question was how to defend the sea control gained at great cost during World War II, and that is the question that may now be arising in east Asia. There is no modern equivalent to the inter-war problem of how to seize control of the sea in the first place. The inter-war games were concerned mainly with seizing sea control, the principal fruit of that control being the ability to choke off an enemy's seaborne trade.

Although the inter-war Navy could certainly bomb and shell enemy territory, such attacks were relatively minor possible consequences of gaining sea control. Bombardment and shelling and the seizure of islands, the techniques developed both in gaming and in full-scale exercises in the 1930s, were a means of supporting a naval campaign, not seen as ends in themselves. No one in the inter-war period imagined a seaborne assault like many in the European theater, as direct strategic attack on an enemy's home territory. It happened that the means conceived to seize island bases for naval purposes provided vital strategic leverage, culminating in D-Day.

The logic of a single concentrated fleet was the logic of battleship-on-battleship combat. During the inter-war period, it was assumed that it would usually take a battleship to sink another battleship.[11] Although other ships could certainly help enforce sea control, nothing but the destruction of the enemy's battleships could ensure it. It was, moreover, generally assumed that numbers would be decisive: A widely accepted mathematical expression was that combat power was proportional to the square of the number of battleships (Lanchester's Square Law).[12] The basic logic predated the mathematics: Rear Admiral Alfred Thayer Mahan had famously told President Theodore Roosevelt never to divide the fleet. The immediate consequence at that time was that U.S. battleships on foreign stations, including those in the Philippines, were brought home and incorporated in a concentrated U.S. battle fleet. After 1918, when U.S. attention shifted from the Atlantic to the Pacific, the logic of concentration implied that the U.S. Navy could either maintain its battle fleet in the Philippines or on the U.S. Pacific coast. Battleships demanded considerable industrial infrastructure to keep them operational. Without much infrastructure in the Far East, that was no choice at all.[13] The inter-war U.S. Navy maintained the small Asiatic Fleet based in the Philippines backed by a battle fleet based at San Pedro (Los Angeles) and San Diego, California. It built up Pearl Harbor into a base capable of accommodating the fleet if need be. Most strategic war games began with the fleet at Pearl Harbor prior to a sortie en route to the western Pacific. The lack of facilities in the Pacific, particularly to repair underwater damage, became a vital point highlighted by war gaming in the mid-1930s.

Thus, the inter-war U.S. Navy was built around a single main fleet, which until 1940 was called the U.S. Fleet (its commander had the unfortunate acronym CinCUS). A scouting element (actually a training fleet built around the oldest battleships) was based in the Atlantic; in wartime it would have joined the main fleet in the Pacific. The Scouting Force, built around modern cruisers, was integral with the fleet in the Pacific.[14] Until the outbreak of World War II, the only other fleet elements were the Asiatic Fleet (which included the Yangtze River Patrol in China) and a small presence force in the Caribbean.

Today's Navy consists of separate carrier and other task forces, partly because the Navy has to be present in many different places at about the same time. That was only beginning to be the case in 1939–40, as events in the Atlantic forced the organization of an increasingly powerful Atlantic Squadron intended to enforce U.S. neutrality. In 1941, it expanded into an Atlantic Fleet. The larger force remaining in the Pacific was renamed the Pacific Fleet. The Atlantic Fleet fought an undeclared naval war against the Germans through much of 1941. It was a convoy-escort force rather than a concentrated battle fleet. When France fell in May 1940, one driver for a massive new U.S. naval building program was the possible need to face an Axis battle fleet in the Atlantic, a scenario that did not feature in inter-war games.

The modern organization is practicable because carriers achieve a considerable degree of concentration by concentrating their aircraft. Several carriers can operate together, but there seems to be a practical limit to how many can do so if they have large numbers of aircraft onboard. The pressure to keep the fleet concentrated is gone, and to some extent the fleet operates more effectively if it is not fully concentrated.

Pre-war concentration of the U.S. Fleet encouraged naval officers to think about their roles in a fleet-on-fleet engagement. As it happened, a wider view of a protracted Pacific campaign required other kinds of operations. The only way that such a view could be disseminated was through campaign-level gaming at the Naval War College. That wider view is a major theme in this book. It turned out that much of the Pacific War corresponded to the build-up to a given decisive battle rather than to the battle itself.

Naval warfare was understood very differently as well. Today, we think of power projected from the sea by aircraft and missiles. That requires reconnaissance to support targeting, but it generally does not require our fleet to locate another fleet in the vastness of the ocean. We are far more interested in preventing an enemy from finding our own fleet in order to attack it. However, in the inter-war years the U.S. fleet was very much concerned with destroying enemy fleets. To do that it had to find them. Scouting was essential, and the broad reaches of the Pacific made it difficult. Naval War College students played games intended specifically to exercise their ability to create effective scouting plans, both for the U.S. (Blue) Navy and for the opposing navies. Large numbers of naval aircraft, including seaplanes, were needed to scout over the vast areas involved. Often, tracking an opponent's fleet would begin with submarines deployed off of its base to pick up the enemy force as it sortied, determining its initial direction. Much attention went into techniques for shaking off scouts, and there was interest in alternative means of scouting such as airships and long-range aircraft. In one game (Operations V, played in April 1934 for the Class of 1934), for example, the Japanese side lost track of the U.S. fleet after it sortied, and then guessed wrong as to the course it might take. It lost the U.S. fleet for some time, wasting considerable effort in a panicked search.

The inter-war Navy was much smaller than the current one, not only in the number of ships but in the number of officers and enlisted personnel. Mainly because the fleet embodied much less advanced technology, the shore establishment supporting it was considerably smaller. The current shore establishment in effect includes large companies that support the fleet. The pre-war equivalent was far smaller. Inter-war Naval War College classes always numbered fewer than 100. In the modern Navy, anything involving fewer than 2,000 officers over two decades would have only limited impact. Such numbers mattered a great deal more in a fleet that included no more than 9,897 officers in 1936. Even in 1940, with the Navy growing rapidly due to the emergency presented by World War II, there were only 17,723 officers.[15] Until the rapid growth beginning just before World War II, all of the officers were Annapolis graduates, which meant that many of them knew each other. An officer returning to the fleet from the War College passed on what he had learned, at least within his wardroom. Given very limited rotation policies, he became a very important window onto the larger naval world, so the experience provided by the War College had a much wider impact.

The structure of the inter-war Navy would also be unfamiliar to us. OPNAV was nowhere near as all-powerful as it has become. It controlled the operational fleet, but the materiel side of the Navy was handled by independent bureaus, such as the Bureau of Aeronautics (BuAer), the Bureau of Construction and Repair (BuC&R), and the Bureau of Ordnance (BuOrd), which were responsible only to the Secretary of the Navy. The situation was complicated because the bureaus' responsibilities conflicted. For example, BuAer bought airplanes (and chose their main features), but BuOrd provided their weapons. Similarly, C&R designed ships, but BuOrd developed their guns and fire controls and armor, and BuAer developed the catapults and arresting gear onboard carriers. A separate Bureau of Engineering (BuEng) developed ships' power plants and also, as it happened, radios and sonars (BuEng and C&R merged in 1940 into the Bureau of Ships). The General Board, compsed of senior officers appointed by the Secretary of the Navy, attempted to coordinate the bureaus, for example in the area of ship design. The Chief of Naval Operations might or might not decide the outline requirements for new ships (which were called "characteristics," and were the province of the General Board). The bureaus of the past are now mostly system commands responsible to the Chief of Naval Operations in a far more unified naval structure, which began to coalesce during World War II.

The Strategic Problem

The inter-war period was a time of considerable U.S. strategic naval development. The emphasis shifted from the Atlantic to the Pacific. Before 1914, the main scenario envisaged by U.S. naval planners was a reaction to a German attempt (actually planned by the Germans) to gain a foothold in the new world by seizing territory in the Caribbean. The main question was whether the U.S. fleet based at Norfolk could get to the Caribbean in time to establish a base of its own. When the Naval War College gamed the operation in 1914, it also raised the question of whether the advancing U.S. fleet would be particularly welcome when it arrived. In 1919, with the Germans defeated, the bulk of the U.S. fleet passed through the Panama Canal to the Pacific. The most likely future naval opponent was now Japan, which was known to covet all foreign possessions in the Far East—including the U.S.-held Philippines. Planners contemplated a surprise Japanese attack on the Philippines or, at times, a preemptive Japanese attack on the U.S. fleet to clear the way for the conquest of the Far East. Surprise attack seemed likely because the Japanese had opened their last major war, in 1904 against Russia, with a surprise attack against the Russian fleet.

Looking toward such a war, the War College always emphasized the end-game, the action it was hoped would force the Japanese to end the war. It was well known that Japan had few natural resources, hence it lived by trade. The favored end-game was therefore blockade—strangulation. Blockade, in turn, required the U.S. Navy to deal with thousands of Japanese and other merchant ships.[16] The unstated implication was that ultimately the U.S. Navy would use large numbers of small warships for this purpose, and that these ships would have to be based on islands near the Japanese Home Islands. Unless it was neutralized, Japan's powerful fleet could sweep aside the blockaders. It followed that in any Pacific War, the U.S. fleet would have to destroy or neutralize the main Japanese fleet as a prerequisite to strangling Japan. None of the war games offered any acceptable alternative to this basic strategy, and it is difficult to imagine one. The game perspective, which emphasized the end-game of blockade as the objective, could not have been gained in a full-scale exercise.

Battleships remained the core of the fleet through the interwar period because it generally took other battleships to sink them. The only aircraft capable of sinking a battleship were torpedo bombers. War College evaluations showed that such aircraft were unlikely to be effective. Early torpedo bombers had low-powered engines, so it took a lot of wing area to lift a heavy torpedo. The aircraft were so unwieldy that for a time the U.S. Navy seriously considered eliminating them altogether. This is a T4M-1 on the deck of the carrier *Lexington*, 8 December 1928. (NHHC NH 51371)

Clearly, blockade required that island bases near Japan be seized, and war planners assumed they would have to do so. However, such seizure was not studied in detail. Island seizures (generally in the face of enemy defenses) became important in the late-1930s context of a new step-by-step strategy requiring intermediate bases.

Through the inter-war period, a key assumption was that the fleet could not be sent to the Philippines during a period of tension that might lead to war; it would be mobilized and dispatched only after war broke out. It followed that the fleet had to cross the Pacific to the Philippines after war broke out, to fight the decisive battle that would open Japan to the end-game of blockade—to seize command of the seas in the Far East. As the think tank of the U.S. Navy, the Naval War College in conjunction with the OPNAV War Plans Division had to find some way of getting there in the face of Japanese attacks. The problem was made considerably worse by the post–World War I settlement that had mandated many of the Pacific islands to Japan. The great question was how (or whether) the fleet could pass relatively unscathed through the mandated islands (generally termed the Mandates) to the Far East.

Until 1934, U.S. naval war planners generally assumed that the fleet could reach Manila, or at least somewhere in the southern Philippines, either before the Japanese did

The Washington Naval Treaty suspended battleship development apart from improvements to existing ships. The treaty envisaged additional protection against bombing and underwater damage. As the first post-war Naval War College President, Admiral Sims used his new precision war-gaming technique to compare the U.S. and British battle lines in simulated battles. He concluded that the British gained an advantage due to the terms of the treaty. That is, reliance on traditional methods of analysis rather than his new one had led to disaster. Later, it turned out that Sims's analysis grossly overestimated the performance of British guns; in fact, the U.S. battle line had the advantage. By that time, however, Sims's analysis had convinced the Navy to press for increased gun elevation (for greater range) in existing ships, a change the British argued violated the treaty. Here the British battleship *Barham*, followed by HMS *Malaya* and the carrier HMS *Argus*, participates in joint exercises of the Atlantic and Mediterranean Fleets about 1930. (NHHC NH 61776)

or in the face of Japanese opposition. The war plan evolved on this basis was called the "through ticket to Manila." Most, but not all, Naval War College strategic games echoed this idea. The college had U.S. Army as well as U.S. Navy students. Many of them apparently told their colleagues that they could not imagine holding the Philippines long enough for the fleet to arrive, but they had no real impact. It appears that the crucial blow to the "through ticket" strategy was experience in a 1933 war game, Operations IV. Once its lessons had been accepted, strategy and the war games changed dramatically. From 1935 on, trans-Pacific war games always included the seizure by one side or the other of an advanced base on an atoll. The fight for the atoll was simulated by a committee of students, as a way of learning how to fight an amphibious battle.

During World War II, the U.S. Navy developed a partial alternative to conventional blockade in the form of unrestricted submarine warfare. The reason this strategy found little favor in inter-war gaming is that the United States of those years lived in what to us was a very unfamiliar world. Having refused to join the League of Nations in 1919–20, the United States rejected any peacetime foreign entanglements (sentiment against the British ran particularly high). In a war against Japan, many of the merchant ships in question would be British, because Britain then had by far the world's largest merchant fleet. War games did show that unrestricted submarine warfare by the United States would inevitably involve attacks on British ships, an unacceptable problem. This sort of conclusion could not, incidentally, have been reached in any way except simulation—gaming. Nor could it have been made obvious that submarines had no potential against Japanese merchant shipping *except* by executing unrestricted submarine warfare (a conclusion obvious from several games).[17]

Once the Germans triumphed in Europe in May–June 1940, the inter-war perspective changed dramatically. The United States found itself pushed toward an unprecedented peacetime alliance with the United Kingdom, and a substantial Atlantic Fleet was created at the expense of the Pacific. However, there was no Atlantic equivalent of the full-war campaign envisaged for the Pacific. To the extent that the United States might expect to fight in Europe, that was much more an Army than a Navy responsibility. The Navy's wartime role would be largely to ensure the build-up of materiel in Europe and to bring the Army ashore there. By this time, too, the Naval War College's advisory role had dwindled. Thus, for most of this book, naval strategy means the strategy of a Pacific war.

Naval Arms Control

The great facts of U.S. naval life during much of the inter-war period were the three arms-limitation treaties, beginning in 1922 with the Washington Naval Treaty; and, for most of the period, congressional reluctance to buy new ships. The Warren G. Harding administration convened the Washington Naval Conference, which wrote the treaty, in 1921 specifically to outflank mounting Congressional pressure to stop the huge expenditures associated with the existing U.S. naval building program. It had a temporary advantage in that the British and Japanese, who had their own expensive programs, were in much worse financial positions.[18] The treaty effectively froze capital ship programs. Among the consequences evident in many of the games was that it left Britain and Japan, but not the United States with fast capital ships (battlecruisers). The cancelled U.S. program had included battlecruisers. War games asked how (or whether) a more powerful but slower U.S. battle fleet could deal with those fast foreign ships, and also how the drastic U.S. deficiency in scouts (cruisers) could or should be addressed.[19]

The treaty had three other important consequences. First, it allowed each signatory to convert two cancelled ships into aircraft carriers. In effect, it jump-started carrier

The shift to the Pacific emphasized the need for various kinds of scouting; war games at Newport emphasized the difficulty of finding and tracking an enemy fleet over vast areas. The U.S. Navy tried a wide variety of scouting platforms, including submarines and airships such as *Akron* (ZRS-4), shown here over Manhattan about 1931–32. War games showed that such craft were too vulnerable to enemy fighters to be worthwhile, although airship advocates continued to press for them as late as 1940. The same games featured a long-range flying boat, the German Dornier Do X, which was more successful. (NHHC NH 43900)

aviation, because the resulting carriers were far larger than any navy would have wanted at the time. Once the two huge U.S. carriers, *Lexington* (CV-2) and *Saratoga* (CV-3), entered service they demonstrated what carrier airpower could achieve. The Naval War College was well aware of the potential of naval air power, and its game rules were often somewhat more optimistic than current reality. Looking back, the major flaws in the rules were excessive confidence in anti-aircraft fire, which was maintained even after fleet experience showed that it was misplaced; excessive confidence in level and torpedo bombing; and perhaps somewhat reduced confidence in dive-bombing. The rules did not capture issues in fighter control, which proved extremely important in the war to come.

The second major consequence was strategic. As part of an ultimately failed attempt to demilitarize the Far East, a Nine-Power Treaty signed alongside the Washington Naval Treaty prohibited further fortification of various Pacific islands. Manila Bay in the U.S.-held Philippines was already fortified to a considerable extent, but now Guam could not be fortified, and war games generally assumed that it would fall at the outset of any Far East war. The question was how long it would take the Japanese to conquer the whole of the Philippines, an assumed target in nearly all the Far East games.

A third consequence of the 1922 treaty and its 1930 successor, the London Naval Treaty, was to set a ratio between U.S. and Japanese strength, measured in the total displacement of capital ships: Japan would have 60 percent of the U.S. limit.[20] Limits on total tonnage were eliminated in the 1936 London Naval Treaty, the last of the series, but not qualitative limits such as maximum ship displacement. The treaties thus raised the question of how weapons not limited by treaty, which included shore-based aircraft, could be used by the Japanese to redress the balance.

Other arms either not limited at all, or less strictly limited, took on far greater significance. For example, it was impossible to limit shore-based aircraft. Until 1930, neither cruisers, nor destroyers, nor submarines were limited, at least in numbers. How could their effects on a naval battle or campaign be estimated? What characteristics should they have? The Japanese might be condemned to inferiority in numbers of battleships, but could other types of ships tip the balance? War gaming, as it was practiced by the War College, offered a way to understand the true balance of forces. That is why the War College was asked to define a "balanced Treaty navy" in 1931. Its conclusions in turn helped OPNAV shape a building program within the limits defined by treaties.

The U.S. Navy's documentation of the inter-war treaties, in the General Board files, seems to show conclusively that the Navy and its think tank had no impact on the U.S. official positions during treaty negotiation. For example, General Board proposals were simply rejected.[21] The War College and others were consulted only after a treaty had been negotiated.

The 1936 London Naval Treaty eliminated limits on total battleship, carrier, cruiser, destroyer, and submarine tonnage, but as of 1941 it was still the old Washington and 1930 London Treaty limits that applied to the U.S. Navy. The reason was political. When U.S. naval rearmament began in the 1930s, most Americans did not want to spend more on the Navy. The country was deep in the Depression, and it seemed inconceivable that it would have to fight in a few years. Naval rearmament was justified as a means of maintaining a "modern Treaty navy," the treaties in question being the Washington Treaty plus the London Naval Treaty of 1930—not the escape from treaty limits that emerged at London in 1936. That language was deliberately adopted in legislation to avoid fights over specific new ships, such as battleships. Having renounced the treaty system in 1934, the Japanese were not similarly bound. Because U.S. naval rearmament was keyed to the Washington and London treaties, it took special congressional authorizations to add carriers, one of which, *Hornet* (CV-8), was authorized in 1938. That was well below what the War College had suggested.[22]

U.S. strategists, including the Newport war gamers, were well aware that the U.S. economy dwarfed Japan's. They assumed correctly, for example, that the U.S. aircraft in-

The fleet needed more effective scouts, so the U.S. Navy invested in large cruisers. To an extent, they replaced the battle cruisers that had been barred at the Washington Naval Conference. At naval arms control conferences, the Navy collided with the British, who badly wanted to limit the size (hence cost) of cruisers to keep them affordable. The War College did not originate the cruiser requirement, but it was extremely important in setting U.S. cruiser policy under the 1930 London Naval Treaty. Here three heavy cruisers of the Scouting Force exercise together in the early 1930s. Note the floatplanes in the foreground, which extended the cruisers' scouting range. (National Archives 80-G-462561)

dustry would grossly out-produce the Japanese. However, they were also painfully aware that building ships would take considerable time. It seemed that destroyers and perhaps cruisers begun when war broke out might be completed in the second or even the third year of a war. Anything larger—a carrier or a battleship—would probably be too late to make much of a difference. As a consequence, when the Naval War College envisaged a future war in, say, 1938, it could not envisage the sort of carrier-heavy U.S. Navy that fought Japan in 1944. In the pre-war context, the only source of tonnage was an ally; as early as 1931, the Naval War College was calling for a wartime alliance with the British.

World War II turned out very differently because the United States began to mobilize in June 1940 when France collapsed and the prospect of a two-ocean war loomed. Congress approved a massive appropriation intended to provide sufficient forces to fight, if need be, in both the Atlantic and the Pacific. Given acceleration after Pearl Harbor, the ships ordered in 1940 entered the war in 1942–43. Without such a source of fresh tonnage, the U.S. Navy would have been in considerable trouble. The 1940 appropriation provided the mass of fleet carriers that won the air-sea battles of 1944–45.

Ships[23]

The ships of the inter-war Navy, whose designs were much affected by war gaming, are relatively unfamiliar to us. We think of carriers as the ultimate capital ships, their aircraft capable of delivering decisive weapons—until fairly recently, including nuclear weapons. The carrier aircraft of the late inter-war Navy were intended mainly to attack ships. Their torpedoes could sink any enemy ship, including battleships, but experiments and games showed that it would be difficult to hit a maneuvering target. The most potent air weapon, the dive bomber, could seriously damage or sink anything *but* a modern battleship. It was widely understood that at some point aircraft would improve to where they would be able to sink battleships at sea, but it was not at all clear when that would happen. World War II did indeed show that aircraft could sink a maneuvering battleship at sea, but in retrospect the main early-war example, the British battleship *Prince of Wales*, fell victim to a lucky hit.[24] The torpedo hit that doomed the German battleship *Bismarck* in May 1941 was possible partly because she had no supporting ships. The Battle of Midway in June 1942 showed that U.S. torpedo bombers were far less effective than had been imagined, and it was not until October 1944 that U.S. carrier aircraft (in numbers unimaginable pre-war) sank the Japanese *Musashi*.

The object of naval warfare, then and now, and very much in the war games, was to gain sea control by sweeping away the enemy's challenge to free U.S. use of the sea. During the inter-war period, the ultimate challenge to free use of the sea was the enemy's battleship fleet, because, if the logic explained above is correct, it usually (if not always) took battleships to sink other battleships. On the other hand, carrier aircraft could reasonably be expected to sink anything short of a battleship, and carrier air reconnaissance would generally keep carriers out of the way of battleships that might otherwise sink them.[25] Carriers were also the only way to project power from the sea against most targets ashore. Such potential targets included an enemy's naval bases and the ships lying in them.

Given the objective of defeating the Japanese fleet in the western Pacific, any U.S. strategist would have understood that the Japanese would seek to wear down (attrite) the U.S. fleet before it got there. It took gaming experience to see exactly how that could be attempted, and what the U.S. Navy could do to maintain its strength. The attrition campaign, which had to precede a full fleet-on-fleet battle, emphasized carrier aircraft and torpedo craft. To fight this campaign, the U.S. Navy had to transform itself from its pre-war character (battleships supported by a few destroyers) into a true all-arms fleet. This force could function effectively even after its pre-war battle line had been destroyed at Pearl Harbor. Gaming simulated all-arms battles. In many of them, the battle line was

Submarines were also valued as long-range scouts, because they could operate off an enemy's bases in the face of his own seapower. The first new U.S. submarine designs of the inter-war period were conceived as long-range cruisers valuable mainly as strategic scouts. Many inter-war games featured equivalent Japanese submarine scouts trying to detect the U.S. fleet as it sortied and also to determine its course (in one game they failed completely, the U.S. fleet went in an unexpected direction, and the Japanese found it nearly impossible to recover). Here the tender *Holland* (AS-3) is shown with four long-range submarines, from alongside to outside: *Nautilus* (SS-168), *Barracuda* (SS-163), *Narwhal* (SC-1), and *Bass* (SF-5), off San Diego, 10 November 1932. (NHHC NH 61589)

secondary or not even present. Full-scale fleet exercises did *not* simulate such battles, in which the battleships played only a minor part.

Aside from carriers, the most visible new type of warship in the inter-war U.S. fleet was the large long-range cruiser. In 1919, the U.S. Navy had three elderly fast cruisers and another ten light cruisers under construction; it also had large obsolete cruisers. The General Board had repeatedly but unsuccessfully asked Congress to approve additional modern cruisers. In a pre-aircraft era, cruisers were vital scouts for a fleet and small cruisers were the best way to beat off enemy destroyer torpedo attacks. In 1920, cruisers were the occasion of the last clear victory of the General Board over Rear Admiral William S. Sims and other rebels who wanted to supplant it with an all-powerful OPNAV. Looking toward the Pacific, the General Board saw a desperate need for long-range scouts, a point which many games would demonstrate. They had to be large, with the most powerful possible guns. Sims, the wartime U.S. naval commander in Europe and the first post-war president of the Naval War College, returned from service in London commanding U.S. naval forces in European waters with much the point of view of his British hosts. They preferred small cruisers because the Royal Navy needed numerous affordable, hence smaller, ships to protect merchant shipping. The British found strategic scouting less important, partly because they had effective long-range submarines, which they thought could operate freely off of an enemy's bases. Moreover, wartime British code-breaking had provided a valuable measure of ocean surveillance. Looking back nearly a century, the cruiser fight seems less than significant, but the later cruiser controversies of the early 1930s offer a useful gauge of War College influence.

The other major surface warship type of this time, the destroyer, seems to have been affected only indirectly by gaming. The U.S. Navy began building modern destroyers in numbers only after the War College and gaming had lost most of their influence. However, even in these ships the impact of gaming can be seen. It was Naval War College games that enforced the view that to survive to fight the main Japanese fleet, the U.S. fleet would have to beat off heavy air and submarine attacks in the Mandates. This idea came up repeatedly when the General Board, for example, discussed what sort of destroyers the Navy needed. It was responsible for the decision to provide U.S. destroyers with a heavy anti-aircraft battery directed by an elaborate fire control system.[26] This choice was unique among the world's navies. It is difficult to dismiss the indirect effect of gaming in causing it. Some of the later inter-war U.S. destroyers were also unique in carrying the heaviest torpedo batteries in the world.

Another important new type was the long-range submarine. During World War I, the Germans built cruiser submarines capable of crossing the Atlantic and attacking merchant ships in U.S. and Canadian waters. Gaming made it obvious that such submarines might perform valuable service as strategic scouts off Japanese bases. During the war, the British built fleet submarines that could operate directly with a battle fleet, an idea that had also attracted the pre-war U.S. Navy. Scout and fleet submarines featured in many war games. The strategic scouting mission, more than the fleet submarine concept, justified U.S. development of the fast long-range submarines that proved so effective during World War II. Because strategic games emphasized the importance of various types of strategic scouting, often by submarines, the characteristics of the fleet submarines of World War II can be seen as consequences of the gaming process.

In the late 1930s, the U.S. Navy and its Marine Corps developed another unique capability: amphibious assault in the face of beach defenses. World War I had featured one major operation of this type, at Gallipoli, and other navies (British and Japanese) seem to have concluded that successful assault could be conducted only away from enemy shore defenses. That made good sense in Europe, with its long shorelines, and in China. The U.S. capability against defended beaches was developed after a crucial 1933 game showed that the fleet could not reach its intended destination unless it had intermediate bases. From 1935 on, games at Newport generally featured attack and defense of fortified islands, and the Navy conducted full-scale Fleet Landing Exercises (FLEXes). The Navy also developed the specialized craft it needed. It planned emergency conversions of merchant ships. They were conceived specifically for assault operations against islands, rather than as conventional port-to-port transports. No other navy seems to have planned this sort of assault force.

2. The Naval War College and Gaming

Before World War I, the Naval War College functioned more as a staff adjunct to the General Board of the Navy—its war planners and policy advisors—than as an educational institution. The General Board itself had been created in March 1900 by Secretary of the Navy John D. Long primarily to resolve planning problems demonstrated by the U.S. Navy during the Spanish-American War.[1] It was deliberately made advisory because Congress strongly resisted the creation of a U.S. general staff. Lacking a large staff, the General Board often turned to War College instructors and attendees for input on policy, strategy, and force structure. It created its Second Committee in 1903 specifically to prepare war plans, and between 1906 and 1910 the committee and the War College worked together. At this time the college ran a "short course" built around the summer conference organized to discuss particular questions posed by the General Board. War College archives include tallies of which officers took which positions on such General Board questions as the best position for a fleet commander in battle and the best tactics. Although there were lectures on various naval and legal issues, answering these General Board questions was at the heart of the War College experience. Students and instructors spent most of their time discussing and gaming those questions, before presenting their "solutions" to the college and the General Board, which usually summered in Newport. The board also made use of the War College outside the conferences. The board valued War College gaming as a way to test new technologies and warship design, sending the college a number of designs and force structures to test. In the decade before World War I, for example, War College analysis played a critical role in the development of "all–big gun" battleships in the Navy.

The "big gun" recommendation exposed a fault line within the Navy. The bureaus that designed and built the ships tended to be conservative. They resisted the General Board's call for a new kind of battleship (which, incidentally, would have predated HMS

Admiral William S. Sims returned to the War College as president in 1919. He transformed the course and sought to make the War College the Navy's think tank. Gaming became the core of the curriculum. Sims is shown after his return from Europe, where he had commanded U.S. naval forces during World War I, on 7 April 1919. (NHHC NH 2839)

Dreadnought, which is usually considered the progenitor of such ships). Reform-minded officers saw bureau resistance as rejection of the legitimate views of those who would have to fight a future naval war. In their view, the fleet rather than the technicians should make the key decisions. The General Board and the War College were steps in that direction. A January 1908 article in *McClure's Magazine* by Henry Reuterdahl, American editor of *Jane's Fighting Ships*, summarized complaints made by the General Board and by many officers, including Commander (later Rear Admiral) William S. Sims, who had gained prominence as a reformer in U.S. naval gunnery.[2] He was serving as naval aide to President Theodore Roosevelt, who was personally interested in the Navy. Roosevelt was not pleased with the article, because it argued that his "Great White Fleet," which was then steaming around the world to showcase U.S. sea power, was much less effective than it looked. The article particularly criticized the new battleship *North Dakota*, the first U.S. dreadnought.[3]

President Roosevelt used the Naval War College summer conference of 1908 to resolve controversy. Because the issue of the design of the *North Dakota* was raised, the conference was called the "battleship conference," although it was more concerned with the way in which design requirements were framed. The President attended. Not surprisingly the conference endorsed the critics' view: Henceforth, the General Board would decide the basic requirements—characteristics—of new U.S. warships. The fleet, not the technicians, would decide what it needed. As the General Board's think tank, the Naval War College would now take a central role in deciding the character of the U.S. fleet as well as its war plans.

It was understood that the critics were attacking the wider administration of the Navy, particularly the independence of the bureaus, which answered only to the Secretary. In 1909 former Secretary of the Navy William H. Moody, now a Supreme Court justice, chaired a commission to study naval administration. He recommended that the Navy Department be divided into functional areas, one of which (operations) would be presided over by a Chief of Naval Operations. As before, Congress was unwilling to adopt a naval general staff. Another board, headed by Rear Admiral William Swift, offered a more modest version of the Moody recommendations, which was adopted in October 1909 by Secretary of the Navy George von Lengerke Meyer. There would be four divisions (Fleet Operations, Material, Inspection, and Personnel), each headed by a rear admiral styled an "aid" to the Secretary. The Aid for Operations issued operational orders to the fleet under the signature of the Secretary of the Navy.

The Naval War College moved from the old Bureau of Navigation to the new Operations Division, which would later become the Office of the Chief of Naval Operations (OPNAV).[4] That formalized its advisory role as the think tank of the operational side of the Navy. In 1912, a British observer likened the long-term students and faculty of the War College to a U.S. naval war staff.[5] This role is easily forgotten if the War College is seen as a purely educational institution.

As—in effect—the staff of the General Board, the War College also became deeply involved in ship characteristics decisions. These roles left little time for any educational function. In 1911, War College President Rear Admiral Raymond P. Rodgers asked for more staff; without it, he could not be both educator and advisor to the Operations Division and the General Board. Instead of providing it, in October 1911, Secretary of the Navy Meyer formally limited Naval War College participation in war planning to work compatible with its educational mission, which meant war gaming. The General Board was given its own staff, but not its own gaming facility. Through the 1930s, successive college presidents insisted that they were not war planners. However, at the same time, the war gaming conducted as the core of the Naval War College course informed the formal war planners, who probably could not have worked effectively without it.

The aid (or aide) system did not last. Woodrow Wilson's Secretary of the Navy Josephus Daniels, who took office in 1913, showed little interest in the naval efficiency for which the system had been designed. Daniels was interested mainly in social issues, such as prohibiting liquor on shipboard. Naval critics satirized him by saying that he was interested mainly in creating a new position, the Aid for Education. The outbreak of World War I in Europe offered Daniels's critics within the service an opening. They thought it entirely possible that the United States might be drawn into the conflict; hence, that naval efficiency was suddenly essential. Daniels's Navy Department seemed not to care; Fiske saw only peacetime thinking. In January 1915, behind Daniels's back, he turned to Representative Richard P. Hobson (D–AL), a naval hero of the Spanish-American War and a member of the House Naval Affairs Committee.[6] Fiske's staff, consisting of officers involved in the previous fight over naval administration, drafted language that Hobson incorporated in the current naval appropriations bill. It called for appointment of a Chief of Naval Operations. Daniels opposed the proposal as "Prussianism," but all he could do was water down the bill. The Chief of Naval Operations was made responsible (under the Secretary) for the operation of the fleet, its readiness for war, and its general direction. He was given a 15-man staff, and the Naval War College was included in his domain (initially the only other naval organization in the Office of Operations was the Office of Naval Intelligence). The CNO could not issue orders in the name of the Secretary. The act was passed on 3 March 1915.

Daniels's revenge against Fiske and the other naval reformers who had come out of the Naval War College was to pass over all the flag officers to appoint Captain William S. Benson to the new post. Benson had attended the short Naval War College course in 1906, and he placed Naval War College graduates in key positions in his office. He also managed to convince Daniels of his value, so that in 1916 new legislation much strengthened his office: All orders he issued were to be taken as coming from the Secretary. His rank and title were raised to admiral.

Freed of much of the war-planning mission by Meyer in 1911, Rodgers turned the War College more toward education for command. He created a new "Long Course" to supplement the usual summertime "Short Course," which had been intended mainly to assist the General Board (the last summer conference was held in 1914). The first Long Course was held in 1911–12 for four students. It was dwarfed by the first and second committees formed for that year's summer conference, with 16 officers each.[7]

The Long Course set the pattern for the war college. That year, the only course was the one beginning in January 1914, with 13 students, five of them captains. In 1915 the War College graduated two classes, one beginning in July 1914 (six officers, only one of them a captain) and one beginning in January 1915 (14 officers, of whom eight were captains).[8]

The star pupil of the first Long Course was Captain William S. Sims, who used that training in his next command, the Atlantic Torpedo (Destroyer) Flotilla. Sims discovered that his new command had no real tactical sea experience. There was, for example, no understanding of how to find and attack heavy ships in the open sea—his command's most basic task (the Royal Navy discovered in exercises at sea that this was an almost impossible problem). Sims compelled his new command to develop practical tactics—to bridge the gap between what could be drawn on paper and what could be done in reality. Sims had been skeptical of the value of the War College course, which included instruction in how to frame tactical decisions and also gaming to test tactical ideas. His experience with the destroyers changed his mind. At Newport in 1914, Sims explained that he used a combination of repeated conferences, including game-board work, before maneuvers with the fleet, followed by thorough discussion afterward, just as the college handled its own tactical games.[9] Sims was rewarded by being appointed president of the Naval War College when that position opened in 1917, and then promoted to rear admiral.[10] He planned to make gaming the core of the curriculum.[11]

Sims did not have long in this post. By this time, it seemed likely that the United States would soon enter the war. The Germans announced unrestricted submarine warfare, which was widely expected to trigger war, on 1 February 1917. The German foreign ministry soon sent the Zimmermann Telegram to Mexico, offering the American southwest in exchange for a Mexican alliance. The combination of the submarine campaign, which would inevitably sink U.S. ships, and the outrageous German offer to Mexico (sent via U.S. diplomatic channels made available as part of a U.S. peace initiative) brought the United States into the war on 6 April 1917. During February and March, as war clearly approached, Sims was often called to Washington for conferences with the General Board and the National Advisory Committee. At the end of March, he was called secretly to Washington and assigned to liaison with the Royal Navy. He sailed on 31 March, just before war was declared. The Naval War College course was suspended. Sims spent the war as commander of U.S. Naval Forces Operating in Europe. He was already considered an anglophile, and during the war he formed close relationships with the Royal Navy. For example, in the absence of the British commander at Queenstown, a vital anti-submarine center, he was given command of all naval forces based there, including those from the Royal Navy.

In London, Sims created the Planning Section, which attempted to apply War College methods to the naval war.[12] It would handle questions of immediate interest raised both by the force in European waters and by the Navy Department in Washington, but it would also raise questions of importance in the longer-term execution of the war. All of this might be considered so obvious as not to be worth mentioning. U.S. officers used to planning and to concepts such as the "Estimate of the Situation" got a glimpse of a very different way of thinking—what might be considered the thinking of a navy without the U.S. type of Naval War College.[13] The Royal Navy created its own plans section (later division) as part of a general reorganization in July 1917. It is not clear to what extent this new group was a direct response to demands by Prime Minister Lloyd George.[14]

The U.S. planners were surprised that the Royal Navy had no real interest in their type of work.[15] The British had a pragmatic approach to everything; they disliked formal planning of any type. They thought in very specific terms, not asking whether some alternative operation might offer a better payoff. The pre-war British decision to create a war staff had had very little impact on British naval thinking. In particular, the British rejected a long-range view of the naval war. They much preferred separate and often unrelated operations, linked (if at all) in the mind of a senior officer. The Naval War College had long advocated a formal planning process, although more for a single operation than for a war as a whole. Sims's London Planning Section tried not only to deal with immediate problems, but also to anticipate problems that might arise. Its most spectacular projection was to point out that the Germans might decide to send some of their battlecruisers into the Atlantic to raid Allied convoys, which had no suitable protection. Its arguments were convincing enough that the U.S. Navy based battleships at Bantry Bay in Ireland specifically to deal with this possibility, and special signals were arranged to warn of a possible break-out.[16]

A few months after Sims arrived, the British naval staff was drastically reorganized. Among its new elements was a plans division, which to Sims's staff should have been the core organization of the staff. American officers later wrote that they were the only ones who took the plans division seriously; Sims's Planning Section took the orphaned plans division under its wing. CNO Admiral Benson was impressed enough with the success of Sims's section to recall one of its members, Captain Harry E. Yarnell, to set up a parallel organization in OPNAV.[17]

The existence of a planning section in OPNAV opened a major question that the legislation of 1915–16 had not resolved. How did the General Board, which was certainly the peacetime planning organization, fit into the evolving Navy? Wartime naturally made the operational naval staff dominant. It alone had the ability to frame and issue orders on a day to day basis. Wartime records show that OPNAV also took most responsibility for the massive wartime building program, designed mainly to fight the war against the German U-boats. This threat had had almost no impact on the pre-war General Board, whose big 1916 building program was designed to provide a U.S. battle fleet "second to none." The logic of the 1916 program was that, at the time, President Woodrow Wilson was trying to protect American shippers from the British blockade imposed on Germany and her allies. In Wilson's view, British naval superiority made it possible for the British to levy unreasonable demands. If the U.S. Navy was as powerful as the Royal Navy, the British might find that much more difficult. The situation changed completely once the United States entered the war. The General Board's sin was not to have foreseen any such shift, at least in framing its building program.

Admiral Sims ended the war as a hero, having presided over the U.S. part in the victory over the German U-boats. Offered his choice of post-war assignments, he chose the Naval War College, which he now saw as the core of the U.S. Navy.[18] He naturally wanted to complete the naval reorganization begun in 1915: The Navy should have one chief, and that should be the Chief of Naval Operations. Sims considered the General Board an anachronism now that OPNAV existed. In response to the demands of war, OPNAV had grown enormously; the question was who would rule afterward.[19] In the wake of the war, CNO Benson reorganized OPNAV into formal divisions. The Naval War College remained in the OPNAV organization, but was not a formal division. The divisions were:

- Policy and Liaison (Central Division)
- War Plans
- Intelligence (Office of Naval Intelligence)
- Naval Districts
- Fleet Training
- Materiel (Fleet Maintenance)
- Ship Movements
- Inspection
- Communications
- Secretarial

Of these divisions, Fleet Training was responsible for tactical development. Inter-war handbooks of tactical practice were generally published as FTPs—Fleet Training Publications (or as USFs, United States Fleet publications).

None of the divisions included any sort of war-gaming capability to act as a laboratory for planners. That role was reserved to the Naval War College. Sims told the December 1919 Naval War College class that their course work and gaming would allow them to "develop new applications of the principles of warfare as applied to modern naval conditions . . . . They will also explain how the tactical game demonstrates the necessity for new or modified types of vessels, or new or modified uses for those already built."[20]

At times, Sims went further, arguing that the organization should not be called a college at all. It was, rather, "a board of practical Fleet officers brought together here to discuss and decide the extremely important question of how we could best conduct naval war under the various conditions that may arise. You should think of this board

As a student, Harris Laning suggested that the War College collect lessons learned during war games and provide them to future students. War College President Sims made him head of the tactics department, where Laning wrote about Captain Joseph M. Reeves's performance as commander of Blue aircraft. Laning returned to the War College as president. Following that assignment, he was made commander of the Battle Force. Admiral Laning is shown (left) in Hawaii in 1935 with Admiral Harry Yarnell, who, as commander of the carrier *Saratoga*, carried out the surprise attack on the Canal Zone in the 1928 Fleet Problem. At this time, Admiral Yarnell commanded the Fourteenth Naval District, which included the critical Pearl Harbor base. (NHHC NH 85318)

as belonging to the Fleet—as being what you might call a Fleet Board on strategy and tactics, frequently making reports to the Fleet and the service upon these vital subjects." Unlike an ordinary college, Newport had no fixed policy, and neither its head nor its staff was permanent. "They [students and faculty] are Fleet Officers and are continually being replaced by other officers from the Fleet. The College is in effect part of the Fleet and it exists solely for the Fleet. The students bring to the War College their practical Fleet knowledge and experience. They are asked to consider this practical knowledge and experience in connection with the principles of warfare [which] are nothing but deductions from the accumulated experience of those who have gone before us . . . . When you finish your course you will carry these principles back to the Fleet, and will, I trust, be guided by them." Students who were assigned to the College Staff would transmit their experience to the new class. Many of the students had never before been involved with strategy or with some forms of tactics.[21]

Sims said that "some officers complain that they do not understand the terms, the strange terms, used by the College. I do not understand any of the strange words used by golfers, because I have never played the game, but I understand that some such words are necessary. They are equally necessary for the game of war. Every art must necessarily have its own rules, principles, and methods, and these must have names if we are to talk about them . . . . This game, like all other games, can be learned only by playing it. The mere study of the art of war, even though very thorough, will no more make you competent in the practice of strategy and tactics than book knowledge of golf and tennis will make you good players. It is for this reason that the College insists not only upon the study of the art of war, but also upon the practice of it in chart maneuvers and on the tactical game board. You will find that by playing these practical games you will gradually acquire confidence in your ability to estimate a situation correctly, reach a logical decision, and write orders that will insure the mission being carried out successfully." Sims added that in the war just ended effort and even many valuable lives had been wasted because of the lack of training needed to reach logical decisions "based upon the well-known immutable principles of war . . . the constant prayer of those who bear great responsibility in time of war is that they may be spared the results of the decisions of the so-called practical officers who are ignorant of the art of war and who have not been trained to think straight—that is, who have not been trained to make a logical estimate of a situation."

Sims also argued that the college was so important to the fleet that flag officers should be appointed to head its two key departments, strategy and tactics.[22] CNO rejected that idea.

To make simulation effective, Sims ordered the preparation of war game rules far more detailed than any yet used, including evaluations of how well various ships in different navies could survive battle damage. While this work was proceeding, the United States convened the Washington Naval Conference, which led to a naval arms-control treaty. At least in theory, the treaty left the United States at parity with the Royal Navy, a result unimaginable before World War I. The General Board recommended approval on that basis. Sims sponsored a study using the War College's new method of fleet comparison, and he publicized its result: The treaty left the British superior (it turned out that when more accurate data were used, the U.S. fleet was superior).[23]

Laning publicized another innovation first tried out in games: circular formations. They were invented by Roscoe MacFall, shown here as a captain in 1945. Initially, their advantage as a cruising formation was that the fleet could respond quickly to an attack from any direction—essential for a U.S. fleet crossing the Pacific in the face of Japanese opposition. During World War II, circular formations turned out to be well-adapted to multi-carrier operations; they were crucial to a fleet concentrating large numbers of carriers, as the U.S. Navy did in the western Pacific. (NHHC NH 48033)

This study, doubtless intended to gain attention for the college as the fleet's think tank, seems to have been the first major post-1918 example of gaming-related arguments used to reach important real-world conclusions. It and successor modified studies became important factors in the first major post–Washington Treaty naval controversy. The Washington Treaty prohibited major modification of the turrets of existing ships. Using the same methodology as the 1922 study of battle line performance, it was possible to show that British battleships could outrange their U.S. counterparts by flooding their bulges intentionally and thus increasing gun elevation. At this time, the U.S. Navy, which was becoming interested in longer-range gunfire as a way of gaining a decisive advantage, wanted authority to increase the elevation of its guns. The British argued that the wording of the treaty prohibited any such modification. Eventually the U.S. Navy rejected British opposition and beginning with the *Nevada*-class (BB-36) ships were modified.[24]

Not surprisingly, Sims became the center of the post-war debate about Navy organization. The pretext for his involvement was differences with Secretary of the Navy Josephus Daniels over U.S. Navy recommendations for the award of the newly created Distinguished Service Medal for wartime service. Sims publicly rejected his own medal, urged others to do so, and charged that Daniels himself had failed to prepare the Navy for war.[25] Sims's attack on the existing system, particularly the General Board, was only partly successful. The General Board retained its central role both in determining ship characteristics and in laying out formal U.S. naval policy. The latter responsibility offered the board at least in theory a central role in the inter-war arms-control negotiations.[26] A quiet struggle between OPNAV and the General Board continued through the inter-war period, the board losing most of its remaining power during World War II.[27] However, it managed to survive through 1950.

In this war, the Naval War College was initially connected to both sides. It advised the General Board and it was part of OPNAV. Attendance at the War College was effectively a prerequisite for service in the OPNAV War Plans Division. War gaming became a key OPNAV laboratory technique. However, only one inter-war War College president, Admiral William V. Pratt, became CNO. OPNAV retained its control over the Office of Naval Intelligence (ONI). The close relationship between ONI and the college is evident in the numerous intelligence reports that survive in the Newport archive (as Record Group 8). To some extent these reports were a necessary way of keeping the war games realistic, but they were also needed for the think tank role.

The Naval War College of the early-to-mid 1920s was essentially that envisaged by Sims: a board of officers developing tactics and strategy by simulating future wars. In accord with this vision, the instructors were all drawn from previous students. At this time, the principal departments were Command, Strategy, and Tactics. Command was headed by the College's chief of staff. The Strategy and Tactics departments ran separate games. As president, Admiral William V. Pratt reorganized the faculty in 1926 to emphasize the vital role of logistics as a separate discipline. The Strategy and Tactics departments were merged into a single Operations Department to emphasize that strategic and tactical considerations could not be separated; to some extent that was due to the reach of aircraft. The Command Department was renamed "Policy and Command." It later became the core of a War Plans/Policy and Plans Department. Pratt added an information office, which later became the Intelligence Department. It received information from the Office of Naval Intelligence, on the basis of which war game rules were constructed.[28]

Probably the most important change, instituted by Admiral Laning as Naval War College president after 1930, was to abandon this practice and appoint long-term instructors (albeit still naval officers, generally Naval War College graduates). The Class of 1933 (1932–33) was the first to work with a new department of research, and that year there was a separate war plans instructor. Intelligence and research merged with the Class of 1935.

Beginning in 1924, the College added a junior class (lieutenant commanders and lieutenants) with its own staff and, in some cases, its own exercises. As in the past, the senior class was intended to teach graduates how to wield a fleet. The junior class was to teach them how to command major units. This book is concerned with the senior class and the games it played, although it includes conclusions from each year's "big game." This was a month-long campaign game played by a larger team combining the two courses, the juniors taking up the more junior commands.

The college also ran a correspondence course. It attempted to teach the decision-making technique employed in the war games, based on the "Estimate of the Situation" (see Chapter 3). Without gaming, this is unlikely to have been very effective. The rise of the correspondence course probably explains why, in the mid-1930s, considerable attention went into producing successive much-expanded versions of the handbook *Sound Military Decision*, sometimes called the "Green Hornet" after the color of its cover. Later editions were roundly criticized as nearly impossible to read (or use), most likely because nearly all readers came out of the correspondence course.[29]

3. War Gaming and War Planning

The "Applicatory System"

The inter-war Naval War College used an "applicatory system" of instruction introduced by War College President Rear Admiral Raymond P. Rodgers about 1910. At the time, it was considered revolutionary: It was learning by doing rather than via instruction. "Applicatory" meant applying principles—having them inculcated by using them. Prior to the applicatory system, much effort went into formalizing the principles of war and teaching them on a formal basis. The students would not apply the principles until they had to fight. The basis of the applicatory system was that memorizing the principles was not nearly enough: A student would not even fully understand what he was being told until he tried to apply them to a realistic situation. That went even more strongly for the more technical aspects of naval warfare, as they might be reflected in maneuvering a fleet against an enemy fleet. A later War College president likened the method to the case method which had been introduced by leading law schools.[1] Students did hear lectures on the law, but they were forced to think through how they would apply their knowledge in a series of cases. Many science students have experienced just this reality, that it is impossible to fully grasp physical laws without solving problems using them. For the War College, the war games were the way students gained experience.

As employed at the War College, the applicatory method incorporated a systematic approach to problems, as reflected in the form used to write orders.[2] At least in theory, this approach helped students apply their knowledge to the particular problems presented. Concentration on practical problems did accelerate a tendency at the War College to discard more academic pursuits; for example, the history chair was eliminated.[3]

Rodgers learned the system from his relative Captain William L. Rodgers, who had learned it at the Army War College. Ultimately, it was derived from practices developed by the German army, who had gained enormous prestige by their victory over France

Admiral Sims' great innovation was to tie the gaming used in the "applicatory" system to realistic data describing U.S. and foreign fleets. The great question, which was not really raised until the War College was in decline as a think tank, was how well foreign navies were described. That applied particularly to the likely adversary, the Imperial Japanese Navy. Japanese counterintelligence was excellent, and we now know that U.S. (and, for that matter, British) knowledge was quite limited. Of the four Japanese capital ships shown in this photo, taken in the 1920s, the *Nagato* (in the foreground) was substantially faster than imagined (26.5 rather than 23 knots). The next ship was the *Kongo*-class battlecruiser *Kirishima*. The U.S. Navy certainly did know that she was fast; the question (which gaming certainly could help answer) was how much of a handicap the United States suffered because it had no fast capital ships. The other two ships, the battleships *Ise* and *Hyuga*, held no great surprises when this photo was taken, but U.S. intelligence was unaware that when they were rebuilt in the 1930s, they were given considerably increased speed. There was also no attempt to credit the Japanese—or any other foreign power—with distinctive tactical style, although war gamers were told about the peculiar Japanese command system in which army and navy were coordinated only through an imperial council. (NHHC NH 111609)

in 1870–71.[4] They seem to have invented the idea of mission-oriented orders in place of detailed orders to subordinates: The subordinate was left to work out how to execute the mission. Mission orientation offered subordinates much more initiative. They had to understand the overall objective so that they could apply their own ideas successfully. Mission-oriented command is often associated with an army's ability to adapt to a rapidly changing tactical situation, with forward commanders understanding what to do without having been formally instructed. Without such orientation, an offensive might well stall because officers who had already reached their initial objectives would not keep going. Exactly that happened during the British amphibious assault at Gallipoli in 1915, when troops facing little resistance stopped at their planned objective line instead of pressing ahead. It might be argued that it was also what prevented British success during the night action after the Battle of Jutland the following year. British CinC (commander in chief, i.e., fleet commander) Admiral Sir John Jellicoe never learned what was happening because no one apparently thought it necessary to report to him. As a consequence, he never knew that the German fleet, which was withdrawing, had passed through his own. Everyone apparently assumed that someone else had reported. After the battle, Jellicoe emphasized the need to report.[5] The idea of mission-oriented orders is still a very active topic. For example, this command technique became an important issue in efforts to reform the U.S. Army during the latter stages of the Cold War.

To support this command technique, the Germans developed a decision-making method beginning with formal determination of the mission—not a trivial problem, since the orders from above would necessarily be much more general than those a subordinate would have to issue. The statement of the mission was followed by an assessment of enemy forces and intentions, assessment of own forces, and then evaluation of alternative courses of action to reach a formal decision. The next step was to write orders using a standard order form based on German practice. Its five parts corresponded to the steps required once the mission had been identified. The Army War College typically evaluated the student's solution using either a map exercise or a staff ride. The latter was a ride to examine the terrain in which the battle would be fought to examine whether planned dispositions and possible enemy actions were practicable.

The Naval War College equivalent of the Army's preparatory document was the "Estimate of the Situation." Its core was a list of possible enemy courses of action followed by the corresponding list—written after working out enemy courses of action—of own courses of action. The student had to guess what the enemy's preferred course of action might be. No one ever described it this way, but comparing courses of action by both sides was not too different from the usual game theory exercise of tabulating possibilities and then evaluating likely payoffs for both sides before choosing a course of action that minimizes risk or seems to maximize payoff. This technique imposes a discipline on whoever uses it because listing what the enemy can do limits wishful thinking.

The inter-war cruiser problem, which will be discussed in detail later, is a good illustration. Asked for advice after the conclusion of the 1930 London Naval Treaty, the Naval War College first laid out courses of action that the other major navies—the British and the Japanese—could and likely would take. It based its advice on those estimates. That might seem obvious, but the British experience at the same time shows what could happen if this process was not followed. The British objective during treaty negotiations was an attempt to end the competition in large expensive cruisers that the Royal Navy was finding unaffordable. The British assumed that, just as they were driven mainly by economic necessity, other navies would welcome the chance to limit themselves to more affordable ships. They surmised further that the cruisers had become so large only because they had to accommodate the largest allowable guns. If only smaller guns were

The war-gaming data system included rules to calculate battle damage. They were as conservative as possible: It was assumed that ships would succumb only slowly to damage. No allowance was made for sudden catastrophic damage, such as that which befell the British battle cruiser (or fast battleship) *Hood* in May 1941, when she was sunk by the German *Bismarck*. *Hood* is shown in Gatun Lake, on the Panama Canal, in July 1924. (NHHC NH 60404)

permitted, surely all the world's navies would build only cruisers comparable to the less expensive ones the Royal Navy planned. British records show no attempt whatever to look at the cruiser issue through others' eyes. They were unpleasantly (and expensively) surprised when neither the U.S. Navy nor the Imperial Japanese Navy followed suit. Both preferred large, expensive cruisers armed with larger numbers of the smaller guns. As a consequence, the British found themselves compelled to build far more expensive ships than those they had previously contemplated.[6]

If an officer understood the mission, he could apply standard tactics. Many of those who espoused the applicatory method saw it as the basis for introducing doctrine into the fleet.[7] For example, as commander of the Atlantic destroyers, Sims developed a menu of tactics. He likened them to the plays of a football team; each member could execute them simply by hearing the number of the play called out by the quarterback. Each member was responsible for applying those plays to a mutually understood objective. Mutual understanding was crucial: It was created by gaming and conferences (reviewing the tactics

Because little was known about Japanese and other foreign aircraft, the War College used the details of current U.S. naval aircraft for both the U.S. Navy and foreign navies (there were some exceptions). These Japanese aircraft are shown on a carrier, about to attack Pearl Harbor. (National Archives 80-G-71198)

used in the games). In modern terms, it might be said that Sims was determined to promulgate commander's intent rather than simply instructions.

World War I Royal Navy experience, particularly at the Battle of Jutland, showed why it was vital that the Navy share a common doctrine based on pre-war exercises. The British seem to have assumed that upon elevation to command, the fleet commander would immediately gain (or quickly develop) the ability to work out the necessary tactics. This belief in the genius of a single senior officer is reflected in the decision, about 1910, not to publish a tactical handbook because that would unduly cramp fleet commanders. Put crudely, officers were not expected to think for themselves until they reached fleet command, at which point they had better think for the fleet as a whole—without much real preparation for that role. At best, a future fleet commander like Jellicoe might enjoy some experience as deputy to a fleet commander (as Jellicoe actually did).

In effect, the Battle of Jutland was a test of the British system of command. Critics of Admiral Jellicoe's performance have argued that his centralized style of leadership failed to develop individual initiative. Probably the worst result was a failure both during and after the battle to report enemy sightings and positions. That was extremely unfortunate, since Jellicoe relied on his subordinates to feed data into the plot he was maintaining. He did not share the nature or importance of the plot with the commanders of his ships, and they in turn often failed to report. The failure of individual initiative has also been blamed for his ships' failure to fire at the Germans when they came into view during the night after the battle. Jellicoe might have argued in the latter case that it was better to withhold fire than to risk fatal errors of identification at night, but this was the sort of question that would have been raised in Sims-type conferences and in war games.

Jellicoe certainly did try to envisage all possible contingencies, often using small model ships his flagship's crew made for him to try different formations and tactics. He encouraged gaming in other ships. However, ultimately his Grand Fleet battle orders were designed to control individual initiative, and he expressed only limited interest in educating subordinates to understand the logic of the orders so that they could function effectively if communication broke down. Neither Jellicoe nor his subordinates had experienced simulated battle situations. For example, no one seems to have understood just how the fleet would be affected by poor visibility. The Royal Navy did conduct many full-scale exercises, but, unlike the U.S. Navy, they do not seem to have provided subordinate officers with a sense of what was happening and of what was needed. Nor, apparently, did the Royal Navy share a common menu of expected tactical moves.

Without much sense of Jellicoe's guiding principles, it might well have seemed that far too much responsibility had rested on him. What would have happened if he had been killed, or had collapsed—as he did the night after the battle? How could continuity of command been assured? The elaborate Grand Fleet battle orders offered little guidance for reactions to unexpected developments. For example, no one had worked out a response to the torpedo attack and retreat the Germans carried out.[8] Prior to the battle, Jellicoe simply counselled that the fleet should not follow the Germans into a submarine and mine trap. He may have been inspired by German tactical instructions warning that the British might do the same thing.

In fairness to Jellicoe, it should be noted that on taking command of the Grand Fleet he wanted to institute decentralized divisional tactics, but did not think that he had enough time before he went to war to train the fleet properly. The kind of decentralized authority that Sims envisaged in his destroyer force required considerable indoctrination and experience on the part of subordinates. When Jellicoe took command, the Royal Navy had no such tradition, and Jellicoe could argue that he needed several years to reverse decades of an attitude that the fleet commander automatically knew everything and therefore ought to be followed blindly.[9]

Every post-war navy studied Jutland intensely, not because such a battle had the slightest chance of recurring, but because it offered so many examples of the way in which the reality of war affected command and tactical performance.[10] The British experience mattered to the U.S. Navy first because the Royal Navy was the largest and most powerful in the world, hence a model for a rapidly-growing U.S. service. It was also the navy

Gaming scenarios were based on a realistic understanding of current political issues, but there was no way to estimate the political impact of particular operations. This did not mean that it was entirely ignored. During the 1930s, a gamer detached a carrier to attack Tokyo. A later summary of lessons learned from gaming mentioned the operation and added that there had been no way to evaluate it (in the game, the detachment of the carrier carried real penalties). When a U.S. carrier force, supplied with Army B-25B bombers, actually did attack Tokyo in April 1942, consequences included the disastrous Japanese decision to attack Midway in hopes of drawing out the U.S. fleet. B-25Bs are shown on the flight deck of the carrier *Hornet* (CV-8), ready for launch against Tokyo. (NHHC NH 53289)

that provided the U.S. Navy with its experience of modern naval warfare during World War I. U.S. cruisers, destroyers, submarines, and sub-chasers were assigned to the anti-submarine war, in many cases integrated with Royal Navy organizations. A U.S. battle squadron was assigned to the Grand Fleet in 1917–18. U.S. officers found the Grand Fleet battle orders (or instructions) far more sophisticated than their own much simpler ones. Overall, exposure to the Royal Navy during World War I was a terrific shock to a U.S. Navy that on the eve of war had considered itself very competent.

However, the unsatisfactory outcome of the Battle of Jutland showed that the British system was limited—that the playbook envisaged by Sims really was necessary. It justified the Naval War College, which was the source of the common command culture the U.S. Navy needed. Command education produced an officer capable of using the

Commentators discussing games often pointed out that new construction would take so long that it probably would not influence the outcome of a war. As it happened, U.S. mobilization began in 1940, and new ships began to appear as early as the end of 1942. In effect, the entire pre-war U.S. fleet was more than replaced, and the mobilization program was responsible for the enormous growth in the U.S. carrier and cruiser forces. All the ships in this photo of Task Force 38, maneuvering off of Japan on 17 August 1945, were completed after war broke out. They include six *Essex*-class carriers, of which *Wasp* is closest to the camera, four light carriers, at least three fast battleships, and cruisers and destroyers. (National Archives 80-G-278815)

playbook—and, incidentally, capable of fitting into an existing tactical organization even if he had not had years of experience within it. A common command culture had to be created, in which all officers understood a common approach to thinking through tactical and strategic issues.

In place of the prescriptive Grand Fleet battle orders, the inter-war U.S. Navy created tactical handbooks best described as playbooks—compendiums of alternative tactics and formations from which a fleet commander could choose.[11]

* * *

War Gaming

William L. Rodgers became War College president in 1911. He combined the applicatory system with war gaming.[12] A student might list an enemy's courses of action, but in reality the enemy had a vote, which could be simulated only by gaming. A game could pit each side's estimates of the other's decisions against the other. It could also tease out surprises an enemy might spring, in a way that no direct analysis could.

The Naval War College seems to have invented the use of gaming to explore the problems of fleet tactics. In all other navies it was apparently assumed that experienced officers would instinctively know what to do when they assumed command. Even full-scale tactical experimentation was a relatively new idea, introduced into the British Mediterranean Fleet by Admiral Sir John Fisher in 1899, and probably not known outside the Royal Navy. The U.S. Navy badly needed the gaming technique, both because it had no recent combat experience (the war against Spain had involved weapons now quite obsolete) but also because its officers had little experience of fleet operations of any kind—the Navy had grown too rapidly for that. Given its ability to conduct naval war games, the Naval War College could use them to develop new tactics against a background of rapid technical change.

Through the inter-war period, the War College emphasized that it was teaching students to outwit capable opponents, who might have equivalent weapons and ships and aircraft. Much is often made of the racism that is said to have contaminated U.S. pre-war thinking about Japan. It is not to be found in gaming. Game documents did include what were supposed to be racial characteristics of the enemy, but in the case of Japan the usual special characteristic was fanatical courage and devotion to duty. It was not stupidity or rigidity. To emphasize that an opponent would be quite as intelligent as the U.S. Navy, the War College rotated its students between the U.S. and enemy fleets. There was no attempt to shape the behavior of an enemy fleet in accord with known foreign doctrine or tactical style.

Note that the War College generally did not refer to its exercises, at least during the inter-war period, as games. They were chart or table maneuvers, the name suggesting that in some way they were comparable to the full-scale maneuvers played out during the Fleet Problems (the term gaming was revived after World War II). Strategic games were chart maneuvers; tactical ones were table maneuvers, because they were typically conducted using a large table marked out in scaled squares. In the 1930s, a checkerboard floor was laid down in Pringle Hall, making it possible to simulate battles over wider areas.

In 1911, the War College already used tactical games to experiment with warship design and also with tactics. For example, the U.S. Navy's Battle Plan No.1 was developed on the basis of gaming at Newport.[13] It was a matter of considerable pride, to the extent that a U.S. naval attaché in Italy commented about 1910 that the Italians had only just progressed to such sophistication. In 1917, Battle Plan No. 1 must have seemed juvenile in comparison to the voluminous British Grand Fleet battle instructions issued

In contrast to ships, the War College assumed correctly that the large U.S. aircraft industry would quickly out-produce the Japanese. Perhaps more importantly, war gaming showed that pilots would suffer a very high wastage rate. The wartime U.S. naval air establishment turned out vast numbers of pilots. Neither of the other two carrier navies seems to have understood the need for pilots in any numbers. That suggests that without gaming, the high wastage rate would not have been imagined. Very rapid production of pilots made it possible to keep large fast carrier forces in action on a sustained basis, pilots and airplanes being replaced while the ships remained in a forward area. This type of replenishment was as important as the more evident development of a fleet train to provide fuel, ammunition, and stores on a sustained basis. The prototype escort carrier *Long Island* (CVE-1) is shown as an aircraft transport, off of the California coast, 10 June 1944. She is carrying 21 Hellcat fighters, 20 Dauntless dive bombers (SBDs), and two J2F utility airplanes on her flight deck, plus other aircraft stowed below; clearly her flight deck is far too packed for her to operate any of them. (National Archives 80-G-236393)

to U.S. battleships joining the Grand Fleet. Battle Plan No. 1 and its relative Battle Plan No. 2 were descriptions of how to fight under a particular idealized set of circumstances. The Grand Fleet battle instructions were designed to respond to the widest likely variety of circumstances. Moreover, they were plans for a combined- arms fleet, in which guns and torpedoes should be used together in complementary fashion. The U.S. battle plans described a gunnery fleet, its destroyers a wholly separate arm. Encountering the Grand Fleet battle instructions was a great shock for U.S. officers attached to the Grand Fleet during World War I.

However, the Battle Plans held the germ of an advantage. They were a corporate U.S. Navy creation, not the fruit of a single tactical genius, if that is what Jellicoe was. They were not identified with the man who happened to be fleet commander when they were adopted. The current fleet commander and other senior officers certainly had a great deal to do with the way in which the battle plan was shaped, but no single officer was responsible for it. It followed that the gap between them and something like the Grand Fleet instructions could be filled by a corporate U.S. naval effort. That effort, in turn, could draw on experimentation, both full-scale and in gaming.

The inter-war U.S. Navy used gaming as the laboratory to test tactical ideas. The full-scale Fleet Problems taught the fleet how to work together, and to some extent they tested the assumptions built into the games. Only the games could simulate full naval campaigns as well as situations, such as massive night battles, which could not safely be exercised by ships and aircraft. Conversely, the game rules were translated into the umpire rules used to evaluate the outcome of the Fleet Problems.

The U.S. Army had its own Army War College. At times, it and the Naval War College conducted joint exercises involving overseas landings. The Army War College seems to have adopted the applicatory system. However, it did not adopt gaming (the Army may have used gaming at its tactical-level school, the Command and Staff College). Thus, its contribution to a joint exercise was an elaborate set of logistical requirements and complementary orders, but not a simulated land campaign after troops went ashore. The Army certainly did become involved in national war planning through its membership on the Joint Board, which ultimately approved U.S. war plans, but its war college seems to have had only limited interest in the strategic level of war that the Naval War College so often examined. That may have been a natural consequence of the difference between the two services. An army ties itself to a particular land area; its job is to seize it or to defend it. Withdrawal is painful and dangerous. A navy is not tied to any geographical area; it can approach or leave as necessary. Particular places are generally expendable, to be revisited later as necessary. This difference was reflected in the later profound disagreement as to the appropriate end-game in the Pacific War.

To the Army, as reflected in its official history, the issue in the Pacific was Japanese seizure of U.S. possessions, particularly the Philippines. The main goal of U.S. strategy was to eject the Japanese from these places. This interpretation may have been affected deeply by the personal identification that the senior Army officer in the Pacific, General Douglas MacArthur, felt with the Philippines, but it seems to have run more widely within the U.S. Army. The Navy's view was that, having attacked the United States, Japan had to be defeated. Once that happened, whatever Japan seized would have to be disgorged. This view was, incidentally, much more compatible with a coalition war than was the Army's. In 1944, the conflict between the two views was reflected in the Navy's preference for seizure of bases on Taiwan rather than a direct assault on the Philippines. At the gaming level, it is obvious from the inter-war games that the objective was generally the Japanese fleet rather than some particular place in the Far East. The unspoken but universal Navy assumption was that once the Japanese fleet was gone, Japan would fall because only the fleet stood between Japan and starvation. Places were valuable to the extent that they made blockade possible or to the extent that their seizure would force the Japanese to expose their fleet to destruction.

Gaming could not and did not make the Japanese fleet an objective. That was fundamental. However, campaign-level gaming could and did give students a sense of what the end-game of blockade and starvation required on the part of the United States. It also avoided making any particular place important beyond its strategic significance.

War Gaming at the Inter-War War College

As Sims intended, war gaming—learning by doing—was the core of the inter-war War College curriculum. The command course did not conduct games; instead its version of learning by doing was to have its students write national policies corresponding to various historical events, particularly the Civil War and World War I. By doing so, they could gain insight into why the various combatants had chosen the policies they actually favored. The games and policy work were supplemented by extensive lectures; non-Navy lecturers were particularly frequent after 1934.[14]

The core curriculum of the War College began with a course in constructing a valid "Estimate of the Situation," which described the mission and then listed courses of action for one's own and opposing forces. Deducing the mission from more general statements was an important part of the exercise. It was not a trivial one, because the statement often could be interpreted in terms of multiple possible missions. If, for example, the fleet was covering an island assault while an enemy fleet was approaching, was the primary mission to cover the assault or to destroy the enemy's fleet in the interests of wider strategic needs? Given the situation, the writer made a formal decision and explained its rationale. The next step was order writing, because poorly written orders could sabotage any operation.

A game began with students receiving papers setting out a strategic or tactical situation. Students turned in their own estimates, which were critiqued by the staff. Since only one estimate on each side could become the basis of war orders and hence of a game, the students of both sides were issued with accepted estimates (incorporating strategic choices) prepared by the staff. Many of these assigned estimates have survived; they are much of the basis of the accounts of particular games in this book.

Once students had learned to write estimates and orders, they were taught to play the war games the War College had developed. Students typically played at least eight strategic and tactical games during their time at the War College after watching one or two demonstrative games to learn how to play. Most of these exercises lasted a few days at most, so they could not simulate the effect of a protracted campaign, in which both sides' ideas of how to fight would probably change due to experience. Each class also played a month-long "big game" representing a protracted campaign (usually, but not always, trans-Pacific). In the 1920s, the Naval War College worked with the Army War College to conduct some joint games involving landings, generally unopposed. Typically, the Army part of the exercise emphasized the logistics of a long-range Army operation, including massive data describing exactly what the fleet had to move.

A history of maneuver was kept for each game, and some have survived. Post-game reviews ("wash-ups") were often written up. They often included analysis by students comparing reality as seen by the fleet with the college's rules.[15] As head of the tactics department in 1923 and 1924, Harris Laning was particularly concerned with tactical lessons the games might teach; his analyses were particularly detailed. That applied particularly to the "big game" played annually by both senior and junior classes. During the period the research department was in operation, it provided a formal post-game analysis, including lessons learned.

The full-scale Fleet Problems did not generally correspond to war games played at Newport, but there were at least two exceptions. One game played by the Class of 1922 was intended to test the scenario for the upcoming Fleet Problem. For that reason, students were asked not to discuss the game with anyone outside the college. It was an unusual scenario, a "bolt from the blue," in which a Japanese fleet visiting the west coast of South America proceeded suddenly to attack the Panama Canal and prevent the Atlantic-based U.S. battle fleet from intervening in the Pacific. The other exception was a partial one. Operations III of 1927 culminated in a battle in Narragansett Bay that was plotted at the War College as it was fought. This also seems to have been the only War College exercise to have received any publicity at all.[16]

To make games realistic, the enemy navy was typically either Red (British) or Orange (Japanese). Each was credited with the ships it actually had, described as accurately as the college's intelligence department could. Its data were the best the Office of Naval Intelligence could provide. Descriptions included ship capabilities and estimated vulnerabilities. Many games simulated parts of the evolving U.S. war plan to fight Orange. It was well known that the U.S. Navy considered Orange the likely future naval opponent; Orange was certainly the object of most naval war planning.

Simulation

After 1919, the War College worked hard to make its simulation more and more realistic. To do that, Sims ordered the preparation of war game rules far more detailed than any yet used, including evaluations of how well various particular ships in different navies could survive battle damage. It was apparently assumed that information about foreign ships and guns was good enough to use directly, but it was understood that information about aircraft was far less reliable than information about ships. The War College generally simply adopted current U.S. aircraft data for both sides. Overall, the rules combined the best estimates by the Naval War College Intelligence Department with actual experience gained by the U.S. Navy in exercises, including the Fleet Problems. War College records (Record Group 8 of the Intelligence and Research departments) include numerous inter-war letters to and from college personnel conducting the annual revision of the rules.

The attempt to improve simulation made for far more complex war game rules than those used before 1917.[17] The new rules required a special staff, substantial documentation, and, to some extent, special facilities. They were very much a War College function. It was no longer possible for the fleet to play the games; they required the staff and the facilities of the Naval War College. Rule books grew from perhaps 20 small pages before 1914 to about a 100 to 150 larger ones from the 1920s on.[18] The rule books embraced so wide a variety of topics that in retrospect they seem to be a good way to learn about the way inter-war technology affected naval warfare. Scenarios included weather, and the rules included estimates of how far away ships and aircraft could be seen at various times and under various conditions. Communications sections included estimates of the time lost in transmitting, coding, and decoding radio messages. Students relied on written ship-to-ship messages, just as in real life they would depend on decoded radio messages rather than voice radio. Students' battle plans generally included frequency plans showing how a fleet's radio resources would be used. From the early 1930s on, the new technology of sonar was included, both for submarine detection and as a means of secure communication.

The effect of imposing increasingly complex rules was to stretch game time well beyond real time, as a great deal of measurement and calculation had to be done after a move that might represent a few minutes in real time. Gaming equipment included calibrated range sticks, so that the ranges between ships could be measured and fed into various fire control and damage calculations. Supplementing the game rules were handbooks describing different fleets and different ships' resistance to various kinds of damage. During simulated combat, ships' remaining combat capabilities were tabulated after each move. Special efforts were made to keep students from seeing more than they would have from the bridges of real ships. For example, the College ran special "quick-decision" games in which students positioned their ships but "fog" simulated by screens kept them from seeing their opponents. The game opened by lifting the screens and the students were timed to see how quickly they could decide what to do.

Generally, the prefaces to student orders listed the objectives of any particular game. The most important were to develop the student's ability to understand situations and to frame decisions and to produce effective written orders. The orders, formalized as deductions from an estimate of the situation, were critically important to the gaming process. Critiques of war games often mention poorly written orders that would have caused huge problems in combat. They cautioned that while students might have a few days to formulate orders for gaming, they might have hours or less to do so in reality, and they would be formulating orders under considerable stress.

Another objective was to familiarize the student with both U.S. and foreign fleets. When they reached the War College, officers were very aware of the capabilities of their own ships and usually of the units, such as destroyer flotillas, in which they operated. They generally did not have much of a sense of how the fleet could or should operate as a combined-arms whole. The War College often provided the only route for developing that feel before being thrust into higher command.

A typical additional objective was to familiarize the student with the waters in which the game was set. Again, a student who had spent much of his career operating in U.S. home waters would not generally have a feeling for the geography of potential theaters of war such as the western Pacific. Fighting notional battles or campaigns in a particular region thus served as a familiarization exercise. Game data included local factors such as normal weather during the period in which the game was set.

The rules were backed by elaborate fire-effect tables giving the effect of various weapons on actual ships—or, at least, on ships whose details were given. They were first constructed in 1922 and revised in 1931.[19] The 1922 publication was the first to attempt to provide a mathematical basis for constructing tables from which the average fire of actual guns could be evaluated in terms of damage to actual target ships in three-minute intervals (representing single moves on the game board). In the past, hypothetical ships equal in attack power and in survivability had been used; now actual navies were being modelled. To enable gamers to compare damage inflicted by different weapons, a penetrating 14-inch hit was taken as the unit of damage and lifetime. Other weapons were assigned capability as proportions of 14-inch hits. The original version of the tables did not take long-range penetration of deck armor into account, because it was believed that no navy had perfected the necessary fuse. In 1924, BuOrd contributed deck armor penetration data, albeit with considerable caveats. The strength of the fire-effects method was that it took into account both offensive and defensive characteristics. It summarized them in a particularly convenient way. A battleship, for example, might have a complex system of protection, but the War College could show how it compared to that of other ships under a particular kind of attack.

The 14-inch gun standard for ship "lifetime" under fire made it possible to compare different fleets and different weapons. The "Estimate of the Situation" generally included tables of standardized firepower and ship lifetime. The idea was widely known in the fleet. When future Rear Admiral Bruce McCandless wrote about his experience as a young officer onboard the heavy cruiser *San Francisco* (CA-38) during the night battles off of Guadalcanal in November 1942, he mentioned his ship's nominal lifetime in 14-inch hits, because she was being shelled by a Japanese battleship armed with just such guns.[20] Despite the tables, *San Francisco* survived many more hits than the War College allowed, partly because they were inflicted at short range, well above the waterline, a scenario not accounted for in the rulebook.

Overall, the fire-effect system was deliberately conservative. The rules credited ships with maximum survivability. Damage was assumed to be cumulative. The rules did not allow for the kind of spectacular results single hits in lucky places sometimes achieved: the explosions that destroyed HMS *Hood* and *Arizona* (BB-39), for example. In the war-game tables of fleet data, HMS *Hood* was what the Royal Navy thought she was: a well-protected fast battleship comparable to her battleship contemporaries. As evidence of that reality, when *Hood* was sunk, the Admiralty warned the commanders of contemporary British battleships that their ships were no better protected.[21] Under the same rules, single dive-bomb hits could not have sunk four Japanese carriers at Midway, although they certainly would have put them out of action by destroying their flight decks. In his memoirs, War College President Rear Admiral Harris Laning wrote that the War College generally described the U.S. Navy as conservatively as possible, but that it allowed considerable latitude to foreign navies.[22] In that way, it prevented over-confidence on the part of students playing the U.S. side.

The War College carefully avoided assigning superiority based on qualities of personnel or leadership to navies. Thus, it assumed that all navies had equally well-trained personnel and comparable guns developing the same rate of fire. Estimates of firing errors, such as dispersion, were based on U.S. performance. It is often assumed that the inter-war U.S. Navy grossly underrated the Japanese on racist grounds. That was certainly the case with the British, who discounted the strength of the Japanese battle fleet on that basis when they estimated what margin of strength they needed in the Pacific. Individual U.S. officers may well have had similar views. For example, one explanation of Admiral Kimmel's failure to provide against a Japanese carrier air threat to Pearl Harbor is said to have been his belief that "the little yellow men" would not dare to mount such an attack. It is therefore all the more striking that no such assumptions appear anywhere in the theory or practice of war gaming at Newport.

The inter-war U.S. Navy had little or no knowledge of foreign tactical practices, because U.S. efforts to collect foreign intelligence were very limited. The best source of tactical information would have been signals intelligence taken from observations of Japanese naval exercises. Although the U.S. Navy did collect such information, its dissemination was very limited for security reasons. It had no discernable impact on the Naval War College. Students were free to choose tactics in simulated battles. The fact that the Japanese players used tactics in games resembling those they actually used during the Pacific War demonstrates that circumstances rather than some form of perceived Japanese military culture determined much of what they did.

The method of analysis developed for war gaming had much wider significance. It made direct comparison of navies possible, based on a combination of offensive and defensive strength. For example, this means of comparison was employed by OPNAV when it advised the Secretary of the Navy in the early 1930s on the appropriate composition of a "balanced Treaty navy."

While the new rules and the fire-effect tables were being developed, the United States convened the Washington Naval Conference, which led to the 1922 Washington naval arms-control treaty. At least in theory, the treaty left the United States at parity with the Royal Navy, a result unimaginable before World War I. The General Board recommended approval on that basis. As touched upon previously, Sims sponsored a study using the War College's new method of fleet comparison, and he publicized its result: The treaty left the British superior.[23]

This study, doubtless intended to gain attention for the college as the fleet's think tank, seems to have been the first major post-1918 example of gaming-related arguments used to reach important real-world conclusions. Sims claimed that he had made two innovations. First, he took into account "the actual types of ships—their actual rates of fire, percentages of hits, and the relative damaging effects of the different calibers [of guns] when fired against the various thicknesses of armor at the various ranges at various angles of impact. In other words the method now in use makes a comparison of fighting strength between ships by taking into consideration their actual structure, armor, and armament. The underlying principles of the present method of comparison are the same as those of the past—that is, if the hitting power of the guns is known and the value of the ships in 'lives' has been determined, then the Fighting Strength is equal to the Aggregate Hitting Power times the Aggregate Life." Second, Sims considered an actual engagement in which factors other than gun power and armor would count. Only the British had battlecruisers. With their superior speed they could concentrate their fire on the head of the U.S. line. The War College also pointed to an overall British one- knot speed advantage. In the War College's analysis, with the two fleets engaged on parallel courses at 15,000 yards, the British speed advantage would allow them to choose the range.

For all the precision Sims claimed, the British never felt particularly superior. They considered their ships far too vulnerable to long-range shellfire, and argued that U.S. ships were better protected. They had nothing remotely like Sims's analysis on which to fall back; it is not at all clear whether they were right. Sims's analysis gives some idea of the advantages the U.S. Navy enjoyed as a result of the War College's new way and far more precise manner of looking at fleets.

At least, that appeared to be the case. However, the apparent precision of the War College technique hid a major problem. Errors in the data could tip the analysis. In this case, the key error was over-rating the most important British gun, the 15-inch/42-caliber which armed ten battleships and three battlecruisers. In Sims's analysis, it was credited with muzzle velocity so high that it outperformed the best U.S. gun, the 16-inch/45-caliber. With a more realistic muzzle velocity, the British gun was midway between the 16-inch/45-caliber and the 14-inch/50-caliber that armed other U.S. battleships. On this basis, overall the British battle line was inferior to the U.S.

Much depended, too, on assumptions about the armor of foreign ships and even about how their shells were fused (because that determined how well shells could penetrate deck armor). The U.S. Navy of this era calculated zones of immunity for ships: the range beyond which side armor could not be penetrated and inside which deck armor could not be penetrated. The fire-effect tables included such data. Bands could be (and were) calculated, inside which one ship could attack another without being endangered. In war games, it was natural for fleet commanders to rely on such calculations as they positioned their ships. In 1937, a Bureau of Ordnance officer pointed out that all of this was far too shaky.[24] The U.S. Navy knew little more about foreign navies than what might be read in open-source publications like *Jane's Fighting Ships*. Some navies, but not the most likely enemy, Japan, did allow gun data like shell weights and muzzle velocities to be published. Even that data was insufficient to estimate how shells would behave once they struck, or how their fuses would act. Most side armor was visible and its thickness (on the outside of a hull) could be estimated, but that certainly did not apply to deck armor. Cruisers might have side armor integral with their hull structure, making observation entirely irrelevant. Worse, by the late 1930s, armor quality was improving dramatically, but the calculations did not take that into account.

The critic missed an important point. Having a way of evaluating and comparing fleets—however flawed—forced War College students to think systematically and comparatively. Moreover, the range-band idea focused officers on seeking combat at greater ranges, with an important if unintended consequence. To fight beyond 20,000 yards, ships had to rely on spotting (to correct aim) by airplanes. Control of the air above a battle line became an important theme well before the striking power of aircraft seemed likely to be decisive. It opened the way for later development of naval air power in War College thinking.

Quite aside from its impact on U.S. tactics, the War College analysis had important policy implications for the Navy. In 1924 the War College reported to the Navy Department that the post–Washington Treaty U.S. fleet would definitely be inferior to its British counterpart. That conclusion fed into an international controversy. The U.S. Navy wanted to overcome British superiority by modifying turrets to increase gun elevation, hence range. However, the Washington Treaty prohibited major modification of the turrets of existing ships. The British argued that increasing elevation was a prohibited improvement. Using the same methodology as the 1922 study of battle-line performance, it was possible to show that British battleships could outrange their U.S. counterparts by flooding their anti-torpedo bulges intentionally and thus increasing gun elevation. Eventually, the U.S. Navy rejected British opposition and beginning with the *Nevada* (BB-36) class ships were modified accordingly.[25]

In 1930, the General Board asked the War College to revisit the question, comparing the U.S. and British fleets after further reductions had been made under the new London Naval Treaty. Now, the War College concluded that the U.S. fleet was clearly superior. Even then, some of the data were not credible. For example, the War College estimated that the British 15-inch/42-caliber guns would penetrate decks the much heavier U.S. 16-inch/45-caliber type could not. Rear Admiral Joseph M. Reeves reviewed and savaged the War College report. He had served as head of the tactics department and he had gone on to revolutionize U.S naval aviation (see Chapter 4); his opinions could not easily be discounted. Reeves found that data for the British gun, which armed nearly all their capital ships, had been handled inconsistently and hence arbitrarily. He was not amused.

How important was all of this? In a naval world apparently ruled by big-gun ships, it certainly mattered politically. In 1922, many U.S. naval officers considered the Washington Treaty a sell-out by politicians because the agreement cancelled a building program that would have given the United States superiority, at least in the most modern ships. Within the Navy, Sims was probably seen as a hero for showing just how bad the treaty had been. Outside the Navy, no one much cared; the pressure for naval arms control was widespread and far too strong to overcome. War College analysis did help justify the aggressive U.S. program of battleship modernization begun after the treaty was signed. How important that ultimately was depends on the reader's view of how important relative U.S. battle-line strength was in 1941.

In 1930, former War College President Admiral William V. Pratt had helped make the London Naval Conference a success. He was soon made CNO. In that job he was clearly determined to sell the treaty. It would have been surprising had analysis at an institution inside OPNAV *not* shown that the London Naval Treaty was a good deal for the U.S. Navy. As it happened, that was true. The U.S. Navy had aggressively modernized its battleships. The British had not, although they had built two very powerful new ships, and they still had the advantage represented by faster ships (the three battlecruisers and five *Queen Elizabeth*–class battleships).

As it happened, U.S. information on the likely future enemy, Japan, was poor. Just how poor is evident in the surviving ONI/naval attaché reports in the U.S. National Archives. Like the U.S. Navy, the Imperial Japanese Navy modernized its battleships. In the 1930s, the Japanese spent very large sums to increase their speed; the Navy became aware of the higher speed of some, but not most, of the Japanese ships only in 1936.[26] The Japanese speed advantage exceeded the known speed advantage enjoyed by the British. At least in theory, it would have enabled the Japanese to choose their fighting range to suit their level of protection. Also apparently unknown to the U.S. Navy, the Japanese were very interested in fighting at maximum range. The U.S. Navy was certainly aware of the advantage higher speed gave the Japanese battle line, and in the 1930s it decided to build fast battleships specifically to deny them that advantage in future.

Assigning detailed capabilities to ships made war games seem realistic. It provided students with a sense of the differences between their fleet and that of their possible opponents. That would have been impossible without assigning details to the various U.S. fleets. It is impossible to say whether the critiques of the fire-effects data mattered in reality, because the United States never fought a World War II battle in which range bands mattered, or in which the balance between the two fleets was close enough that details of relative firepower and protection would have affected tactical decisions. That was true even of the bitter cruiser-on-cruiser battles in the Solomons. The chief tactical issues there had to do with the problems of night combat, which the fire-effects data did not touch.

War gaming had a much more practical impact on U.S. war plans, fleet composition, and tactics outside the battle line.

* * *

Some Limits of Gamed Reality

The game rules missed important facets of naval combat, many of them connected with the subtleties of command/control and air tactics. For example, although many games featured night combat, it is evident in hindsight that the problems presented by night battle were not properly simulated. Thus it is clear from the games and from tactical handbooks that the inter-war navy thought it understood how to fight at night. The problem seems to have been that the inter-war U.S. Navy generally did not exercise at night, for fear of fatal accidents. The Japanese and the British did practice to prepare for night combat in the inter-war years. Their efforts paid off in successful night engagements such as Savo Island and Cape Matapan.

Game designers admitted that air tactics were poorly simulated; they wrote as much in the war game rules. That applied to both air-to-air combat and to attacks on ships. The main implication of air-to-air rules in actual games was that fighter defense was unlikely to be effective. Combat air patrols were generally ineffective in the games (as indeed they would be until the advent of radar-based fighter patrols). Some games did hint at the need to reserve fighters to deal with multiple waves of attackers, or with attackers approaching simultaneously at different altitudes, in the sense that multiple attacks tended to overwhelm the defenders. Overall, the games offered no hint of the need for, or the complexity of, fighter control, only pessimism as to fighters' ability to beat off attacks, which was realistic in a pre-radar world. Air combat could really only be simulated in full-scale engagements in Fleet Problems, not on a game board.

The exceptions to the conservative approach were attempts to test new weapons and new types of ships. This meant aircraft carriers in the years before the U.S. Navy had experience with them; an abortive new type of cruiser called a flight-deck cruiser; fleet and cruiser submarines before these types had been developed; and the radio-controlled Hammond torpedo in 1920. Games also employed drifting mines laid during naval battles and gas in various forms. Both were in various navies' weaponry through the inter-war period, but neither figured in World War II naval combat.

Reality could never be fully simulated, and it is important to keep the limitations of gaming in mind. The number of players was always limited. Players notionally commanded units such as destroyer squadrons, but not individual ships. War gaming was unsuited to modeling the lowest levels of war, and could not reflect the operational reality of individual pilots or smaller warships. To some extent, the full-scale Fleet Problems acted as a corrective.

The games also missed features of foreign practice that might be significant tactically. The inter-war U.S. Navy observed major Japanese fleet exercises, largely by intercepting and decoding radio signals. No tactical insights gained in this way seem to have been provided to the intelligence department at Newport. The entire code-breaking operation was presumably far too secret to be risked in this way. Unfortunately, the records of the signals intelligence collection operation are too vague to show whether it might have been applied to the Orange side in the war games. However, it seems notable that there was apparently no attempt to apply specific British practices, either, even though the Royal Navy was far easier to observe. That may be because the games against Red (the British) were intended mainly to test U.S. tactics against a fully equal enemy—as the Japanese would be if they succeeded in destroying U.S. ships early in a war.

Although students at Newport spent considerable time evaluating national policy, the games were never political, in the sense that they did not evaluate the political side of any future war. They did involve U.S. politics, but only to the extent that, for example, students might be told that the scenario was placed well after the outbreak of war, and that the U.S. population was impatient for offensive action. There was certainly no attempt to evaluate Japanese reactions to various U.S. national strategies. It was simply assumed, not only by Newport, but also generally by U.S. war planners that a Japanese government would come to rational terms if it faced strangulation by blockade. There is no evidence that anyone imagined unconditional Japanese surrender as a possible outcome of a Pacific War. This war was never envisaged as the fight to the death that U.S. war planners came to foresee after bloody island battles beginning with the conquest of Saipan in June 1944. Nor, for that matter, did the gamers ever envisage an invasion of Japan. U.S. war planners assumed that the United States could never create a mass army to fight in the Pacific. The only alternative they imagined to strangulation was a bombing campaign against Japanese cities, which it was assumed were flimsy and vulnerable. Because the islands needed to support the bombing campaign were the same as those needed to support a blockade, this possibility did not affect the logic of the Pacific War as fought in games at Newport.

Using War Gaming

Toward the end of Sims's tenure, one of the War College students, Commander Harris Laning, pointed out that the war games were yielding valuable lessons.[27] Surely it would make sense to collate them as a source for future work. Laning was planning to spend the next year at the Army War College; such cross-education was not uncommon. Sims convinced him to stay on at the War College as head of the tactics department, and to implement his idea. When Laning returned as president in 1930, he formalized the system, creating a Research Department headed by Captain Wilbur van Auken.[28] The research department function of maintaining files of lessons learned seems to have lapsed when van Auken left the War College in 1936.

Van Auken was outspoken in his War College analyses, which were embodied in reports and in letters for the president of the War College—presumably for onward transmission to CNO and the War Plans Division. These analyses survive mainly in the files of the OPNAV War Plans Division rather than those of the War College. Van Auken was well aware of the artificialities of the games, since he was also responsible for keeping the game rules up to date (they were revised annually). In 1931, he concluded that the United States had neither the ships nor the bases (with repair facilities) it would need for a war against Japan. In a remarkable postscript to a 1931 Blue-Red (i.e., against the British) game (Operations III for the Class of 1932), he wrote that it was essential to game and plan a cooperative war by both countries against Japan and its allies. Van Auken referred both to a shortage of cruisers and to the impossibility of replacing losses during a war. Van Auken repeated his comment the following year in commentary on a Blue-Orange game (Operations II for the Class of 1933).

It says a great deal about the college's ability to think freely and to avoid publicity that van Auken was able to draw the conclusion he did in a document that circulated freely through the college, and also through the War Plans Division of OPNAV, without any ill effect. It is difficult to exaggerate how radical his proposal was. Popular U.S. sentiment at the time, and for years afterwards, was squarely opposed to any sort of peacetime alliance with the British. The belief was widespread that the United States had been lured into World War I by the British in combination with U.S. East Coast bankers, who had lent them vast sums and did not want to lose their investment. This explanation of recent history had real impact; it led Congress to pass a neutrality act that precluded future loans and required that any arms buyer pay in cash ("cash and carry"). This act badly hobbled British and French arms purchases in 1939–40. Moreover, van Auken was writing two years after the big triumph of American arms control diplomacy (at the London Naval Conference, which wrote the 1930 London Naval Treaty) had been to force the British to accept cuts in cruiser numbers so that they would no longer be overwhelmingly superior to the U.S. Navy. Given Van Auken's analysis, this triumph had been disastrous: The more cruisers the British had, the better they could join the United States as an ally against Japan. Van Auken himself had written the college's analysis of what sort of ships the U.S. Navy should build under the London treaty. One of his assumptions had been that U.S. politics would demand some sort of parity with the British in cruiser strength. The unspoken assumption as the United States entered treaty negotiations was that the goal was to cut overall British cruiser strength to what the United States could match.

Another point van Auken made struck home. The fleet had to have a mobile support force far beyond what already existed. To some extent the desired auxiliary ships were included in the building programs of the late 1930s. To a much greater extent, they were built by the Maritime Commission created in 1938 specifically to provide both new U.S. merchant ships and potential wartime auxiliaries. It seems fair to argue that it was the war games that showed the need for many of those auxiliaries, particularly the ones that constituted a mobile base and repair organization.[29]

In several cases, the War College staff applied gaming experience to vital current questions. The cases developed in detail in this book are aircraft carrier and cruiser characteristics and associated recommendations for the U.S. Navy's building program. There is also evidence that War College experience was reflected in the battleship program.[30]

On the strategic side, the OPNAV War Plans Division, which developed War Plan Orange used against Japan during World War II, received and used Naval War College gaming material: The surviving files include substantial amounts. Exactly how it was used is less clear than in the case of the building program, however.

The historical record of lessons learned and transmitted is somewhat obscured for the 1920s because both OPNAV and the General Board had direct informal access to War College experience: The Naval War College was an OPNAV division and its president was an ex officio member of the General Board. Only after 1930, when the president was removed from the General Board, did the War College have to provide written comments on various questions.[31]

War Gaming and War Planning

The Naval War College maintained a connection with the OPNAV War Plans Division, but in the absence of correspondence files on either side it is difficult to say exactly how close. The War Plans Division files include considerable Naval War College war game material, much of it not duplicated in Newport.[32] The large quantity strongly suggests that the college was the think tank for the War Plans Division. Conversely, each year the Naval War College played a month-long "big game." At least the 1927 version was described explicitly as a simulation of the war plan as then understood. In this connection, it seems significant that attendance at the college, which meant experience in its gaming technique, was considered a prerequisite for selection to the War Plans Division. At the very least, that made the war planners sensitive to whatever surprises games simulating their plans might reveal.

The "war plan" means the war plan against Japan. That Japan was considered the most likely enemy throughout the inter-war period is evident in General Board hearings on new types of warships, in which the question "how will this ship operate in the Orange war" almost always arises. However, some war games, including some "big games," examined a Red-Blue scenario: war against the United Kingdom. A reader gains the impression that no one seriously believed that such a war could break out.[33] The usual stated cause for such a war was economic rivalry, which in the 1930s was widely believed to have triggered World War I. By way of contrast, Blue-Orange wars were generally triggered by U.S. resistance (via the "Open Door" policy in China) to a Japanese attempt to gain control of the Far East.

It seems likely that war games against Red were intended to teach students how to fight a defensive war against a peer competitor. They would also familiarize students with areas outside the usual U.S. operating sphere, the Pacific. For example, a defensive scenario might be set in the Caribbean. It would give students insight into the strategic significance of various islands and potential bases, very valuable experience once the Atlantic sea war broke out in 1939. Most of the Red-Blue games were fought in U.S. or

Canadian waters. The two typical scenarios were either U.S. resistance to a British expedition to deal with a U.S. threat to Nova Scotia or U.S. resistance to a British attempt to relieve bases in the Caribbean. In effect, the first reversed the usual U.S. problem of relieving the Philippines in the event of a Japanese attack. The Caribbean scenario offered a sense of which British Caribbean bases were worth trying to acquire, as the United States actually did in 1940 in the destroyers-for-bases deal.

For three consecutive years, a very different Red-Blue scenario was played out. Like Japan, Britain lived largely by her seaborne trade. The three "big games" focused on a U.S. attempt to defeat Britain by choking it off. They included a U.S. attempt to establish a base off of the West African coast as a preliminary to trying to block the Straits of Gibraltar. The games showed how difficult it would be to set up and support such a base without any chain of intermediate bases. By analogy, it would be difficult to do so in the Far East, if for some reason the fleet could not go directly to an existing base in the Philippines.

This game showed how rapidly Blue's cruiser force would be expended. That raised the more general question of how wartime losses of major warships could be made up. Even the huge U.S. industrial base would need time to build ships. The more complex the ships, the longer it would take to build them. A cruiser ordered at the moment war was declared might not be completed for two years. Battleships and carriers would take even longer. Looking at the sheer scale of losses in the big trans-Atlantic trade war game, van Auken wrote that the only way to gain sufficient cruisers to fight a sustained war against Japan would be to ally with a country with a large cruiser force. The only candidate was the United Kingdom.[34]

The United States did fight the Pacific War in alliance with the British, but the Royal Navy was so fully occupied in European waters that it could not provide the U.S. Navy with additional cruisers. What saved the U.S. Navy was that the European war, and particularly the fall of France, inspired a massive building program begun in 1940 (under the "Two-Ocean Navy Act"). The 18-month head start, coupled with much more intense working conditions once the war began, provided numerous new ships far earlier in the war than pre-war analysts had imagined.[35]

The Blue-Orange war was of far more direct significance, and it accounted for most of the "big games." Students correctly saw the "big game" as a version of the current Orange War Plan. Introductory material for the 1927 "big game" shows that this was exactly what was intended.[36] Until the mid-1930s, the game generally reflected the prevailing concept that the fleet would begin the war by heading directly from its war base in Hawaii to the Philippines. The latter would be the forward base from which the fleet could fight and destroy the Japanese fleet, and thus open Japan to blockade. Under the Washington Treaty, even the forward base in the Philippines was strictly limited. Further fortification was banned, although existing forts were not destroyed. These already controlled the entrance to Manila Bay. Manila's repair facilities were limited to the floating dry dock *Dewey*, which had been designed to handle battleships in service in 1907. The ships in service during the inter-war period were about twice as large. Congress was unwilling to buy a replacement dry dock or a permanent graving dock at Manila. Thus, even if the fleet reached Manila, it had no way of making good underwater damage to its capital ships. The battle-damage issue was exactly the sort of problem gaming emphasized, because games envisaged the entire campaign leading up to the entry of the fleet into a Philippine port.

It was generally assumed that the Japanese would open the war with an attack on the Philippines. Just like the Americans, they would realize that they were subject to strangulation by blockade. Because the War College's method of analysis including examining a situation from the enemy's point of view, assessments generally began with Japan's need to assure control of vital overseas sources of supply, mainly in East Asia but also farther afield. The Japanese would have to secure the sea routes to the south. Because the Philippines lay astride such vital routes, it seemed obvious that any initial moves by Japan would include the islands' capture as well as the seizure of relatively nearby undefended Guam. Conversely, unless a Japanese position in the Philippines was either overthrown or neutralized, the desired blockade of Japan would be difficult.

The significance of the Philippines exemplified the difference between the strategic view developed at the Naval War College and in the War Plans Division of OPNAV and the Army view that the issue in a Pacific War was the recovery of a U.S. possession (the Philippines) which the Japanese might seize. The strategic view was that whatever the Japanese seized would inevitably be disgorged once Japan surrendered or came to terms. The Philippines became an important strategic objective not because it had to be recovered from Japan, but because it was an obvious part of the only end game that could be envisaged.[37] It seems arguable that this kind of strategic analysis was prompted by war gaming and by the thinking it encouraged.

The Japanese actually had an alternative supply route. If they could make their peace with the Russians, they could rely instead on the short shipping route across the Sea of Japan from Siberia; goods could come from Europe via the Trans-Siberian Railway. The (simulated) Japanese tried this in a few games, and it was not at all clear how the United States could deal with this possibility. This possibility did not arise in reality because even though Japan signed a non-aggression treaty with the Soviet Union, when war broke out in the Pacific, access to goods was cut off by the war on the Eastern Front in Europe.

The question was how long the Philippines, or at least Manila Bay, could hold out against a Japanese attack. In theory, the fleet could convoy sufficient troops to beat back a Japanese invasion force—if it arrived in time. The war plan therefore envisaged the quickest possible run from Hawaii to the Philippines—the "through ticket to Manila."

To get to Manila, the fleet had to pass through the Mandates. It was widely suspected that the Japanese had violated mandate agreements forbidding fortification of the islands. If anything, this suspicion was reinforced by the Japanese themselves, who successfully prevented U.S. visits to the islands.[38] Even without fortifying them, the Japanese could base submarines, surface torpedo craft, and seaplanes there. This possibility transformed the War College simulations. Instead of concentrating simply on the hoped-for decisive battle against the Japanese battle fleet, the War College had to simulate the fight between the U.S. fleet (or a massive convoy) and Japanese forces based in the Mandates.

The college correctly credited the Japanese with a strategy of attrition to balance the odds for the decisive fight. The Washington and London naval treaties had left the United States with a 5:3 superiority over the Japanese in numbers of battleships and in total carrier tonnage. It was understood that in the ultimate battleship fight, the balance of forces would be proportional to the square of the numbers. The Japanese could use aircraft, which were not limited, to improve their situation. Thus, they could use unlimited aircraft based in the Mandates to make up for limited numbers of carriers.

A subtle consequence of the assumption of a fight in the Mandates was that the carrier-based aircraft defending a U.S. fleet had to be able to counter the highest-performance land-based aircraft. Had it not been for the Mandate problem, the U.S. Navy might have accepted instead that its aircraft only had to be good enough to deal with other carrier-based aircraft. Other navies, particularly the Royal Navy, accepted the idea that carrier operation in itself limited aircraft performance. Through much of the inter-war period, the British accepted that limitation on the theory that they would be fighting only carrier-based aircraft. That left them in considerable difficulty when they faced land-based Axis aircraft in the North Sea and the Mediterranean during World War II. They seem to have been rather fortunate in that they had access to U.S. naval aircraft whose designers had assumed from the outset that their fighters would face land-based attackers. It also turned out that Japanese carrier-based aircraft were quite competitive with land-based types.[39]

It took war gaming to force the fleet and the larger navy to understand the importance of the highest possible aircraft performance. The U.S. Navy was also fortunate that the particular scenario its war planners envisaged required the fleet to fight its way against large numbers of enemy land-based aircraft.

It was not universally agreed that the "through ticket" strategy would work, but the "big games" reflecting expectations in the Far East generally involved something like it.[40] It was by no means clear that Manila Bay and its defending positions on the Bataan peninsula and at Corregidor could hold out, but it seemed that it would take the Japanese time to occupy the whole of the islands. The assumed destination of the fleet was shifted well to the southern part of the archipelago, to Tawi Tawi or Malampaya Sound or Dumanquilas Bay.[41]

The 1933 "big game" (Operations IV) tested this idea. As usual, the Japanese mounted attrition attacks as the fleet steamed toward its southern Philippine haven. The southern Philippine harbors were entirely unprepared, with no facilities for repairs (Manila was not too much better, given the limited capacity of its one floating dry dock). The U.S.

fleet emerged from a battle still superior to the Japanese, but with most of its battleships damaged by torpedoes. These ships were certainly designed to withstand such damage without sinking, but they had to keep pumping to stay afloat. That would not have been a problem had they arrived at a fully equipped base with dry docks, hence with the ability to repair underwater damage. Van Auken pointed out what had to happen: As soon as the ships stopped pumping, they would settle onto the bottom. They could not be restored until something like a first-class base could be improvised.

Worse, the battle had not been conclusive. Japanese battleships also still remained afloat, albeit damaged. Unlike the U.S. force, the Japanese fleet had fully equipped shipyards within steaming distance. Unlike the U.S. fleet that limped into Dumanquilas, it could be restored to full capability. Neither side could complete new battleships in time to affect the balance of sea power. Japan would emerge superior in the Far East, as though the Orange fleet had won the big battle outright.

This conclusion was so important that War College President Rear Admiral Luke McNamee, Laning's successor, sent van Auken's analysis and a summary of the game to the Chief of Naval Operations. As far as the records of the War Plans Division show, this was the only war game material ever sent to the CNO. The entire package survives in the files of the War Plans Division.[42] The game carried two implications, both of them fatal to the "thrusting" concept. First, it showed that underwater damage would likely trump whatever the fleet could do in the western Pacific at the outset of a war. Ships would almost inevitably be torpedoed, and they would have to go somewhere other than the Philippines for repairs. If enough of them had to be repaired at a rear base, the fleet would no longer be superior to Orange's. Given the agreements barring fortification of Far Eastern bases, only Pearl Harbor could repair U.S. capital ships. Sending them back to Hawaii would forfeit the Far East to the Japanese. Yet the U.S. fleet had to retain superiority to win the engagement making the blockade—the end game against Japan—possible.

Second, in the Mandates the Japanese had exactly the resources—aircraft and submarines—they would need to inflict the type of damage that would send the U.S. fleet back to Pearl Harbor for repairs. Gaming showed that the Japanese would almost certainly be able to inflict sufficient underwater damage on enough ships. The U.S. war plan changed. The westward advance would have to be step by step. The United States would need an advanced base west of Hawaii with substantial repair facilities. Ultimately, that meant development of floating dry docks and other auxiliaries that could turn an empty lagoon into a major fleet base. Initially, it meant rapidly developing one or more permanent bases.

A step-by-step advance could also change the terms of the air threat to the fleet. Once an initial base had been seized, aircraft operating from it could strike Japanese air power deeper in the Mandates, reducing the air threat the fleet would face as it seized further islands. In effect this would reverse the expected Japanese plan to shuttle aircraft between islands so that they could be concentrated against the U.S. fleet wherever it penetrated the area Japan controlled. To a lesser extent, long-range U.S. aircraft based on one island could also, in theory, attack other Japanese assets in the Mandates, such as submarines.

From 1935 on, the War College's "big games" always entailed the seizure of island bases by one side or the other. Gaming material always included a special module involving such an operation. Specially assigned Marine Corps and Army officers conducted this phase of the exercise. In 1935, when this was first done, War College President Rear Admiral Edward C. Kalbfus told students that this time the "big game" did not reflect the war plan. That was probably to protect the security of the radical change, from the "through ticket to Manila" to a more gradual advance requiring the capture of islands. That year, Army Chief of Staff General MacArthur signed a new version of the joint war plan in which the relief of the Philippines was no longer contemplated.[43]

The 1935 strategic shift was seismic, with visible consequences. The Marines adopted amphibious assault as their primary future fleet role, and they sponsored development of specialized landing craft (ultimately the "Higgins boat" and the amphibious tractor). The Navy began to conduct Fleet Landing Exercises. Among other things, these taught the fleet how to conduct bombardments in direct support of landing Marines. It is difficult to avoid associating the changes with Operations IV of 1933, particularly since the OPNAV War Plans Division was using gaming as a means of testing elements of its plan. It does not appear to have had any other means of simulation.[44]

All of the Orange games showed how vital scouting would be. Game solutions always included scouting plans, generally employing long-range aircraft. The penalties of failed scouting were demonstrated. In the most spectacular case, the Japanese failed to find the U.S. fleet before it had reached the Far East, and the game showed their panicked reaction. Various alternative long-range scouts were tried. For example, in some games the fleet was provided with long-endurance dirigibles, sometimes equipped (as two U.S. dirigibles actually were) with fighters for self-defense. Long-range flying boats, notionally represented by the German Dornier Do-X, were also tried, and they were more satisfactory. By 1933, the Navy could buy a U.S. equivalent, which became the PBY Catalina. Unlike carrier aircraft, such long-range scouts were not limited by the arms control treaties. They could move across the Pacific, supported by tenders that could operate from unimproved lagoons.

It seems significant that by 1937 the fleet's patrol aircraft had been reorganized. Before that, they were assigned to the Base Force at Pearl Harbor. Mobility meant that they could move with the fleet to the Philippines, where they would be the fleet's scouts providing support prior to the big battle against the main Japanese fleet. In 1937, the patrol aircraft were made part of the Scouting Force. The building programs of the late 1930s included new seaplane tenders as well as ships described as seaplane carriers. There was also a program of small seaplane tenders (AVPs) designed specifically with draft shallow enough to enable them to enter small lagoons.[45] The big tenders made the seaplanes far more mobile. Given the step-by-step strategy, they could move with the fleet. That would not have been possible in the context of the "through ticket," because the big seaplanes could not be handled or launched or landed in the open sea. They needed temporary bases in sheltered water from which they could hopscotch with the fleet in the step-by-step advance across the Pacific. No one seems to have imagined that the U.S. Navy would gain the ability to construct air bases ashore as rapidly as it did during World War II. Games did not include any sort of rapid base construction.

Gaming explicitly avoided political considerations, except for an assumption that in some cases U.S. public opinion would demand offensive action after any lengthy period of consolidation. Nor did the Japanese players in any of the "big games" develop a Japanese end game that might cause the United States to abandon a war. They concentrated on resisting the U.S. strategic attack. This was realistic. Once Japan had seized the Philippines and gained a free hand on the Asian mainland, the best it could do would be an offensive-defensive, the goal of which was destruction of a U.S. fleet steaming west. The Japanese had to hope that the United States would quit at this point. The gaming system never took into account the possibility that the United States, its population enraged, would refuse to quit while creating a new fleet, as happened in World War II. Conversely, because it strenuously avoided special assumptions about the enemy, the war-gaming system never raised the possibility that the Japanese would refuse to come to terms when they saw that they had been defeated—as happened in 1945.

It is difficult to see how the U.S. naval war planners would have reached the conclusions they did without using simulation—gaming—to tease out the implications of their work. It was easy enough, for example, to say that the Japanese would try something to wear down the fleet as it passed through the Mandates, but gaming offered insight into just what that probably would be. The Navy's Fleet Problems were necessarily concentrated on the fleet-on-fleet battle that was expected to be fought after the U.S. fleet managed to reach the Far East. It was gaming that offered insights into the rest of the war, or at least into possible events prior to that big battle.

The alternative to gaming was to apply what would later be called military judgment. It was a composite of wartime and exercise experience. As simulation, gaming often offered results contrary to military judgment. They were not always at all acceptable within the wider Navy, and some of the War College's judgments were rejected outright in the late 1930s. Perhaps the fairest evaluation of military judgment versus gaming would be that in areas in which considerable full-scale experience had been accumulated, military judgment was much more likely to be accurate. Gaming offered insight into wars that had not yet been fought, involving new weapons—particularly aircraft. As we look back, the shift in War Plan Orange—a shift which proved extremely beneficial—was one of those areas.

Strategic gaming offered a kind of evidence about such warfare. It was particularly important as the U.S. Navy shifted its preferred war plan after 1933. Whatever the war planners might have said, it seems unlikely that their shift would have been accepted without the weight of that 1933 "big game" behind it. This was a major change, with large political implications. If the fleet did not rush immediately to the Philippines, their loss and the loss of their garrison would have to be accepted. The Philippines was the Army's important stake in the world outside the United States. The service could not easily accept loss of the islands, even in the context of a strategy designed to defeat Japan and thus force that country to disgorge the islands later on. Junior Army officers in the Philippines might be aware that the islands could not be held, but what counted in the mid-1930s was the view from Washington. That was where it mattered that the president of the War College provided OPNAV war planners with a solid forecast of what would happen if the fleet tried to cash in the "through ticket."

4. War Gaming and Carrier Aviation

Carrier aviation was both the greatest triumph and the greatest test of inter-war gaming. It was the most radical technology that had come out of World War I, the one for which previous experience was least relevant. War experience had certainly shown that aircraft were important and had great potential, but they had not been involved in battles. No one knew how they would perform in large-scale naval combat.[1] The triumph of gaming was both to show the U.S. Navy a way forward and to educate non-aviators in what aircraft could and should accomplish. During World War II, U.S. non-aviators such as Admiral Raymond F. Spruance and Admiral Frank Jack Fletcher (both War College graduates) successfully wielded carrier forces. Non-aviators in other navies do not seem to have done nearly so well.

In the immediate aftermath of the war, the closest the U.S. Navy could come to simulating future air-sea warfare was the gaming facility at Newport. The Navy was converting the slow collier *Langley* into an experimental prototype carrier (CV-1), but she could hardly be considered a viable fleet unit. No airplane landed onboard her at sea until October 1922, and she did not officially join the fleet until the end of 1924. She was classified as an experimental ship until 1926, and her initial air complement was a paltry eight aircraft. In 1919–21, there was active U.S. Navy interest in building a fast fleet carrier. What actually provided the U.S. Navy with such ships was an accident: The terms of the Washington Naval Treaty signed in 1922. As noted above, it allowed each signatory to convert up to two capital ships, which would otherwise be scrapped, into carriers. As it happened, the U.S. candidates were huge incomplete battlecruisers, *Lexington* and *Saratoga*. As a result, these first U.S. fleet carriers were large and capacious. Before they were completed, many thought them far too large, with too many aircraft to operate effectively. The Japanese converted two more or less equivalent ships (*Akagi* and *Kaga*); the British converted smaller and flimsier "large light cruisers" built during

The carrier *Ranger* (CV-4), shown here in April 1938, was conceived before the War College could formulate any concept of carrier characteristics based on war-game experience. When she was designed, it seemed clear that the main requirement was for the carrier force to generate maximum immediate striking power, which meant providing the maximum number of strike aircraft on deck for a first strike. Calculation showed that the best compromise between flight deck area (which set the number of aircraft that could warm up at the same time) and seakeeping within the limit set by the Washington Treaty was a relatively small ship with a flush flight deck. By the time *Ranger* was being completed, trials of the two big carriers *Lexington* and *Saratoga* had shown that it was best to provide an island for ship control; nothing could be done about the awkward arrangement of tilting funnels (shown here in the "up" position). Limited displacement allowed no protection and also favored a simple lightweight flight deck separate from the hull (the British and the Japanese used flight decks integral with their carriers' hulls). *Ranger* is shown in April 1938. (NHHC)

World War I (*Courageous* and *Glorious*). Until well after the two big U.S. ships entered service late in 1927, no one in the U.S. Navy had anything but War College games as a basis for deciding how they might be used. Of the foreign navies only the Royal Navy had any operational carrier force at all, and the U.S. Navy had no access to its experience. Americans were thus forced to debate the challenges and future of naval aviation without much empirical evidence.[2]

* * *

Guessing What Aircraft Could Do

When he returned to Newport as War College president in 1919, Admiral Sims was well aware of the need to understand aircraft and almost immediately ordered the preparation of new war game rules to reflect their potential.[3] By 1923, the aircraft rules were quite elaborate.[4] They indicate what the War College (and the fleet) thought that carrier aircraft would be able to do once the carriers entered service. As such, they were a basis for simulating naval war in that near-term future. Like other sections of the game rules, the rules for aircraft were constantly revised on the basis of feedback from the fleet and from other Navy experts. Asked to comment on the current rules in 1935, Chief of the Bureau of Aeronautics (and later CNO) Rear Admiral E. J. King pointed out that aircraft were the only naval weapon not yet really tested in war, so to some extent the rules reflected opinion as well as experience.[5]

The 1923 rules distinguished between torpedo bombers (VT), fighters (VF), Scouts (VS), patrol planes (VP), two- and three-seat observation planes (VO), and a projected submarine-borne one-man scout floatplane (VSS). Now that the United States was building carriers, most types could be produced as either land or seaplanes. If it could hit its target, a torpedo bomber could deliver a significant blow using either a 1,650-pound torpedo or a 1,500-pound bomb (or three 520-pound bombs). Fighters could carry 25-pound bombs, which were significant because they could wreck a carrier's flight deck. Aircraft speeds were considerably more impressive than in 1919: 95 knots for a torpedo bomber, 135 knots for a carrier fighter. It was assumed that carriers (and other ships) would stow their aircraft disassembled—it would take two hours to assemble a fighter and six to assemble a torpedo bomber (a carrier would have six assembly crews). From 1926 on, the rules embodied the actual characteristics of current U.S. naval aircraft, attributing the same performance to equivalent aircraft in other navies unless there was no U.S. equivalent.[6]

The game rules envisioned four different classes of carrier: a large offensive carrier (CVT), a small offensive carrier (CVO), a second-line carrier (OCV), and an auxiliary carrier (XOCV). *Lexington* and *Saratoga* were CVTs; *Langley* was an OCV. A CVT was an armored ship credited with a lifetime of twelve 14-inch hits, however they might be delivered. She might easily have her flight deck wrecked, but she would be difficult to sink. The CVO displaced about 23,000 tons (which happened to be one third of the tonnage available to the U.S. Navy after the two *Lexington*s were completed), with a speed of 27 knots and a life of ten 14-inch hits. An OCV was a converted merchant ship with a life of four and a half 14-inch hits. Finally, an XOCV was a large passenger steamer temporarily converted into a carrier.[7]

At this point the CVT was credited with 18 VF and 6 VT on her flight deck plus 18 VF and 30 VT in her hangar and 18 of each stowed in her hold as disassembled replacements. Other ships had proportionately weaker air complements. However, neither OCV nor XOCV carried any torpedoes. The rules assumed that half of the torpe-

The War College played a very different role when a new carrier was planned in 1930. War College arguments, based on game experience, led the Navy to reject the idea of building a small carrier specifically to work with the battleships. Instead, a heavy fleet carrier was built. War-game experience also showed that carriers often lost their aviation capacity to flight deck damage. The new carrier *Yorktown* (CV-5), shown on trials in May 1937, had several special features to keep her in action despite attacks. The light wooden flight deck (on a steel supporting structure) conceived to save weight in *Ranger* was retained because, unlike the usual heavy steel structure, it could be repaired rapidly by the ship's force. Arrester gear was installed at both ends of the flight deck so that the ship could recover aircraft even if one end of the deck was disabled. The roller blinds shown in the nearly down position over the forward part of the ship's hangar covered one end of an athwartships catapult, which could launch aircraft even if the flight deck was entirely destroyed. Flight deck survivability features kept *Yorktown*'s sister ship *Enterprise* (CV-6) in action well enough that she fought in every carrier battle of World War II. (NHHC NH 42341)

does aircraft dropped would run properly. The courses of those judged to run properly were plotted to see whether hits would be made. By this time, real warships were being attacked with bombs on an experimental basis, but it was difficult to say what the chance was that a bomb or torpedo would hit a moving, maneuvering warship. Estimated bomb-hitting probability declined as experience showed how difficult it was to target a maneuvering ship.[8]

Through the inter-war period, the most salient reality of bomb damage was that it would take very little to neutralize a carrier by wrecking her flight deck. According to the 1923 rules, two 100-pound bombs would destroy a carrier flight deck. Ten 25-pound bombs were credited with the same effect. The effectiveness of light bombs was important, because from 1926 onward the U.S. Navy practiced dive-bombing, initially using fighters limited to light bombs. This technique offered vastly better accuracy than conventional

The standard fleet carrier of World War II, the *Essex* class, was essentially an enlarged *Yorktown*. *Essex* (CV-9) herself is shown soon after completion. (National Archives 80-G-68097)

(level) bombing. No single bomb was expected to sink a carrier (as actually happened at Midway), but a carrier whose flight deck had been wrecked—a much easier goal—would be completely neutralized.

In combat, much would depend on how quickly aircraft could take off and land. The rules stated that a fighter could take off every 15 seconds, a torpedo plane or scout or observation plane every minute. The assumed elevator cycle was two minutes. An elevator could bring two fighters or one torpedo bomber to the flight deck at the same time. To recover aircraft, a carrier had to steam into the wind. She might need six minutes to steady on that course. Airplanes could land every two minutes (every four minutes at night). That was dangerous: The chance of wrecking an airplane while it landed was 5 percent (10 percent at night). Flying was dangerous in general: There was a 4 percent chance that an airplane would not make it back to the carrier at all.

The rules included considerably more detail, such as ranges at which aircraft could see other aircraft and ships, and rules for catapult aircraft and seaplanes. Rules were also given for air-to-air combat.

Looking back, the central questions for carriers were how many aircraft they could operate and how efficiently; how much of a threat those aircraft represented both to other aircraft and to enemy warships; and how vulnerable their carriers were to air attack (all other forms of attack were easily taken care of because there was considerable experience of them). Before 1928 and the completion of the two big carriers, all of this was guesswork. There was very little experience on which to base rules.

By 1925, the experimental carrier *Langley* was in commission and the designs of the two big fleet carriers were relatively mature. Operations with *Langley* provided initial experience in bringing airplanes up to the flight deck and launching them. The interval between arrival on the flight deck and take-off (to manhandle the airplane into place, start and test its engine, and make last-minute adjustments) might be four minutes for a fighter and five for something larger such as a VT. Once the airplanes on deck were ready, a fighter could fly off every 15 seconds and a torpedo bomber every two minutes. With proper preparations, a limited number of aircraft could be flown off twice as quickly. The maximum size of such a quick launch (under the 1925 rules) was nine fighters or seven VT. These figures explain why the U.S. Navy thought in terms of deck-load strikes, bringing a mass of aircraft onto the deck, starting them up together, and then launching them as quickly as possible. By this time, aircraft were considered sturdier, so the rules offered a better chance that a landing airplane would not be wrecked.[9]

When carrier characteristics were finally included in the pamphlets describing the fleets fighting in the games in 1926, they were standardized. Each carrier was assumed to have the five-squadron organization then planned for the two big U.S. carriers: two fighter squadrons (36 operating aircraft), two torpedo-scout squadrons (32 operating aircraft), and one observation squadron (12 operating aircraft), with further knocked-down aircraft in reserve (18, 16, and 6 respectively). A ship would also carry spares equivalent to another 25 percent of the total number of aircraft (120).

Gaming and Early Carriers

The first major impact of war gaming on U.S. naval aviation thinking came in 1924. The Washington Treaty allowed the U.S. Navy 69,000 tons of carriers beyond the two big carriers, assuming that the experimental *Langley* was scrapped or converted to non-carrier use. How should that tonnage be used? Led by Rear Admiral William A. Moffet, the Bureau of Aeronautics argued that the key issue was how to take the largest possible total number of airplanes to sea. It would take a carrier to accommodate the larger naval aircraft: the spotters, scouts, and torpedo bombers. However, although it would take a carrier to recover them, the smaller fighters could take off from short platforms onboard other types of ships. During World War I, the British had placed take-off platforms onboard capital ships (on turret tops) and cruisers. Moffett pointed out that similar platforms could be built onboard tankers and even destroyers. The new Washington Treaty

The special features of the U.S. fleet carriers were deduced from gaming—from simulated war— experience. Other navies took very different approaches. The Royal Navy adopted armored hangars (the armor did not cover the whole flight deck) on the theory that, when faced with air attack, carrier aircraft should shelter in the hangar (U.S. carriers were expected to mount an active defense using fighters). Flight deck armor impressed U.S. naval officers who inspected HMS *Illustrious* when she was repaired at Norfolk, but they apparently did not realize that she was spending a lengthy period in the U.S. yard because any bomb which actually did penetrate that armor would put the ship out of action for many months. U.S. carriers survived serious damage because they had considerable deck armor below their flight decks. On the other hand, British armored decks were certainly enough to keep out kamikazes, which badly damaged U.S. carriers by causing massive fires in their hangars. HMS *Victorious* is shown about 1941, an Albacore torpedo bomber landing on board. The Albacore, with its limited performance, seems to have been accepted because the Royal Navy did not expect to face high-performance enemy fighters. Again, war gaming forced the U.S. Navy to see things very differently. (NHHC NH 73690)

imposed no limits on the numbers of cruisers, destroyers, and tankers. Later, the turret-top flying-off platforms would give way to catapults, and they would typically launch scouts and spotters, but that was not yet the case in the early 1920s.

Gaming offered a way of envisaging a war involving carriers. In January–February 1924 the War College played the "Battle of Siargao" (actually the Surigao Strait) as Tactical Problem III.[10] It was a convoy problem. Blue had secured most of the Philippines and also held Guam as a forward base. The fleet in the Philippines badly needed supplies, including oil, carried by a convoy from Guam.

Each side had a carrier (with some fighters onboard) plus fighters on turret-top platforms onboard capital ships. Each of Blue's 19 tankers had fighters onboard: The Blue fighter force included 50 from the tankers and 12 from the six battleships escorting the convoy in addition to fighters onboard the carrier. Blue's cruisers carried only observation planes. Blue chose to place its carrier in the center of his formation for protection. This ship had to steam away at high speed when it launched aircraft. By way of contrast, Orange chose to keep its carrier well away from his main force after it launched aircraft. As a consequence, it was not seen by Blue scouts who spotted the rest of the Orange fleet. Blue used carrier aircraft as scouts. Initially, Orange used scouts that were launched by light cruisers.

Once they found the Blue fleet, the Orange scouts were to maneuver near the Blue carrier in hopes of drawing off the defending fighters. Orange could therefore hope that its bombers would reach the Blue carrier unopposed. Initially, the Blue fighters did chase the Orange scouts, but they noticed that they were unarmed. Recognizing that they were decoys, most Blue fighters did not chase them.[11] The Orange bombers had no trouble finding the convoy, as the weather was clear. However, the convoy was defended by numerous fighters, including those onboard the tankers and the capital ships.

The Orange attempt to keep its carrier from being seen was frustrated because Blue scouts saw the high-altitude fighter barrier that had been set up specifically to shield the Orange fleet and carrier. Worse, although the barrier protected the Orange fleet to some extent, Blue attackers could get at the carrier by flying around it.[12] The barrier was too far from the carrier to shield it, and it offered the rest of Orange's fleet only limited protection.

As usual in a game, the element of chance was represented. In this case, Orange's strike approached the convoy just as Blue's carrier headed into the wind to recover aircraft to refuel them. The convoy still had its own fighter cover provided by airplanes from the tankers. When Orange's bombers headed for the carrier, fighters waiting to land (to refuel) joined the defense. They shot down eight of 12 Orange attackers, and the others made no hits. Meanwhile, Orange launched a second attack (6 bombers covered by 18 fighters). It met 29 (of the original 32) Blue fighters, which were chasing the surviving 4 bombers of the first strike back to their carrier. That forced the Orange bombers back, but as soon as the Blue fighters flew back to their carrier the Orange strike re-formed.

Meanwhile, Blue was unable to find the Orange carrier. Faced with a possible air attack, Blue chose to orbit a ready strike force over its carrier while more airplanes were assembled on deck.[13] When an Orange cruiser group caused Blue too much trouble, eight of those bombers were sent out to attack. They were level bombers, so the targets largely frustrated their attacks by zig-zagging. Even so, two hits sank one cruiser and another inflicted 50 percent damage on a second. Blue submarines working with the convoy sank another light cruiser.

The re-formed Orange strike was spotted by outlying Blue ships, giving the force time to launch fighters. When the attack arrived, the carrier launched 18 fighters, which joined 22 more waiting to land. The tankers contributed another ten and the Blue battleships another four. This large fighter force drove off the attackers, who did not manage to damage the Blue carrier. Two of the five Orange bombers that got near the Blue carrier were shot down.

War gaming showed again and again that the U.S. Navy could never have enough carriers to fight a Pacific War. Through the mid-1930s, the solution was to convert fast liners into carriers. The projected conversions were elaborate, and they seemed to require much more time than would be available. They may also have been dropped as plans shifted toward a step-by-step advance through the Mandated Islands held by the Japanese. Such an advance would require numerous amphibious assaults and troop transport capacity would become much more valuable. Once war broke out, the need for numerous carriers re-asserted itself. The main solution was a series of simple conversions of existing merchant or auxiliary hulls (later some escort carriers were built as such from the keel up, but their hulls were essentially those of merchant ships). The escort carrier *Chenango* (CVE-28), shown in 1944, was built on a tanker hull. (NHHC NH 95703)

Blue finally found the Orange carrier the next day.[14] Blue had already launched two strikes against the Orange cruisers. Before a third could be assembled to attack the Orange carrier, Orange launched a third strike against the Blue carrier. It sighted the Blue force about the same time that a Blue battleship sighted Orange. Blue had 41 fighters overhead, including aircraft waiting to land to refuel. They sufficed to drive off the Orange attack, shooting down two bombers and protecting the carrier from any hits. However, many of the fighters were low on fuel. They could not land back onboard the Blue carrier quickly enough. Seventeen of them had to ditch (their pilots were saved).

During the day, the Orange commander realized that he could not prevent the Blue convoy from reaching Surigao. He was reduced to bombing it and also trying to drive off Blue bombers holding down his submarines. He had to give up any hope of attaining air superiority by attacking the Blue carrier. The attack on the convoy was more successful,

To meet a perceived emergency need for fast fleet carriers, nine light cruiser hulls were converted on much the same lines as the escort carriers (a proposal for a more elaborate conversion was rejected). The first of the class, *Independence* (CV-22), is shown in San Francisco Bay, 15 July 1943. Comparison with *Essex* gives some idea of just how small her flight deck was. This was very much not a concept developed by the War College, but it can be traced back to the War College's conclusion from gaming: Carriers were essential, in large numbers. (National Archives 80-G-74433)

because the convoy had very little antiaircraft protection, and fighters assigned to protect the convoy had gone off chasing Orange scouts.

Blue now launched an attack on the Orange carrier. It was not seen until it was nearly overhead. Orange fighters already in the air attacked, but were not successful; Blue planes managed to drop 66 bombs, of which eight hit. They demolished the ship's flight deck. Under the rules then in force, such damage could not be repaired outside a shipyard. Blue now had air superiority. Moreover, Orange aircraft in the air when the carrier was struck had to ditch.

Much of Blue's success was attributed to the care the Blue air commander took to conserve his airplanes by arranging to refuel them constantly. Unfortunately, only the carrier could refuel these aircraft. Because it took so long for them to land on the carrier, Blue lost a significant number of fighters, which had to ditch. Thus, the game emphasized the need to be able to recover aircraft very quickly.

In his analysis, Captain Laning, the head of the Tactics Department, pointed out that aircraft could not easily be replaced during a campaign in the western Pacific. Heavy losses were unacceptable. Conversely, to be viable late in a campaign, a fleet steaming west from Hawaii would need to take very large numbers of aircraft with it.

In retrospect this was a remarkable game. It was not conceived as a test of carrier aviation. At the outset, Orange's main tactical plan was to mount a night torpedo attack on the convoy. The players were not aviation fanatics; most likely none of them was even a pilot. The players' reasoning led them to see that air superiority was key to anything else they might want to do. Both commanders in the game sought it; the carriers became primary targets. Only in a game could the large numbers of aircraft that fought the battle have existed, at least at the time. Captain Laning was a surface officer without any aviation experience. The Blue air commander, whose efforts he considered excellent, was almost certainly Captain Joseph M. Reeves Jr.[15] He would soon graduate and be chosen as Laning's successor as head of Tactics.

Surviving game material shows that, as head of the Tactics Department, Reeves received another lesson in carrier operations while at the Naval War College. In Tactical Problem II of the Class of 1925, two large U.S. carriers faced six much smaller British ones, total air strength on both sides being about equal.[16] Both of the big U.S. carriers had their flight decks destroyed while only four of the six British carriers had their flight operations similarly put out of action. The other two would have enforced British control of the air (although that proved more difficult than many had imagined), with devastating consequences for the U.S. fleet.

In game after game, the lesson was that carriers had to achieve as much as possible with their initial strikes, because they might be unable to launch further ones. At the very least, the fleet with fewer carriers had to strike all of its enemy's carriers. To do that, it needed as many aircraft as possible. Much the same could be said of any attempt by the carrier's own fighters to defend her. Overall, games were ambiguous as to whether fighter defense was likely to be effective.

At this time, the U.S. Navy had a single experimental carrier, *Langley*. *Lexington* and *Saratoga* were still being completed (they would be commissioned late in 1927). Assuming that the *Langley* would be replaced, the United States could build three 23,000-ton carriers or five 13,800-tonners, or some combination of these ships, none of them displacing more than 27,000 tons.[17]

There was no limit on carriers displacing less than 10,000 tons. The Bureau of Construction and Repair prepared sketch designs of all possible carriers. It pointed out that although larger ships would find it easier to handle more aircraft, a smaller carrier had about 15 percent more deck area per airplane. The total number of aircraft a carrier could accommodate would depend on deck area (hangar and flight deck). A larger number of smaller carriers would offer more total deck area. The bureau estimated that five 13,800-tonners could accommodate 6 percent more airplanes than three 23,000-tonners. It rejected the small carrier as useless for anything larger than a fighter, hence not worth pursuing.

In 1925, Secretary of the Navy Curtis Wilbur asked the General Board whether it would make sense to use the hulls of some of the recently authorized 10,000-ton cruisers to build small carriers not limited by the Washington Treaty instead. The Naval War College provided its own analysis, based on tactical games.[18] The games showed that carriers were generally damaged by bombs dropped on their flight decks; it took only a few hits to wreck a flight deck and eliminate a ship as a carrier. This type of damage could be repaired only by a shipyard. Control of the air "has nearly always been obtained by destroying the flying-on decks of enemy carriers." This was more or less obvious, but the lesson of the games was not: It took no greater effect to disable a large carrier than a small one. Numbers were vital.

The War College President quoted the result of Tactical Problem II. Blue did about twice as much offensive bombing as Red, so it knocked out twice as many carriers. Unfortunately, that was only four of the six Red carriers—and Red knocked out both Blue carriers. Once Red had control of the air, its spotting aircraft represented a crucial advantage. Using spotting from the air, Red guns could hit Blue at ranges Blue could not match without deploying its own air spotters. When carriers were dispersed singly, attacking each one required the same search effort, whether the target was large or small. It also seemed that each carrier, large or small, could launch aircraft at about the same rate. A larger number of smaller carriers could place more aircraft in the air more quickly. That was likely to be a considerable tactical advantage. Against that, for the greater number of aircraft to concentrate, the carriers launching them had to operate close together. In that case a successful enemy search would locate all of them, and they could all be attacked by the same enemy force.

Which was better, then—concentration or dispersion? Concentration increased the number of aircraft that could work together. Dispersion made it more difficult for an enemy to find all the carriers. This question was important through the whole inter-war period and into World War II. At the Battle of Midway in 1942, the U.S. carriers were dispersed to some extent. The Japanese carriers were concentrated to focus their striking power. To some extent that concentration was their downfall, as the dispersed U.S. carriers managed to strike three of the four Japanese carriers in very quick succession.

In 1925, it seemed that the most important aircraft role in a major battle would be to enhance the gunnery of the battleships inflicting serious damage on the enemy's battleships. That meant seizing and using control of the air above the battleship fight. The side that controlled the air could spot for its battleships, gaining them maximum range. It seemed that small carriers could provide both the fighters maintaining air control and the spotters exploiting it. It also appeared that small carriers could operate with the battle line while the two new carriers *Lexington* and *Saratoga* mounted long-range strikes, operating far from the battleships. Smaller carriers operating directly with the battleships

War gaming certainly affected the U.S. Navy's idea of what it needed in its carrier aircraft. War games depicted campaigns in which carrier-based U.S. fighters would face land-based Japanese bombers; the fighters had to have the highest possible performance despite the limits a carrier might impose. The Vought Corsair, which won the 1938 fighter competition held by the Navy's Bureau of Aeronautics, had the highest performance in the world (in terms of speed) when it flew in 1940. An F4U-1A flown by the Navy's then highest-scoring ace, LTJG Ira Kepford of VF-17, is shown over the Solomons in March 1944. (National Archives 80-G-217819)

would be easier for an enemy to find and destroy, but no such success would deny the U.S. fleet all its airpower, because the two big carriers would still be available.

The War College was not, of course, equipped to say whether the small carrier was technically feasible. For the time being it was dead because the ship designers of the Bureau of Construction and Repair rejected the concept altogether. When the General Board submitted a five-year building program in March 1926, it favored the 23,000-tonner.[19] The board subsequently reversed its position based on an April 1927 Naval War College submission emphasizing the need for a maximum number of aircraft.[20]

The CNO asked the War College to comment on the General Board plan, based on game experience. In reviewing past experience, War College President Rear Admiral

William V. Pratt decided to include the results of the just-conducted Operations Problem II He also took into account a new study of how deck handling would affect the outcome of a carrier battle. Carrier deck handling was simulated using model flight and hangar decks and cardboard airplanes. The detailed analysis was carried out by then-Lieutenant Forrest Sherman, a future CNO. Pratt wrote that "in view of the present quiescent status of the carrier question in the United States Navy," the delay involved in completing this detailed study seemed warranted.

How the U.S. Navy should allocate its carrier tonnage was controversial. Of five officers at the War College who were queried, two favored a large number of small carriers, one a small number of large ones, and two a medium size (20,000 to 23,000 tons). Clearly the large carrier (like the *Lexington*) offered convenience and coordination of air operations, but no such ships could be built under the treaty rules. The intermediate size offered similar advantages. Sherman wrote that it "seems to be a crystallization of British experience." Pratt observed, however, that the smaller carrier offered important advantages: greater scouting area (each carrier could scout a given area, so more smaller ones could scout a larger area); better overall security (due to more ships, as had been argued before); mutual support increased by operating in pairs; and, not least, more available aircraft. The last point could not be emphasized too strongly. Pratt wrote that

> ...one of the outstanding lessons of the overseas problems played each year is that to advance into a hostile zone the fleet must carry with it an air force that will insure *beyond a doubt* [emphasis in the original] command of the air.... [T]he only way to secure it is by large numbers of carriers and planes. This means not only superiority to enemy fleet aircraft, but also to his fleet and shore-based aircraft combined. This is a large order, and the only way to secure it is by a large number of carriers and planes. If small carriers will put more planes in the air than the same tonnage of large ones, we must have the smaller carriers, regardless of the small extra cost per plane put in the air. This means quantity production, to which a smaller type is much better adapted than a large one. Our shipyards (such remnants as we have left) and our national engineering bent favor quantity production.... Too, the smaller the carrier the quicker it could be built: we may need to build carriers in a hurry.

Pratt added that seaplanes would be extremely useful in a Pacific campaign, a recommendation that the Navy certainly heeded later.

Pratt reported that the games showed that no allotment of carrier tonnage within the available tonnage would suffice to gain and retain air control. It would be unwise to experiment with that tonnage, so he supposed that three 23,000-tonners might be best. He hoped that numbers could be made up out of the unlimited 10,000-ton (or smaller) class.

War gamers struggled to evaluate the dive bomber, the great attack development of the inter-war period. It offered extraordinary accuracy; the question was how that would be affected by anti-aircraft fire and other combat factors. The other important issue was lethality. The war game rules accepted that dive bombers could severely damage or sink anything short of a battleship, but that typical battleship deck armor would defeat its weapon. Thus, the carrier's best weapon could affect the outcome of a surface battle—which it was assumed would be decisive—by stripping the enemy's battle line of all auxiliary ships, such as destroyers and cruisers. Since torpedoes delivered by destroyers certainly could sink any battleship, this sort of action could well be decisive in itself. War experience showed that this was a reasonable evaluation of dive bombing; it took torpedoes to sink large battleships. Here, SBD-3 Dauntless dive bombers of VB-6 prepare for takeoff from *Enterprise* for the Wake Island raid of 24 February 1942. Bombs are not very visible because they were semi-recessed under the fuselage. (National Archives 80-G-66037)

The games showed that speed was essential for a carrier. Whether a carrier could be as fast as a cruiser depended on delicate design issues. Games did show that even when operating with the fleet, carriers soon became detached as they turned repeatedly into the wind to launch and recover aircraft. In this case they needed high speed to regain the protection offered by the battle line. A carrier should be able to maintain air patrols that would enable her to evade surface attack, and to send up fighters to counter a developing air attack.

Gaming experience already showed the shape of a Pacific war in which the U.S. fleet protected a necessarily massive convoy steaming west. Pratt wrote that

>...we were more often forced to expose a vital bombing focus than was the enemy and it was usually easier to locate a vulnerable bombing area in our floating forces than it was in those of the enemy, for he had the power of a wide initiative and was less circumscribed in his movements. Further due to the great sea efforts we were forced to undertake, the quantity of targets vulnerable to bombing attacks was much greater in our case than was the case of the enemy. By proper strategical and tactical dispositions the enemy was enabled to dispose his forces so that a bombing attack delivered by us need not usually fall upon a force massed, a contingency we could not avoid. Also it was usually found that since the enemy was initially inferior in gun power he was forced to develop his bombing tactics in order to neutralize this gun preponderance as much as possible; that is, that the bombing attacks assumed much of the character of the [torpedo] attacks delivered by destroyers, namely, as an attempt to equalize the fighting strength (measured in gun fire) of the two fleets, and to destroy that very vulnerable portion of the fleet, the convoy.

Therefore, the enemy air objectives were usually first the U.S. carriers, then the battleships, and finally the convoy. Given limited U.S. air resources, the main U.S. concerns were to protect the carriers' facilities and U.S. fighting strength (which included the aircraft needed to control gunfire); protection of the convoy was relegated to third place. If these requirements could be met, the U.S. force could counterattack. If U.S. information was good enough, the attack could be made first. However, the enemy was usually able to take the initiative and to avoid great vulnerability to air attack.

All of these considerations led to concern to protect the fleet during the air battle that usually preceded a gun battle, conserving those aircraft that would be needed during the latter. Hence, the fleet tended to hold bombing power in reserve until the enemy massed and made itself vulnerable, thus augmenting the gun superiority of the U.S. fleet. Not all games went this way, but Pratt thought that a review of campaign games would show that he was right.

Pratt concluded that his light mass-production carriers should be equipped mainly with fighters, the larger carriers accommodating mainly bombers. Observation and patrol aircraft should be limited to non-carriers (cruisers and battleships). If possible, destroyers and submarines should be equipped with observation planes (this included the gunfire-spotting role).

A key point was that a carrier's arresting gear could handle one airplane about every two minutes, which was already faster than had previously been possible.[21] If airplanes landed one after the other, and they had a two-hour endurance, a carrier could accommodate and operate 60. However, in most cases an airplane's wings had to be folded before it could be struck below to clear the deck for the next landing. More time was needed to

In war gaming, the chances of torpedo bombers were problematic. In 1942, U.S. aerial torpedoes had to be released near their targets at relatively low altitude and at low speed, making the bombers very vulnerable and allowing the targets a good chance of evasion—as happened at Midway. The situation changed radically later in the war, as torpedoes were modified so that they could be dropped at high speed from a much greater altitude. This TBD-1 Devastator from *Enterprise* is shown dropping its torpedo on 26 October 1941. War gaming showed the consequences of the reality that the U.S. fleet had to deal not only with the Japanese fleet, but also with Japanese aircraft (and, probably, submarines and surface torpedo craft) based on the Mandated Islands. The fleet therefore needed heavy bombers as well as anti-ship aircraft, and its torpedo bombers had a dual role. Devastators performed well as bombers during the early carrier raids against the Mandated Islands, but they were wiped out at Midway. (National Archives 80-G-19229)

maneuver an airplane after it had been stopped on deck. Some of that crucial time could be saved if the airplane did not have to go below to be serviced and rearmed. That implied that the flight deck had to be large enough for airplanes to be parked while the ship launched and recovered aircraft. Gaming also showed the need, which Sherman pointed out, for a new type of flight deck that could quickly be repaired after damage from light bombs.[22]

Summarizing carrier-design considerations for Admiral Pratt, Sherman looked to British experience. The British had reconverted their 10,000-ton carrier *Vindictive* to a cruiser. Sherman saw this as proof that 10,000 tons was too small. Sherman interpreted the British choice to convert the 20,000-ton *Courageous* and *Glorious* to carriers as evidence that this was about the right size for a future U.S. carrier.[23] Apparently, he did not

For a time in the 1930s, it seemed that big flying boats like this PBY-4 could be effective bombers, with performance similar to that of land-based types. Many war games featured mass attacks by such aircraft, which were attractive because they could move forward with the fleet, based on mobile tenders. The PBY was given its bomber (B) designation when it was fitted with the Norden precision bombsight in recognition of its high performance (including extraordinary endurance). By World War II, the PBY was no longer competitive with land-based bombers, but higher-performance seaplanes were about to enter service. This PBY-4 is shown dropping a heavy bomb, probably before the war. (National Archives 80-G-10550)

realize that these ships were the only viable British candidates for conversion under the Washington Treaty. The British had no equivalents of the big but incomplete U.S. and Japanese capital ships to convert. Sherman argued that a small number of larger ships enjoyed important advantages. They would absorb fewer escorts, they would not require so much communication to mount a large air attack, and, most importantly, they could more easily coordinate their air efforts. He recognized the argument for dispersion, but thought that three 23,000-tonners plus the two huge carriers offered enough.

For Pratt, the key argument favoring the 23,000-tonner was that it could be supplemented by a large number of 10,000-ton carriers. However, by this time the Bureau of Construction and Repair had decided that no such ship was practicable. In this light, Pratt's advocacy of numbers translated into a preference for the 13,800-tonner. It seems to have been particularly significant that such a ship would not carry too many fewer aircraft than a 23,000-tonner—aircraft capacity was proportional to the areas of the flight and hangar decks, which would not decrease very rapidly with displacement.[24] This seems to have been the basis for a reversal of policy by the General Board. In September 1927, its building program included five 13,800-tonners rather than three 23,000-tonners.

In its submission to the Secretary of the Navy, the General Board echoed the War College's 1925 reasoning: slightly greater aircraft capacity, but much greater air effectiveness. Each carrier could launch the same number of aircraft at one time, so more carriers meant putting many more aircraft into the air more rapidly. The larger number of smaller carriers could also recover their aircraft much more quickly, and thus could sustain an air effort for longer. Too, the loss of any one carrier would be far less devastating to the overall air effort.[25]

The new carrier became *Ranger* (CV-4). She embodied Pratt's thinking: Aircraft numbers mattered more than anything else. The ship carried the maximum number of aircraft at minimum cost. As an indication that aircraft numbers trumped other considerations, the designers traded off power (hence speed) for more aircraft. Instead of the 33 knots of the converted battlecruisers, *Ranger* was rated at only 29.6 knots. Moreover, in order to squeeze in more aircraft, *Ranger* was not armored, although she did have torpedo protection. The logic was simple: In games, and therefore probably in reality, carriers would be able to evade surface attackers. Their main enemies would be fast airplanes and invisible submarines. It was impossible to provide enough armor to protect the flight deck against bombing; more airplanes were a better bargain. To provide maximum stowage space, the constructors followed Sherman's advice, and used nearly the whole of the upper deck as a hangar. Above it they erected a light superstructure topped by a flight deck. Accounts of the design typically justified the lightness of this structure as a way to minimize top weight. However, but it also satisfied Sherman's requirement that it be quick and easy to repair. This type of structure was repeated in all U.S. carriers that fought in World War II. Neither of the other carrier navies, the British or the Japanese, adopted this type of construction. Neither seems to have contemplated the sort of drawn-out air-sea war that made it desirable.

Just after the General Board made its recommendation, Captain F. J. Horne (later to become the first CNO in World War II), acting director of the War Plans Division, summarized the points that seemed most important to him. Although he did not cite war gaming, his organization used war gaming as its laboratory.[26] Horne's memo reiterated points Pratt had made, but in more striking form. It argued that a carrier, no matter what her size, could advantageously operate only a limited number of aircraft. Although a larger carrier could accommodate more aircraft, she could not operate them very efficiently. Efficient operation demanded the maximum number of smaller carriers. They could launch more aircraft in a given time, and they could also recover more of them. On the other hand, it was easier to build a large fast carrier than a small fast carrier, because machinery would take up less of the available space in the larger ship. War Plans pointed

Seaplanes were attractive as part of the island-by-island strategy developed after the War College showed (by gaming) that the earlier straight advance across the Pacific would not succeed. They could relocate quickly, their tenders moving with or after the fleet. Once World War II began, the U.S. Navy supplemented its specially designed tenders with converted merchant ships. Here tender *Currituck* (AV-7) lies alongside the converted merchant ship *Tangier* (AV-8) at Morotai in October 1944 to handle seaplanes supporting the Leyte Gulf operation. The aircraft are small OS2U Kingfishers, but both ships had enough space aft to take on board full-size seaplanes. (National Archives 80-G-1022364)

to the importance of the other air-capable ships in the fleet, those with catapults. The key role of spotters in a battle-line engagement demanded that a carrier be included in or near the battle line specifically to refuel spotters.

The episode is remarkable. *Ranger* was not laid down until 1931, but she was conceived and designed well before *Lexington* and *Saratoga* had been completed.[27] The reasoning leading to the design was based on war gaming—on the only way of visualizing the entire campaign the U.S. Navy contemplated. It took war gaming to convince Admiral Pratt that the United States could never have enough carriers to fight an Orange war, hence that something new was needed. That might be converted liners or small mass-produced fleet carriers—or a better way of operating carriers themselves.

Above all, the campaign orientation offered by gaming showed that any Pacific war had to involve huge numbers of aircraft. Wastage would be rapid. Ultimately that meant that the U.S. Navy needed large reserves of aircraft and pilots. The two other carrier navies seem not to have reached any such conclusion. In the case of the British, it might be claimed that competition with the Royal Air Force for scarce resources made it impossible to build up reserves.[28] This argument does not, however, apply to the Imperial Japanese Navy, which did value its air arm—but never built up the reserves needed to fight a protracted war.

After leaving the college, Admiral Pratt headed the U.S. delegation to the 1929 London Naval Conference. Pratt was well aware that naval arms control was grossly unpopular in the Navy. When he returned from London, he argued that he had extracted an important concession. It was a clause that allowed the construction of a limited number of cruisers with flight decks, which Pratt saw as a way of adding more fleet aircraft.[29] Flight-deck cruisers became a fixture in war games and were retained in the building program as possible new type of ship until Pratt was relieved as CNO (see Chapter 5).

The other possibility, the converted liner, also became a fixture in U.S. war planning. Such ships had figured in war gaming as early as 1923, but it seems likely that in their earliest incarnation they were inspired by the British *Argus*, converted in 1916–18 from an incomplete Italian liner. By 1929, the converted carrier (XCV) figured in lists of ships to be taken up from the U.S. merchant fleet upon mobilization. The Bureau of Construction and Repair produced conversion drawings, and ships were earmarked. The concept died because there were too few suitable fast ships in the U.S. merchant fleet and also because it seemed that conversion would take far too long. These ideas prefigured the escort carriers of World War II, but those were far more austere and conversion was far quicker.

Reeves and Operating Practices

While all this was going on, another gaming-related development was affecting the way in which U.S. carriers would operate. Captain Reeves, who had done so well in the 1924 game, left the War College in the summer of 1925. His reactions to the two games described here are not recorded, but what he soon did suggests that he digested their lessons. At the War College, Reeves had shown no special interest in aviation; his student thesis concerned battle-line gun tactics.[30] He then went to Pensacola to take the aviation observers course, which qualified him to command an aviation unit. Given that credential, in September 1925 he was appointed Commander, Aircraft Squadrons, Battle Force, a billet from which he could put his Naval War College experience into practice. In later years, many U.S. naval officers with non-aviation backgrounds took the course at Pensacola specifically because it opened important naval air commands to them. That Reeves took the course before there was any such rule suggests that his college experience—essentially gaming—had convinced him of the future importance of naval aviation. When Reeves arrived in the battle force, his command amounted mainly to the aircraft onboard the new *Langley*. His main assignment was to develop the tactics

The inter-war Naval War College missed the potential for quick development of air and other bases on seized islands. The wartime Navy included construction battalions (SeaBees) who could rapidly create substantial airfields. As a consequence, the fleet was able to adopt high-performance land-based patrol bombers like these Privateers (PB4Y-2s) from VP-23, shown over Miami in July 1949. The Navy retained seaplanes until 1965; it valued their inherent mobility, gained by using tenders. (National Archives 80-G-440193)

they should employ. At the least, that meant air tactics. Reeves saw something more. He would be responsible for the tactics of his force, meaning the way it was launched, the way it was recovered, and the way it was used against an enemy. He had, in effect, a deadline. His new tactics had to be ready by the time really large numbers of aircraft reached the fleet onboard the two huge carriers then being completed.

For Reeves, the lesson of the war games was obvious: *Langley* had to be able to take aircraft onboard much more rapidly.[31] That would also allow her to operate many more aircraft. The two-minute landing interval enshrined in the war game rules was unacceptable. Reeves was anxious to get the most out of his new command, so he formulated a questionnaire for his pilots: the "Thousand and One Questions." Among many other things, he wanted to know how many aircraft his ship could support beyond the eight she was assigned. Reeves's war college experience showed that much larger numbers were needed.

Quite aside from the need for numbers to make carrier aircraft tactically meaningful, Reeves needed them to devise the tactics and strategy the very large numbers planned for the big carriers would need. Early in November 1925, he assembled his pilots and crew in an auditorium at the naval air station at North Island, San Diego.[32] He told them that he planned to increase the air complement of their ship from eight to fourteen and then to more. To do that, he had to increase the tempo of takeoffs and landings. Reeves's chief of staff later wrote that the pilots had learned that it was dangerous enough to land on the small flight deck. They resented having a non-pilot tell them to take even worse risks. Making takeoffs more rapid was not difficult, but cutting landing intervals was.

When Reeves took over, *Langley* typically recovered an airplane and then struck it below before recovering another. The great innovation that made it possible to shrink the landing interval was to push the airplane forward into a deck park instead of taking it below into the hangar. Aircraft parked forward were protected by a wire barrier, a key innovation.[33] Arresting gear stopped a landing aircraft short of the barrier. It took far less time to stop an airplane and roll it forward than to winch it down into the carrier's primitive hangar. Once all the aircraft were recovered, they could be rolled back to the after end of the flight deck and set up for another flight. The new technique was not necessarily popular with pilots; it required them to land into a space closed off by the barrier, rather than onto an open runway-deck. It required a much greater degree of control of landings, hence the institution of landing signal officers and, ultimately, a very fast tempo of landings and takeoffs. All of these seem so natural now that it is easy to forget that during the inter-war period they were unique to the U.S. Navy.

To an extent the pilots were right: The new technique was much more dangerous than the past one. Anyone watching movies of U.S. carrier operations well into the jet age will remember numerous clips of real accidents, many of them fatal. By way of comparison, the Royal Navy had no landing signal officers; pilots landed using their own judgment, just as they did ashore. An account of inter-war British carrier aviation was titled *It's Really Quite Safe*. No account of U.S. inter-war carrier operations would have had such a title. It took someone like Reeves to understand that the increased risk was worth the gain in combat power. It took gaming to inspire him in the first place.

In December 1925, it took *Langley* 35 minutes to recover and stow ten airplanes without a barrier—and that was considered an excellent performance. *Langley* had a barrier by June 1926. In August 1926, recovery time per airplane was 90 seconds, and Reeves was sure it could be improved much further.[34] He recommended that *Langley* be reclassified as a full combatant rather than an experimental carrier, and he expected to double her air group to 28 aircraft. Reeves's figures were presumably based primarily on airplane endurance versus recovery time—at most a carrier could operate only as many aircraft as she could recover before they ran out of fuel. The idea of using the flight deck for servicing aircraft (as in Lieutenant Forrest Sherman's December 1926 memo on the new carrier) was a natural consequence of the new landing technique.

Reeves's barrier made it possible to stow aircraft permanently on the flight deck, not just on the hangar deck, because aircraft could land while the former remained in place. That must have been particularly important for the small *Langley*, which had a tiny hangar and very slow elevators. For the future, it offered the U.S. Navy many more aircraft per ship than foreign navies, which counted hangar capacity as ship capacity.[35] Reeves's barrier and the consequent innovation of the deck park were why three U.S. carriers at Midway had onboard about as many aircraft as four Japanese carriers. In this battle, the U.S. fleet was outnumbered in the number of carriers, but not in the number of its aircraft.

Reeves's innovation—born of game experience at the Naval War College—had a subtler consequence. For most of the inter-war period, the Royal Navy had more carrier tonnage than the U.S. Navy. However, the U.S. Navy of the inter-war period operated many more carrier aircraft. That made for a much larger and livelier market in naval aircraft. In Britain, the market for naval aircraft was tiny compared to the market for land-based aircraft for the Royal Air Force. The Royal Navy could not demand the investment that would have been needed to produce really high-performance carrier aircraft. The much larger U.S. market encouraged manufacturers to innovate in order to meet demanding naval requirements. The effect of that larger market is apparent in the high performance achieved by operational U.S. naval aircraft before and during World War II.

It was already well understood that very large numbers of aircraft were needed. The lesson of the 1924 war game was that an air effort could not be sustained unless aircraft could be recovered and serviced quickly. Aircraft could already be launched in quick succession. It was no surprise that a carrier could already launch a substantial strike. Doing so more than once demanded quick landing, too. That was also important for a carrier's security. A carrier had to steam steadily into the wind to recover her aircraft. The longer that took, the better the chance that she would be caught by a submarine. If the carrier did operate with the battleships, the longer she steamed into the wind, if that took her away from the battleships, the farther she got from their protection.

Reeves's innovations were reflected in the 1929 rules. At least a quarter of a carrier's planes would normally ride the flight deck, and half could be accommodated there "in such a manner that either flying off or flying on, but not both simultaneously, can be carried on." A fighter could be launched every ten rather than every 15 seconds, and a torpedo bomber every 15 seconds rather than every two minutes, a vast difference. Crucially, it was no longer necessary that the flight deck be cleared before an airplane could be allowed to land. After they landed airplanes were moved forward. The 1933 rules cut the day landing interval to half a minute.

The rules changed in 1936 to allow for launch *and* recovery with aircraft on deck. All aircraft could be parked on deck, but in that case there was only enough space for aircraft to take off, but not to land. With three quarters of the aircraft on deck, airplanes could take off (if the aircraft were all aft) or land (with all of them forward), but not both simultaneously. However, with half the airplanes on deck, presumably all amidships, aircraft could take off and land simultaneously. No more than half the aircraft could be in the hangar at any one time. Such practices differed completely from those in other carrier navies. Launch interval was given as ten seconds, and landing interval as 30 (intervals would double in darkness). These figures corresponded to current U.S. practice.

Putting It Together—the *Yorktown* Class

Ranger having been authorized in the FY29 program, the question for 1930 was whether she should be repeated. It was assumed that an additional carrier would soon be built. The U.S. Navy still had 55,200 tons left in the carrier category, and the new London Naval Treaty did not change that. How should that be allocated? There was still very little operational experience to inform any such decision; gaming still offered the best guidance. *Lexington* and *Saratoga* became fully operational only in the fall of 1928, just before Fleet Problem IX. That game, in which *Saratoga* successfully mounted a surprise attack against the Panama Canal despite the numerical superiority of the defending fighters, was the first demonstration of what a fast carrier operating more or less independently could do.[36] *Ranger* had sacrificed speed and protection for aircraft capacity. The spectacular *Saratoga* operation suggested that this had been the wrong choice. If a carrier needed the highest possible speed, comparable to that of the cruisers working with her, then cutting size would also entail a considerable sacrifice in aircraft capacity. However, it could be argued that the *Saratoga* spectacular was only a single exercise. The War College had been running carrier battles for several years.

It was generally accepted, as during Admiral Pratt's tenure at the War College, that the U.S. Navy would need every seaborne airplane it could get. Both enemies in the games (Japan and the United Kingdom) had large numbers of land-based aircraft unlimited by treaty. Game after game showed not only how important aircraft were, but also how rapidly they were shot down. Current War College President Rear Admiral Harris Laning made this point, for example, when he provided the General Board with the War College's version of the composition of a balanced treaty navy in December 1931.

Ranger was not so much an ideal carrier as the best that could be done to maximize the number of aircraft at sea. Admiral Pratt brought a partial solution back from the London Naval Conference. A flight-deck cruiser could accommodate a squadron of aircraft; a full-size carrier could handle four squadrons. The eight flight-deck cruisers allowable under the new treaty could be considered equivalent to two carriers. Games played in 1931 showed that the flight-deck cruiser could be extremely valuable in scouting and screening. In reporting this success, the War College admitted that it could not judge the platform's feasibility. Given the prospect of aircraft onboard flight-deck cruisers, it could be argued that the Navy should build larger but less numerous carriers, perhaps three somewhat larger ones instead of four more *Rangers*. As it happened, no flight-deck cruisers were built, but by the time that project was dead, the vital decisions shaping the new carriers had already been taken.[37]

By this time, no one liked the small *Ranger*, which had not yet been completed. Available alternatives in 1931 were four more *Rangers*; or two heavy general-purpose carriers (20,000 or 20,700 tons each) and one smaller carrier; or three smaller general-purpose carriers (18,400 tons each). Each of the alternatives offered about the same total number of aircraft and the same number of flight decks.

This rethinking, in which gaming played an important role, turned out to be very important. The U.S. Navy built two more fast heavy carriers, *Yorktown* (CV-5) and *Enterprise* (CV-6). Not only were both very successful in the Pacific War, they were in effect the prototypes of the *Essex* class fleet carriers that served so effectively in the Pacific during World War II. The argument favoring two distinct types was that a ship somewhat larger than *Ranger* might yet prove attractive. CNO Admiral Pratt liked the combination of two large and one smaller carrier, and the General Board agreed.

There was real interest in what was then called a battle-line carrier. Such a ship would operate near the battleships, and she would keep their crucial spotters in the air both by providing fighter protection and by servicing them. The War College's fire-effect rules showed how valuable long-range fire could be. It required the support of spotting aircraft. In 1931, Commander of Aircraft, Battle Force considered a battle-line carrier as the fleet's most pressing need. He thought the new *Ranger*, which was being built, could fill it. There was a real possibility that the next carrier would be a relief battle-line ship.

Naval War College President Rear Admiral Harris Laning used war-game experience to argue persuasively against a second battle-line carrier. He pointed out that both full-scale experience and gaming showed that carriers would and should often operate more or less independently, subject mainly to attack by enemy cruisers. They needed high speed and a degree of hull protection, both of which would cost tonnage. When the General Board revisited the carrier characteristics question in mid-1931, it demanded both. Hence, the choice to build two heavy carriers was made. A third would be built using the tonnage left over.

On 20 July 1931, as a preliminary to developing characteristics and setting carrier policy and a building program, the General Board asked CNO to conduct a study of the uses of individual carriers and of U.S. carrier-based aircraft that were permitted under the London Naval Treaty. Asking CNO meant asking his think tank, the Naval War College.[38] The college study has not survived, but it was almost certainly the 5 August 1931 letter cited and heavily quoted in a 30 July 1932 letter to the General Board, which has survived.[39]

The Naval War College observed at the outset that the more realistic the games became, thanks to inputs based on fleet experience, the more important it was to increase the number of fleet aircraft. The two great unanswered questions, then and through the inter-war period, were the effect of antiaircraft gunfire and the effect of air-to-air combat. Laning wrote that not until both questions had been resolved, "can we be sure of what we ourselves will be able to accomplish with aircraft or what an enemy may accomplish with his aircraft."

Even with those questions unanswered, Laning could be confident that "in all forms of naval warfare, aircraft will exert a decisive influence, not necessarily by their direct action and hitting power, but rather by the cumulative effect of initial advantage they often gain for a fleet." During the 1931–32 year in the War College's problems and games, "aircraft have continued to exert a decisive influence." In Blue-Orange problems,

> the information obtained by aircraft to both Blue and Orange, and the effect of aircraft bombs and torpedoes, was invaluable to each fleet. The games again emphasized the necessity of the United States' possessing an adequate number of carriers of various types to neutralize the advantages of Japan's carriers and shore based aircraft in well established bases near the probable theater of war operations (i.e., in the Mandate Islands, islands of the Far East, and through the Philippines). In these games, Orange demonstrated how serious a blow she could strike against Blue in a strategy of "attrition" using her aircraft, light and submarine forces, either independently or united. And, against a Blue trans-Pacific slow-speed expedition, it appears that an overwhelming Orange air force in the vicinity of the Marshall, Caroline, and Marianas Islands operating from carriers and bases can exert a great influence against the Blue expedition and add greatly to the damage already inflicted by previous submarine attacks. In such a Blue-Orange Campaign, before the possibility of a *fleet* engagement, the games show that the United States must have a *great* preponderance of carriers and aircraft strength. It is vital that we have as many carriers, 'flying deck cruisers,' and converted auxiliary merchant carriers as possible, and all the aircraft practicable aboard each vessel, to use not only for all purposes prior to a main battle line action, but to ensure that we maintain *air superiority* during the *approach* and the *action* itself [emphasis in original].

Because it was so difficult to locate enemy air forces once they were aloft, it was generally impossible to gain "command of the air" by air combat alone. It could be attained only by destroying or rendering inoperable the sources of enemy air operations—hitting the enemy's carriers as early as possible. Moreover, U.S. carrier flight decks had to be protected in all possible ways. Game rules and game experience emphasized how vulnerable flight decks were to air attack, and how much could be lost when they were destroyed. At this time, it was assumed that it would take many days for a ship's company to effect temporary flight deck repairs. All of the 1931–32 games emphasized the vulnerability of carrier flight decks and flight-deck cruisers and the lack of suitable repair facilities in the western Pacific.

The need for enough carriers and aircraft for both the attrition phase of the war and the decisive battle affected the choice of the types and numbers of ships carrying aircraft, the types of aircraft, their armament, and the plans for the use of the material and personnel of this combined force. The treaty allowance of carriers was hardly sufficient; the War College was looking for end-runs like the flight-deck cruiser and converted merchant ships. The other point emphasized by the games was that everything had to exist on the outbreak of war "with our most probable enemy, Japan."[40]

The War College warned against trying to categorize carriers for particular duties. This applied particularly to pressure being exerted at the time to build battle-line carriers on the theory that the fleet already had two scouting carriers in the form of the two *Lexingtons*.[41] Unfortunately, battle-line carriers would probably be the main objective of any enemy air attack prior to a battle line engagement. The 1931–32 games offered only a single case of the use of battle-line carriers. In this case, the Red fleet had three 31-knot carriers and two slower ones. The latter were assigned to the Red battle line. They exchanged their bombers for scouts and spotters. During the game, all five Red carriers were surprised near their battle line by Blue scouting forces, which damaged all five Red flight decks before the action began. "In general, no carrier which has *remained within sight of their own battle line* has come through without destruction of her flight deck even though not assigned to strictly battle line duties [emphasis in original]." All that the college could suggest was that such a carrier should be small enough to hide, that she should have the maximum possible flight-deck protection, maximum antiaircraft protection, stowage for at least some of the battleship spotting aircraft, and possibly few or no main battery (antiship) guns. The War College planned to assign carriers to the battle line during the 1932–33 games. Among the students were the former captains of the three existing U.S. carriers, who were expected to provide insights into the use of carriers and aircraft.

Carrier antiaircraft guns offered little protection against bombers when the ship's aircraft were away. Battleship officers were certainly acutely aware of their need for air services. During the wash-up after Fleet Problem XX (1939), they accused Rear Admiral King, who was commanding the carriers on their side, of fighting a "private war" against the enemy's carriers. King replied that unless he dealt with the enemy carriers at the outset, they would get no air services at all. Other Fleet Problems showed that a carrier's best hope of survival lay in evasive movements at high speed. A ship tied down to a slow formation was likely to be hit. The War College noted that carriers had found high speed very valuable in scouting, in evading submarines, and in dodging torpedoes and shifting position during operations.

All carriers could and should operate all types of aircraft. All of their flight decks were vital, particularly given their vulnerability. Each carrier or flight-deck cruiser should carry the maximum number of aircraft practicable, compatible with maintaining all its military characteristics.

Comments by the Bureau of Aeronautics dated July and August 1931 survive in the General Board file on the carrier issue.[42] They appear to show the influence of war game experience. BuAer quoted former War College President (and then-current CNO) Pratt:

"[O]ne of the outstanding problems played each year is that to advance into a hostile zone the fleet must carry with it an air force that will assure, beyond a doubt, command of the air. That means not only superiority to enemy fleet aircraft, but also to his fleet and shore-based aircraft combined. This is a large order, and the only way to secure it is by large numbers of carriers and planes." BuAer pointed out that given treaty limits, the only way to add much to the fighting value of the fleet was by adding aircraft. Even large numbers of additional cruisers (which were the main type the Navy was then considering building) would add little to U.S. ability to destroy enemy capital ships or bases.

In November 1931, BuAer pointed out that its request for an estimate of how many carriers and flight decks would be required had never been answered. Its own studies, assisted by individual officers in the War Plans Division, indicated that at least 14 would be needed. Moreover, the war games showed that aircraft attrition would also be enormous. Initially it would be impossible to grow the carrier air force since, for the first nine months of the war, the entire U.S. aeronautical industry would be making up attrition losses.[43] BuAer cautioned that the six fleet carriers—all the United States could have, under the treaties—and the eight flight-deck cruisers it envisaged "by no means represent our final requirements in any war with a first-class power." BuAer warned, moreover, that carriers would inevitably be lost during the early stages of a war. Only merchant ship conversions might make up some of the difference—but unfortunately there were few suitable U.S. merchant ships, and it turned out that the conversions envisaged would have taken far too long. Perhaps in hopes of keeping the carrier program popular in a battleship-oriented naval establishment, BuAer accepted the idea of the battle-line carrier. Admiral Moffett must have been aware of the danger of tying a carrier to battleships.

In 1931, the General Board took the War College's advice. It could not do anything about *Ranger*, so it characterized her as a battle-line carrier and argued that she could benefit from the cover offered by the battleships. The Naval War College had already pointed out that such protection was delusionary. For the future, the War College bought two fast protected carriers. The third, whose tonnage was limited by the Washington treaty, would be a slightly inferior general-purpose carrier. Unlike *Ranger*, this ship—*Wasp* (CV-7)—was ultimately used in the Pacific.[44] The General Board finessed the question of whether she was worth building by saying that she was a point in the carrier spectrum that could be compared with the larger *Yorktowns*. Relative value was indicated by the order in which the ships were built: The two bigger and more useful carriers came first.

The Naval War College emphasized the need to make the flight deck survivable. In games, carriers were most often put out of action by bombing, not by attacks by surface ships or submarines. In its July 1931 memo, the college proposed that various flight-deck facilities be duplicated to make them more difficult to destroy. This idea did not appear in the formal characteristics reported by the General Board, but it was embodied in the ships as built. The previous month, BuAer had already called for arresting gear at both ends of the flight deck, so that aircraft could land on at either end. The hangar side was to have an open section forward from which aircraft could be launched using a fixed athwartships catapult. This idea had been raised by Admiral Reeves as early as December 1927.[45] He saw it as a way of maintaining carrier capability despite flight-deck damage. Reeves also preferred the open hangar (which was introduced in *Ranger*) because it made for maximum aircraft capacity, offering the maximum proportion of hangar space to the upper deck. He wrote at the time that "it is evident that the fleet which takes to the scene of battle and is able to launch into the air the greatest number of aircraft should win the air battle, other things being equal." Airplanes in an open hangar could be gassed and warmed up before they were brought up to the flight deck, hence could be launched far more quickly. All of these features were included in the new carriers and in the follow-on *Essex* class.[46]

In describing improvements embodied in the new carrier, the Bureau of Construction and Repair responded to a specific request from Air Squadrons Battle Force for onboard facilities to repair flight-deck damage.[47] The last item in the C&R letter describing features of the new design stated that "it is assumed that the flight deck damage...refers to local damage to planking and plating occasioned primarily by bombing. It is believed that the only practicable method of making quick repairs to such minor damage on the part of the Forces Afloat will be by spiking steel plating over the holes in the decks. Such plates can be taken from the regular stock carried by the vessel. A repair party station is being provided just under the flight deck accessible to the galleries on the *Ranger* and CV 5 [*Yorktown*]." This was a rather casual reference to something very important—something the War College had seen in its games. It harked back to the *Ranger* design and to Forrest Sherman's call for a light flight deck that could be repaired quickly after it was bombed.

The *Ranger* flight deck certainly need not have been repeated. Other navies did not follow suit. British and Japanese carriers had flight decks integral with their hulls, their hangars closed in at the sides to form structural supports. *Ranger* and her light flight deck and open hangar had been a distinct departure from the steel flight decks and fully-enclosed hangar of the *Lexingtons*. Prodded to think about how the flight decks atop open-sided hangars could be repaired, C&R noticed something important: The light wood over steel structures could be fixed surprisingly quickly.[48] The rather casual way the issue was addressed suggests that it had not been raised before; it was certainly not included in the formal characteristics of the next carrier, *Wasp*. By far, the bulk of correspondence concerning the new designs concentrated on whether (and by how much) the flight deck could be extended to increase aircraft capacity—another important issue raised by war games.

Once *Yorktown* was about to enter service in 1937, the war game rules were rewritten to allow for relatively quick repairs by a ship's force. The time required to repair flight-deck damage was still very conservative, but now that the ship's own crew could make the repairs, the ship could remain in action despite damage. The reality of such repairs was demonstrated again and again during the Pacific War. That was particularly important early in the war, when the U.S. Navy had few carriers available in the Pacific. The reason *Enterprise* was able to fight in nearly every Pacific battle was that her crew could patch her immediately after she was hit. Without the ability to make quick repairs and thus keep its few carriers in action, the U.S. Navy would have been far less effective in 1942–43, before the mass of new carriers appeared.

This type of construction was an alternative to flight-deck armor. The choice was raised when U.S. naval officers inspected the British armored-deck carrier *Illustrious* as she was being repaired at Norfolk Naval Shipyard in the fall of 1941.[49] They were very impressed, and said so at a General Board hearing. The response was that the U.S. Navy had opted for more than twice as many aircraft and for the ability to make quick repairs after battle damage. The officers were told that U.S. practice had been designed for a Pacific campaign rather than the Mediterranean or European operations for which the British had conceived their carriers. The use of the word "campaign" reflects an understanding of the whole of a Pacific war gained by gaming. *Illustrious* could certainly resist some bombs, but she was being repaired because her deck had been penetrated by much heavier ordnance dropped by dive bombers. As in the war game rules, she needed a first-class shipyard to repair that kind of damage. *Enterprise* and her sisters were able to patch their own flight decks and keep fighting. Even so, they were just short of fully exploiting the light flight deck. That was left to the *Essex* class, unlimited by treaty and hence considerably larger.[50]

The preliminary design of the new 20,000-ton carrier was formally submitted to the Secretary of the Navy on 28 December 1931.[51] In submitting initial characteristics in May 1931, the General Board had listed roles in order of importance support of the battle line; support of battle forces (e.g., contact scouting, both tactical and strategic, and attacks on the enemy air force); support of fleet operations and movements, including security of train and convoy; and operations with task forces or independently. It is a measure of the War College's success that the first priority changed dramatically in favor of independent operations of various kinds. Despite some experience with the big *Lexington*s, the General Board stated in a covering letter to the October 1931 version of the characteristics that "the present broad fields of usefulness of carriers have not yet been revealed in a practical way." Gaming provided a lot of the needed experience.

The perceived urgency of carrier construction shows in the recommendation to build two such ships in the FY33 budget and the third in FY34: The General Board clearly understood that carriers and their aircraft were essential.[52]

Aftermath

After 1934, the Naval War College never again cited war-game evidence in any advice it offered. The terms of the 1930 London Naval Treaty closed off further carrier construction, leaving the U.S. Navy with two options to provide the aircraft it needed. One was merchant ship conversion, but, as noted, there were too few suitable ships and conversion, as then envisaged, would take far too long. The other was to seek a source of aircraft not limited by treaty: seaplanes. Particularly after the war plan strategy changed to a step-by-step advance, the atolls of the Pacific offered numerous potential seaplane bases. In war games, patrol planes—seaplanes not limited by treaty—often added to naval air power not only for scouting but also for attack. Beginning in 1934, the U.S. Navy started work on seaplane tenders that could expand the fleet's air reach and its numbers. In a 1937 reorganization, the fleet's seaplanes were moved from the Base Force at Pearl Harbor to a new Patrol Wings organization as part of the Scouting Force. The implication was that they would move forward with the fleet.

The evidence of interest in patrol planes as attackers, outside the war games, is the decision to designate the Consolidated Catalina as a patrol *bomber* (PBY) rather than in the pure patrol series of the past. Previous flying boats had been capable of carrying heavy bombs, but the PBY added the Navy-developed Norden precision bombsight. The attack role of the seaplanes is obscured because, by 1941, they could no longer compete effectively with modern fighters (although PBYs did turn out to be effective night bombers, at least for a time). However, the conceptual effort was not wasted; the wartime Navy effectively deployed large numbers of land-based patrol bombers in the Pacific. These aircraft never figured in war games, because until 1942 the U.S. Navy was prohibited from operating them under an agreement between CNO Pratt and Army Chief of Staff Douglas MacArthur.

In 1936, the new London Naval Treaty abandoned limits on total tonnage in all its categories. Everything said and written in 1931–32 would have pointed to a large U.S. carrier program in parallel with the substantial battleship program that then began. Instead, only one carrier was authorized in FY38 (*Hornet*, effectively a repeat *Yorktown*), another (*Essex*) not following until FY41. The design of the latter ship was greatly enlarged once World War II began and all remaining treaty limits were abandoned. There seems to have been a sense that seven or eight fleet carriers were quite enough. The kind of campaign reasoning which demanded many more was not brought up. The absence of carriers in the U.S. building programs of this time can be taken as evidence that the war-gaming point of view was no longer of interest, at least outside the OPNAV War Plans Division. The emphasis on battleships suggests that attention was concentrated on the (hopefully) decisive battle against the Orange battle line rather than on the entire campaign against Orange. That is somewhat ironic, given that the "through ticket to Manila" had been abandoned.

On the other hand, war games had also shown that carriers were terribly vulnerable to air attack. It was widely accepted that, once enemy carriers were nearby, there was no security at all until they had been neutralized. The same applied to an enemy's view of U.S. carriers. It was entirely possible that all carriers on both sides would be put out of action early in a war. The greater combat survivability bought by the new type of flight deck seems not to have been widely understood. Carriers did not really gain survivability until the development of radar-based fighter direction, which was only dimly in prospect in 1940.

Any interpretation of what happened is muddied by the character of the legislation under which new ships were authorized. In 1934, Congress passed the Vinson-Trammell Act, under which sufficient tonnage was authorized to build up to a "modern Treaty navy." In practice, that meant replacement of over-age ships. When the total tonnage limits were abandoned, it meant that over-age ships no longer had to be discarded. Since there were no over-age carriers in the U.S. fleet, the question of new carrier construction did not arise. All of the battleships were far older, hence were automatically candidates for replacement. In 1938, a second Vinson-Trammell Act added new tonnage, and the question of new carriers certainly did arise. The new act simply enlarged each tonnage category by a set percentage. Thus, the authorization continued to reflect the balance between different types of ships reflected in the Washington and London treaties. In the case of carriers, it added 40,000 tons, enough for two ships. The use of a set percentage avoided controversy; even in 1938, naval rearmament was a very contentious issue. The FY38 carrier, *Hornet*, was a special authorization outside the tonnage increases in the 1938 act.[53] Thus, it was a direct reaction to the opportunity opened by the 1936 treaty.

No carriers were included in the FY39 and FY40 programs. When the Secretary of the Navy asked the General Board in July 1938 for relative priorities for design work for the FY39 and FY40 programs, a new aircraft carrier came last out of five design projects. The first priority was the new 45,000-ton battleship, which became the *Iowa* class.[54] Even in this case, the situation is muddied, since the new carrier might be a repeat *Yorktown*, in which case no design work would be needed, and no characteristics for an entirely new carrier design had yet been laid down. In 1938, the General Board expected to include two 20,000-ton carriers in the FY41 program. Formal General Board policy set out in October 1934 was not to reduce the number of carriers (seven) "in view of the need for aircraft with the Fleet and of the large wastage factor which, due to the vulnerability of their decks, may be expected of carriers in an active campaign." In a 1940 memo, the General Board observed that the combination of expected wastage and the "increasing relative performance of land based planes to carrier based planes" had resulted in a "problematical appraisal of carrier value. Before this question gets in the open, and due to the present popularity of aviation in Congress, it is believed wise to include one, if not two, carriers in the 1941 program. Unless a war occurs in which the role or value of carriers is definitely determined, it is believed that before carriers are scrapped, improvements in carrier technique and landing and take-off facilities, will enable them to handle maximum performance planes of the fighting and pursuit type, which would warrant their retention in the fleet"[55]

Presumably, this language reflected fleet skepticism as to the potential of carriers, which the General Board warned would be taken poorly by the public and by President Roosevelt, who was a strong advocate of air power. On the other hand, the General Board should have been aware that better carrier aircraft were coming. At this time, BuAer was testing the F4U Corsair, which it thought was the fastest fighter in the world—and which was intended for carrier operation.

At about the same time, the bureaus proposed a ten-year (FY39–48) program including six carriers (two each in FY41, 45, and 46) and 16 battleships.[56] The carriers were the two new ones Congress had provided plus replacements to be laid down when existing carriers became over-age. Not even BuAer seems to have advocated a massive carrier-building program. None of the General Board policy papers of this time refers back to the drastic need for aircraft in an Orange war, as war gaming had demonstrated. Since 1931, Pacific War strategy had changed, and it is possible that the new seaplane bombers were considered an effective counter to Japanese aircraft based in the Mandates. The theory would have been that these bombers would provide effective air cover to the fleet by destroying Japanese land-based air power at source. The fleet would need much less air cover as it steamed west, at least until it encountered Japanese carriers and their aircraft. In that case, it might be reasonable to relate needed U.S. carrier strength to Japanese carrier strength rather than to the much larger total Japanese air threat including land-based aircraft.

On the other hand, the step-by-step strategy demanded the seizure of island after island in the face of Japanese resistance, including aircraft. Carriers were needed for these seizures; analyses conducted in 1938 made it clear that long-range aircraft from other islands could not effectively support a landing. Every such seizure, which tied the carriers to a particular area, left them vulnerable to submarine attack. A rational peacetime program should have allowed for such losses. There was some interest in providing the Marines with floatplane bombers that could be launched by a seaplane tender, but even if such aircraft were effective, they could not operate until U.S. forces had seized control of sheltered areas of water. It is not at all clear to what extent the OPNAV War Plans Division estimated how many carriers would be needed for this phase of a war.

Even if the big seaplanes were envisaged as alternatives to carrier attack aircraft, that turned out to be a poor choice. The gap between seaplane and land-plane performance opened rather than diminished as war approached, although there was some hope for improvement in seaplane performance as engines became more powerful. The U.S. Navy found itself operating versions of Army Air Force land-based bombers from the islands it conquered. Even their ability to project air power forward with the fleet was limited. It was not at all clear that such aircraft, based on the initial islands seized, could neutralize Japanese air bases deeper in the Mandates. In the games, neutralization was achieved by periodically gassing bypassed airfields after they had been struck. In reality, gas warfare was a political issue, and it was never permitted. Among other things, that meant that airfields could never be permanently neutralized. None of this would have been obvious before the outbreak of war, but it does seem useful to point out that the carrier role changed substantially.

The General Board record suggests no wide-ranging reappraisal of overall U.S. requirements in view of a Pacific (or other) strategy. The question of when carriers would largely replace battleships was raised, but it seems to have been assumed that would lie in the distant future. Nothing in the General Board record suggests that the new kind of repairable flight deck had made an appreciable difference in thinking. There was no allowance for possible war losses because there was no estimate of how many carriers the strategy would require.

Changing U.S. interest in carriers was illustrated by the FY41 program, which ultimately added 11 of them. When this program was first considered in June 1939, it included a single new carrier, CV-9, which was conceived as a slightly modified *Yorktown* displacing at most 20,400 tons.[57] In October, war having broken out, OPNAV was considering emergency expansion of the U.S. Navy on the basis of a 25 percent increase in authorized tonnage. This time, it seems that tonnage of carriers, cruisers, destroyers, and submarines could be swapped. The program showed four new carriers. A fifth new carrier (*Hornet*) was already under construction, to make an ultimate total of 11. In addition, of 14 new light cruisers and ten new heavy cruisers, at least four were to have flight decks (as flight-deck cruisers). Additional seaplane tenders (and seaplanes) were included. This program was based on a "Two-Ocean Navy" program that had been developed inside OPNAV. It was based largely on an inventory of shipyard capacity. That approach was reasonable: Gaming had shown that the U.S. Navy would need vastly more ships. All that could be done was to make maximum use of existing resources.[58] As of late May 1940 the projected FY41 program was headed by two battleships and four carriers. Another eight battleships had recently been authorized, plus a single carrier.

The building program was totally revamped after France fell. Until that point, the United States might plan to assist the allies in the Atlantic, but the bulk of the fleet could still be concentrated in the Pacific. The U.S. Navy could fight a one-ocean war while participating, if need be, in the other ocean. After France fell, it was no longer so certain that Britain would survive. The United States might find itself fighting full-scale naval wars in both oceans, in both cases against opponents with powerful battle fleets. That is not obvious now, because the worst case scenario in the Atlantic never came to pass. The British never surrendered, and the Germans never gained control of the French fleet.

The U.S. reaction to the fall of France included the massive new Two-Ocean Navy Act calling for a 70 percent increase in authorized tonnage. A BuShips estimate dated 27 July 1940 showed seven more battleships and seven more carriers, to make a total of 11 FY41 carriers. The carrier figure seems to have been based on nothing more than a ratio between the original plan for a 25 percent increase (4 carriers) and the new 70 percent increase. The carrier program was by no means universally supported. Priorities swung to carriers only in 1942, when the Navy was forced to make choices due to a steel shortage. New battleships and large battleship-like cruisers were cancelled, but the carriers were not. The way those decisions were taken is outside the scope of this study. What is striking is the shift from, in effect, "we can't have enough carriers" on the basis of simulation by gaming to an emphasis on more or less meaningless ratios of numbers or tonnages of carriers divorced altogether from any sort of campaign analysis.

Gaming shaped the carriers that proved so effective in the Pacific, because it forced their designers to think in terms of the entire campaign they would fight. Gaming forced the Navy to think about the campaign that had to be fought prior to the expected decisive battle. That it would have been easy to disregard the campaign prior to the battle is suggested by the analysis—or lack of analysis—of the late 1930s, when the United States was rearming. The force ratios involved made sense, if at all, only in the context of a single battle. The reality was that the Pacific War resembled the pre-battle campaign. In this campaign, the force ratio was not carrier to carrier, but U.S. carriers to full Japanese naval air forces.

5. The War College and Cruisers

Cruisers, which may not seem very important in retrospect, were the largest surface warships that could be built during the "battleship-building holiday" imposed by the Washington Treaty. They were also the class in which the United States was weakest, Congress having rejected new cruiser construction after approving three small scout cruisers before World War I (the big 1916 building program included ten more scouts). Cruisers became a major concern for the inter-war War College and a major application of its gaming technique.

For inter-war navies, cruisers had three distinct roles. One was as scouts, acting either independently or in the van of the fleet. A second was as the fleet's defense against enemy destroyers attacking with torpedoes. Despite their name, destroyers were not considered an effective defense against enemy destroyer attacks because they were poor gun platforms. In effect, they were ocean-going torpedo attack craft. A third was to deal with enemy commerce raiders. At this time, the law of war prohibited "sink-on-sight" attacks; merchant ships had to be stopped and examined, and only then sunk or seized. The safety of their crews had to be ensured, which often meant that they had to be taken onboard the raider. These legal conditions generally required that a raider be a surface ship with enough speed and firepower to stop a fast merchant ship. An enemy would deploy cruisers or, more likely, armed merchant ships. The first two roles were the most important to the inter-war U.S. Navy, but for the Royal Navy trade protection was at least as important.

Gaming demonstrated what everyone in the U.S. Navy already knew: The Navy had far too few cruisers. Worse, it was impossible to place an upper limit on what it needed, just as it was impossible to provide enough airplanes to support an Orange war. The inter-war arms-control agreements functioned as justification for particular numbers of cruisers under the 5:5:3 ratio (Britain: United States: Japan) applied to total battleship and carrier tonnage (not numbers) at Washington in 1921. From 1922 on, the General Board was responsible for a formal U.S. naval policy. It included the important political requirement that the Navy be "second to none," meaning parity with the Royal Navy. The General Board also called for the United States to build the maximum possible number of the largest permitted cruisers armed with the most powerful possible guns. That meant 10,000-ton ships armed with 8-inch guns. The numbers that the General Board proposed were based on the 5:5:3 ratio, applied to the total tonnage of British and Japanese cruisers. It seems to have been generally assumed that further arms-control negotiation would extend the Washington Naval Treaty to cruisers and even to destroyers and submarines.

Gaming offered only limited insights. It could certainly demonstrate the consequences of the weakness of the U.S. cruiser force, particularly in a fleet-on-fleet battle. Campaign games could show that the United States had far too few cruisers overall. However, it was impossible to say how many would be enough. The games of this era did not, moreover, provide insight into the range of cruiser types the U.S. Navy needed, or to the proportions between types. In effect the ten "scouts" of the 1916 program provided the U.S. Navy with all the smaller fleet cruisers it was likely to get, the General Board pressing only for construction of the much larger ships armed with 8-inch guns. Here, fleet cruisers meant ships capable of dealing with (or supporting) destroyer torpedo attacks.

The big cruisers favored by the General Board were well-adapted to function as scouts, operating either independently or in the van of the fleet. To some extent they offered a capability the United States had been denied when the Washington Treaty had cancelled the U.S. program to build battlecruisers. However, big cruisers with heavy guns were poorly equipped for other cruiser functions, particularly beating off destroyer attacks on the battleships, since the powerful, 8-inch guns fired too slowly to be sure of hitting rapidly maneuvering destroyers.[1]

U.S. Navy goals were not necessarily the goals of the U.S. government. The Washington Treaty came to be seen as the first step toward all-embracing naval arms control, which was seen by successive U.S. administrations as a way both of avoiding war and of reducing the crushing cost of armaments. U.S. President Calvin Coolidge convened a follow-on conference at Geneva in 1927 specifically to extend the Washington Treaty to smaller warships, specifically cruisers. The major navies, particularly the British and the Japanese, were already building substantial numbers of large cruisers, and there was a general perception that this new naval arms race had to be stopped. The General Board's hopes that Congress would provide enough such ships to make up for the gross U.S. deficiency had been disappointed. Congress had authorized eight ships in 1924, far short of what the General Board's policy required. Any attempt at negotiation was complicated by British insistence that the Royal Navy needed more small cruisers than the U.S. Navy in order to protect the much more massive British seaborne trade. For their part, U.S.

The War College did not originate the big cruisers the U.S. and other navies built in the 1920s. Nor did it evaluate them, based on their flimsy protection and their heavy armament. It certainly did appreciate the need for fleet scouts. This is *Houston* (CA-30), a first-generation heavy ("treaty") cruiser armed with 8-inch guns, but very lightly protected to achieve high speed. In the background is the far better protected *New Orleans* (CA-32). Both were photographed off San Pedro on 18 April 1935. (National Archives 80-CF-21337-1)

naval officers badly wanted sufficient numbers of modern cruisers before they were willing to accept any sort of limitation. At Geneva, as at the Washington Conference, naval officers were among the delegates. U.S. naval officers were widely credited with breaking up the conference, infuriating President Coolidge.[2]

The Washington Treaty required that a new conference be called in 1930, when the suspension of battleship construction was due to end. The conference met in London. The sort of naval engagement that had crippled the 1927 conference was avoided by barring naval officers as delegates.[3] They did attend as naval advisors, however. Rear Admiral William V. Pratt, recently president of the War College and now commander of the U.S. Fleet, was chief advisor.[4] He was widely credited with resolving problems and thus making the 1930 conference successful. In the past, he had been the Navy's expert on arms control. His key role in the 1930 conference made him extremely unpopular within the service.[5] Pratt had served as Admiral Sims's chief of staff when Sims had commanded U.S. naval forces in Europe during World War I, and the British remembered him warmly. His role at the conference was much appreciated by President Hoover, who was well aware of the failure of his predecessor's 1927 Geneva conference.

In the wake of the London Naval Conference, President Hoover made Pratt Chief of Naval Operations. To the extent that this appointment was seen as a reward for service at the conference, it cannot have helped Pratt's popularity within the Navy. Presidential favor gave Pratt more power than he might otherwise have had. To a substantial extent, he could decide what the U.S. Navy would build. As CNO, Pratt removed the president of the War College from his former ex officio position on the General Board. The War College president retained his position within OPNAV, as a direct advisor to the CNO.

The War College itself appears to have had no direct impact on the U.S. negotiating position.[6] That is no surprise; the U.S. Navy did not change its preference for the most powerful cruisers. The demand for parity was a political method of justifying construction of the maximum possible number of such ships. Conversely, the Coolidge and Hoover administrations wanted to shrink the British cruiser fleet so that they could limit U.S. construction of expensive cruisers within a demand for parity. Both U.S. administrations doubted that the U.S. public would stand for an agreement leaving the British with a significant superiority in cruiser numbers, even though it was obvious that the British needed more cruisers to protect their huge merchant fleet (the threat to which, it was assumed, was surface raiders, not submarines). The British had built their own cruisers armed with 8-inch guns, and they found such massive ships unaffordable in the desired numbers. Since they associated the heavy gun with large (expensive) cruisers, their single main objective at the conference was to preclude further construction of cruisers armed with 8-inch guns. For trade protection, a much smaller gun, of 6-inch caliber, was large enough, because no converted surface raider could be armed with anything more powerful.[7]

Probably the key factor in success was a very friendly 1929 summit conference between President Hoover and British Prime Minister Ramsay McDonald, during which both agreed that differences would be resolved. British naval officers saw this as a surrender to the Americans. McDonald agreed to clauses, such as a dramatic reduction in destroyer tonnage, that the Admiralty specifically rejected. He was anxious to reach agreement both for idealistic reasons and because his country was already in severe economic trouble, even before the onset of the Great Depression. McDonald found himself accepting the American pressure to force down the size of the British cruiser force to the point where the U.S. Navy could match it. The Admiralty admitted that it could not quickly replace the numerous ageing cruisers left over from their World War I building program.[8] McDonald told his critics that numbers could be built back up after the treaty expired. This time, too, the U.S. Navy was more favorably inclined toward cruiser limitation because Congress had authorized 15 more cruisers in February 1929, of which five were to be ordered each year from FY29 on.[9]

Given the new Congressional authorization, the United States might build 23 cruisers armed with 8-inch guns. The British had stopped building such ships after completion of the fifteenth unit. Initially the U.S. delegation offered either parity with the British (15 each) or 18 8-inch gun cruisers, in each case with a tonnage allowance for 6-inch cruisers.[10] As ratified, the treaty allowed the U.S. Navy to build its last three 8-inch gun cruisers, but their construction was to be delayed. As an incentive to the United States not to build those ships, it could substitute 15,166 tons of 6-inch cruisers for each 8-inch cruiser not built. Parties could also transfer up to 10 percent of their allowed destroyer tonnage into cruisers armed with 6-inch guns. All of this left exactly the sort of questions war gaming was designed to investigate. Were the last three 8-inch gun ships worth more than the larger tonnage that could go into 6-inch gun cruisers? What sort of 6-inch gun cruisers should the U.S. Navy build?

The high status of aircraft in the U.S. Navy was evidenced by the appointment of Rear Admiral Moffett as a member of the naval staff at London. Because the initial U.S. proposals were drafted by the State Department without naval input, they did not reflect his thinking. Thus, the only carrier proposal initially offered by the United States was to abandon the existing provision that anything under 10,000 tons would not count in the carrier tonnage total. The Japanese had taken advantage of this clause by building the 8,000-ton carrier *Ryujo*. She had not yet been completed, so it was not yet evident that she was too small and would have to be rebuilt. Once that had been done, she no longer displaced less than 10,000 tons, although the Japanese chose not to admit as much. The British agreed to ban small carriers despite their potential value in protecting British trade. They were unlikely to build many such ships, given their economic problems, and if anyone else did, such carriers would be a considerable threat to British seaborne trade.

The British took carriers far less seriously than did the U.S. Navy. The Washington Conference had allowed guns as large as 8-inch caliber onboard carriers. The British suddenly decided that a country could build air-capable cruisers instead of true carriers. Such ships could be somewhat larger than the 10,000-ton cruisers allowable under the treaty, and they might overwhelm existing cruisers. To prevent a cruiser from being built in the guise of a carrier, the maximum caliber of carrier guns was reduced to 6.1 inches. The British apparently did not imagine that anyone would build a large cruiser armed with such weapons.

At the upper end of the scale, the British wanted to reduce the 27,000-ton limit agreed at Washington to 25,000 tons. On that basis, the 135,000-ton total agreed at Washington, equivalent to five 27,000-tonners, could be cut to 125,000 tons (five 25,000-tonners). This was anathema to the U.S. Navy. The two large carriers had taken up 66,000 tons, and the remainder was just enough for five 13,800-ton *Rangers*. It seemed most unlikely that the two huge carriers would be reclassified as experimental, hence instantly replaceable. The British proposal translated into one fewer U.S. carrier. The U.S. Navy needed more, not fewer carrier-based aircraft.

The carrier provision mattered in the context of cruisers because Moffett was vitally interested in an end-run: the flight-deck cruiser. He wanted to use some of the allowed

It proved possible to improve cruiser protection (survivability) dramatically. Here, the second-generation heavy cruiser *Minneapolis* (CA-36) fires in battle practice, 29 March 1939. It seems to have taken War College analysis to demonstrate clearly how superior such ships were to contemporary foreign (particularly British) heavy cruisers. Without the analysis of the type developed for war gaming, it was certainly clear that ships like this were better than their predecessors; but only the analysis could show how much better they were. (National Archives 80-CF-21343-2)

cruiser tonnage to build, in effect, fractional aircraft carriers. His chief, Admiral Pratt, certainly understood: As Naval War College president, he had pointed to the desperate need for aircraft in an Orange war—as demonstrated by gaming—and to the impossibility of providing enough within the terms of the earlier Washington Treaty.[11] Moffett and Pratt managed to insert the clauses they wanted in the final treaty.[12] Up to a quarter of allowable cruisers could be fitted with flight decks, and they would not count toward the allowed total of carrier tonnage. That left another question that gaming could help solve: would it be worthwhile to build flight-deck cruisers?

As noted above, the flight-deck cruiser had been conceived some years before in connection with the 1924 Congressional authorization of the first eight 10,000-ton cruisers. At about the same time, air power advocates were arguing that aircraft could and should replace virtually all major warships. Cruisers became a battleground. The Washington Treaty imposed no limit on the total tonnage of surface ships displacing 10,000 tons or less. On 31 March 1925, Secretary of the Navy Curtis Wilbur asked the General Board whether, "in view of the propaganda now going on in favor of aircraft," some of the 10,000-ton cruisers should be completed either as small carriers or as combination cruiser-carriers.[13] The General Board rejected completion of any of the ships as carriers. Not only was the board reluctant to sacrifice cruisers, but preliminary studies by the the Bureau of Construction and Repair showed that 10,000 tons was too small.[14] The bureau similarly rejected the idea of a combination cruiser-carrier. The Secretary agreed, and no such ships were built—for the moment. However, the idea of a combination ship survived.

The proposal occasioned a formal comment from the War College, based on experience in tactical games.[15] The college pointed out that the two functions of light cruiser and carrier were fundamentally incompatible. To function effectively, a cruiser had to place herself within gun and torpedo range of the enemy. A carrier had to avoid exactly those weapons because she was inherently vulnerable. She would probably be unable to function as a carrier once she had been involved in a gun action, because it would not take much gun damage to disable her flight deck and to wipe out any aircraft on it. Her aircraft could not affect the outcome, because planes would be unable to fly off or on, or to survive on deck, during a gun battle. On the other hand, tactical exercises at Newport certainly had shown that a larger number of smaller carriers offered advantages.

The situation changed once total cruiser tonnage was limited. Pratt returned from London pointing to the treaty provision allowing construction of flight-deck cruisers as a major opportunity for the United States; there was enough tonnage for up to eight such ships. Each could accommodate a squadron of aircraft, so together the cruisers might equate to two aircraft carriers beyond the limit assigned by the Washington Treaty. As noted in the previous chapter, this possibility affected the choice of the next carrier design, justifying the choice of larger (hence less numerous) ships.

Evaluating Alternatives

In the aftermath of the London Naval Treaty, War College President Rear Admiral Harris Laning wrote the General Board in September 1930 stating that for the past several months the War College had been working on a table showing what it considered the most effective fleet the United States could have when the treaty expired in 1936—what might later be called a "balanced treaty navy."[16] The letter is interesting because it led to an exercise showing how the war-game style of fleet comparison (and indeed of thinking about problems) translated into proposals for construction. Laning was pleased that the General Board's program corresponded roughly with what he had in mind. Figures for carriers (at that time four more *Rangers*) and submarines were fixed. The question was how to use available cruiser tonnage. Laning wanted to use all the available cruiser tonnage and also to build destroyer leaders (a type not yet in the U.S. fleet) for five of the seven squadrons of destroyers allowed under the treaty. He was particularly concerned that the General Board did not intend to use all the allowable 6-inch gun cruiser tonnage. The London and Washington treaties were due to expire in 1936. A replacement treaty would be negotiated at a 1935 conference. It might well further reduce allowed U.S. cruiser strength. Anyone trying to make such a cut would find it easier if the United States failed to build all the cruisers it already could. Whatever was built or under construction in 1935 might define the new limit. Laning recommended keeping decisions as to details of 6-inch cruiser tonnage as fluid as possible, yet at the same time obtaining authorization (and laying down all permissible ships) before 1936. The General Board clearly regretted the loss of 8-inch cruisers at the London Naval Conference. It was not about to accept the treaty provision offering nearly 46,000 tons of 6-inch gun cruisers in exchange for the last three 8-inch gun cruisers. Without this tonnage, the treaty allowed the United States a total of 143,500 tons of 6-inch cruisers. The ten *Omaha*-class "scout cruisers" that had been ordered in 1916 accounted for 75,000 tons. However, because they had been laid down from 1918 on, they would be eligible for replacement beginning in 1938.

The value of the powerful carriers had recently been demonstrated. The dramatic success of the carrier *Saratoga* during the 1928 Fleet Problem showed that they could usefully operate independently of the battle line. War games emphasized that point: They should stay as far as possible from the battleships to avoid being caught by enemy aircraft. A fast carrier had only one type of surface ship to fear: a fast cruiser. Conversely, only cruisers were fast enough and powerful enough to be effective carrier escorts. To protect a carrier, at the least they had to match enemy firepower. In 1930, the U.S. Navy contemplated ultimately operating seven fast carriers. If each required a four-ship cruiser division as escorts, the needs of carriers alone would account for 28 8-inch gun cruisers.[17]

Laning's letter to the General Board was a new departure. In the past, since the War College President had been an ex officio member of the General Board, most communication was informal. Admiral Pratt had just cut that connection. Probably as a result of Laning's September letter, General Board Chairman Rear Admiral Mark Bristol wrote Laning on 14 October asking that the college work out tactical and strategic problems using proposed types of ships, either those recommended by the department or those suggested by the college itself—the latter being preferable.[18] Bristol emphasized that he wanted the War College to work out what was needed on the basis of its gaming; If games demonstrated the worth of a proposed type, "the Department and the General Board would then have a very sound foundation for preparing plans and developing such a type of vessel and in the same way avoid working on types which the College has demonstrated are not the best for the Navy.... [T]he College should suggest new types and improved types by utilizing them in your game. Specifically such new types have been suggested recently." The latter presumably referred to the flight-deck cruiser.

As for the choice between 6-inch and 8-inch guns, the U.S. Navy saw the former very differently from the British. To the British, 6-inch guns would match the heaviest guns any merchant ship converted for commerce raiding could mount, making a cruiser

The Naval War College did not have much impact on the agreements reached during the 1930 London Naval Conference, but it certainly did affect what the U.S. Navy decided to build afterward. In convening the conference, the British were concerned mainly to kill off the heavy cruisers, which they considered ruinously expensive. They assumed that any navy building cruisers armed only with 6-inch rather than 8-inch guns would adopt ships roughly equivalent to their own 6-inch gun cruisers, displacing about 7,000 rather than 10,000 tons (cost was about proportional to tonnage). The principal U.S. naval expert at the conference was Admiral Pratt, who returned home having accepted the British view. It says much about the strength of the analysis the War College was able to bring to bear that he reversed his position and accepted that the cruiser the U.S. fleet needed was much larger. The result was the *Brooklyn* class. *Honolulu* (CL-48) is shown here in a 9 February 1939 photo. (NHHC NH 53562)

armed with such weapons an effective trade-protection ship. To the U.S. Navy, on which the 6-inch gun had been imposed, its main virtue was that it could fire much more rapidly than the 8-inch gun arming its large cruisers. That made the 6-inch gun most valuable as a means of protecting the battle line from enemy destroyer torpedo attacks during, before, or after a fleet action.

At this time, the General Board liked the idea of small cruisers, so it wanted five ships displacing about 7,000 or 8,000 tons. Laning wanted the remaining 30,000 tons devoted to three large 6-inch gun cruisers of "special type," to be laid down in 1932, 1935, and 1936. The first might well be a CLV, the other two of an unspecified type to be chosen in a year or two, when U.S. needs would be clearer. The CLV should be tested before more were built. Laning would use gaming to test the CLV, a well-protected 6-inch gun cruiser, and other possibilities. He was careful to point out that the college was making no claim as to the feasibility of the CLV. Assuming it was feasible, its tactical value could be tested in a war game.

Laning's September 1930 letter seems to have had some effect. The General Board dropped its demand that no light (6-inch gun) cruisers be built until all the approved heavy cruisers (8-inch guns) were completed. In December, it approved design requirements for a 10,000-ton cruiser armed with 6-inch guns, and CNO Pratt approved them in January. Pratt also favored a small (6,000-ton) cruiser for distant scouting and for fleet work. He may well have been influenced by the British, who had assumed during the run-up to the treaty conference that 6-inch gun cruisers would be scaled down, hence much less expensive, versions of the unaffordable 8-inch gun cruiser. It is not clear why they thought so. British internal documents reveal no idea whatever of the logic of the other maritime powers. It seems to have been assumed that they, like the British, would be driven by a desire to reduce the cost of ships. The British were content to limit guns but not cruiser size. They were soon unpleasantly surprised that both the Americans and the Japanese built rather large 6-inch gun cruisers, the *Brooklyn* and *Mogami* classes, ruining what the British clearly imagined had been a diplomatic victory.

In contrast to British reasoning, which was based on self-interest and a tacit assumption that all governments shared the same rationale, the Naval War College examined the logic likely to drive the other major powers, particularly Japan. That was the method the college had long been teaching to students who had to prepare an "Estimate of the Situation" before deciding what to do. The first stage was not what they wanted out of a negotiation or problem. The first stage was to guess what the enemy wanted, and what the enemy's logic might be. Gaming might show that the guesses embodied in the "Estimate of the Situation" were flawed, but simply writing an estimate offered great advantages.

The initial War College report (6 December 1930) focused on the proposed flight-deck cruiser. It concluded that the ship should have the same speed and endurance as existing heavy cruisers (32.5 knots, 10,000 nautical miles at 15 knots) and should be protected at least against 6-inch (preferably against 8-inch) fire. On 18 December, CNO asked for a formal comparison between an 8-inch gun cruiser and a flight-deck cruiser armed with 6-inch guns. The War College submitted results on 2 January. They included a comparison in which both types of ship tried to beat off a destroyer attack on the battle line during a general engagement. A current 8-inch gun cruiser (*Northampton*) was unable to sink all four attacking destroyers before one of them managed to fire her torpedoes at the battle line. The CLV, with or without aircraft, did far better, because its nine 9-inch guns fired much faster than the heavy cruiser's nine 8-inchers. Either gun would kill an approaching destroyer with a single hit. In a simulated night action, in which the enemy destroyers were illuminated by U.S. destroyers at 8,000 yards, the CLV sank all four attacking destroyers in three minutes, the 8-inch gun cruiser in three to four minutes. In offensive screening, the CLV was far more effective. However, for screening purposes, the CLV needed protection against 8-inch fire, even at the cost of some aircraft.

At this point, the War College examined the much wider issue of what sort of cruisers the U.S. Navy needed. CNO and the General Board envisaged building a variety of cruiser types, as foreign navies did. Gaming showed that this would be folly: No U.S. commander could be sure that his ships would encounter comparable foreign ships. What a foreign navy used depended on what it was trying to do. "For this reason, the development of special type cruisers based solely on the opposition they are to encounter on a particular duty is *not* practicable [emphasis in the original]." The type of cruiser to be used in a particular role should depend entirely on what the U.S. commander was trying to accomplish. "We should design our cruisers primarily to accomplish a *purpose* for us, but at the same time giving to each the maximum power for offense and defense." Numbers were valuable, but in wartime, weak cruisers could quickly be created by converting merchant ships. However, under the existing treaty there was no way to quickly increase fighting strength. Ammunition allowances should be as large as possible, and gaming indicated that radius of action should be at least 10,000 nautical miles at 15 knots. All of this seems to have been an indirect way of killing Pratt's 6000-ton cruiser.

On 11 December 1930, Admiral Bristol of the General Board formally requested a War College study of the characteristics of future fleet cruisers, based on a list of duties.[19] The eight 8-inch gun cruisers the Navy had already laid down (the *Pensacola* and *Northampton* classes) were lightly protected, as were their foreign equivalents. However, the Bureau of Construction and Repair had managed to provide far better protection in a new design, designated CL-38 at this time.[20]

C&R pressed for a version of the latest 8-inch gun cruiser (CL-37–38), armed with 15 6-inch guns (three quadruple and one triple mount) instead of nine 8-inch guns. It argued that when employed with the fleet, she would be "worth half a dozen lightly armored, lightly gunned 6-inch gun cruisers such as are the prevailing style today in foreign navies as well as our own." Added displacement provided vital advantages not always included in military characteristics: seaworthiness, sustained sea speed, and habitability. Compared to a 6,000-ton ship, a 10,000-ton ship had every advantage at sea, in anything but nearly smooth water, in maintaining her speed. Moreover, a larger ship could carry more guns per ton. That would make each gun less expensive, since ship cost was roughly proportional to tonnage.

If the three allowed 8-inch gun cruisers were built, 73,000 tons were left. It could be distributed in various ways, assuming that all of it went on the same type of ship. How could these ships be compared? The designers thought that the largest type (10,000 tons) could be protected to the point that she could get close enough to an 8-inch gun ship for her 6-inch guns to become effective. The War College's gaming techniques were centered on ideas of relative fighting strength. Its initial approach was to graph the relative fighting strengths of the different possibilities at various ranges, and to compare them with the existing U.S. heavy and light cruisers (*Pensacola* and *Omaha*) and with the latest proposed 8-inch gun cruiser (CL-38, which became *San Francisco*). Fighting strength took rate of fire into account, so for the 6-inch cruiser the War College tried both the rate of fire realized in the *Omaha*s and an increased rate of fire that it was already using in other cruiser comparisons. All of this entailed an enormous amount of work, and as of 6 January 1931 the results had not yet gone to the General Board. There had not been enough time to digest them.

First, the War College compared U.S. cruisers to U.S. cruisers. The 6-inch gun cruisers performed surprisingly well inside 10,000 yards, where they could penetrate the armor of an 8-inch gun cruiser. However, there was an important caveat: At a target angle of 45 rather than 90 degrees (i.e., when the ships were not on parallel courses), which might not be unusual in a cruiser-on-cruiser engagement, the new heavy cruiser (CL-38) could penetrate the 6-inch gun cruiser without being penetrated. It was pointless to continue to pursue all the possibilities, so the analysts reduced the study to two 6-inch gun schemes (one armored against 8-inch guns with twelve guns, and one with protection against 6-inch fire and nine guns), plus the existing *Omaha*, and *Pensacola*. A comparison with Orange cruisers showed how much better the new CL-38 was, compared to the lightly protected (but more heavily armed) *Pensacola*. This seems to have been a distinct surprise; the Naval War College officer wrote that "Lord, if I had only had this a year and a half ago [mid-1929], when I sat at a meeting of the General Board, considering our new cruisers, and I could find no sympathy there for my prefixed opinion that we needed more protection on our cruisers." A similar comparison with the Red (British) cruiser *Sussex* again showed how effective better-protected ships (CL-38 and the 12-gun ship) were at "nice cruiser ranges" of about 13 to 20,000 yards.[21]

So much for single ships; what of the numbers that could be built, given the N-square law (combat power is proportional to the square of the number)? The War College analysts compared three ships with 8-inch protection with four with 6-inch protection (a total of 28,000 tons in each case). They were surprised to discover that better protection overcame the N-square law at cruiser ranges of 11,000 to 17,000 yards. When the *Pensacola* was replaced by the slightly better-protected *Northampton*, the value of superior protection was particularly obvious outside 18,000 yards. The increased rate of 6-inch fire would devastate a poorly protected ship. When the two 6-inch ships were put up against an 8-inch cruiser with good protection (CL-38) the result was very different; at a 45-degree target angle, it would have been even more pronounced in favor of CL-38. As a further experiment, the two cruisers and an *Omaha* were pitted against a division of six destroyers. *Omaha* looked rather weak. It was obvious that numbers were attractive, but fewer well-protected ships were a much better bargain. That effectively killed the small fleet cruiser. The graphs were included in a formal reply to the General Board dated 10 January 1931.

The War College was so effective in setting U.S. cruiser policy that cruiser design might be considered a measure of its success. In that sense, the outcome of the London Naval Conference of 1935 (which led to the 1936 London Naval Treaty) might be considered a measure of the downfall of the War College as a think tank basing its advice on gaming and game-related analysis. This time, the British succeeded in reducing the maximum size of new cruisers to 8,000 tons, even though the War College had advised U.S. delegates to avoid exactly such an outcome. Their argument was based on what war games tested: how well the fleet could carry out the national strategy. Cruiser size does not really figure in accounts of the negotiation; the U.S. delegation was far more concerned to beat down a British attempt to reduce the size of battleships and to develop a means of controlling Japanese construction (the Japanese removed themselves from the treaty system). The War College was not, it seems, even consulted as the design for a new small cruiser was developed. This is the outcome: USS *San Juan* (CL-54) of the *Atlanta* class, photographed on 3 June 1942. (National Archives 19-N-31525)

All of this calculation, which employed methods developed specifically for war gaming, had real consequences. In 1931 the U.S. Navy had eight completed 8-inch cruisers and was building another seven. The War College analysis revealed just how weak those ships were. The ideal solution was the new CL-38 design, but it was difficult to change the designs of ships already under construction. Even so, that was done. Ships to be built in Navy yards were reordered to the new design. The two ordered from private yards could not be totally redesigned, but they were given considerably better protection.

The War College's initial formal answer to the cruiser query was a study by the Department of Intelligence, delivered in February 1931.[22] It reflected gaming techniques, both in its attention to what other countries were likely to do and in its methods of comparison. As the War College saw it, the question was how the United States might best distribute the available cruiser and destroyer tonnages *based on what the other two major naval powers already had and were likely to build*. To get to this point, the War College looked at the situation from the points of view of Orange and Red (primarily Orange) in terms of the likely tasks of their fleets, as derived from war-gaming experience. To ask the question about the policy a likely enemy would adopt before setting U.S. policy was the core of the decision-making process the War College taught. This approach may seem obvious, but it is striking that when the British looked at the same problem they made no such analysis.

Orange's vital tasks would be to maintain command of the western Pacific and eastern Asiatic seas and to maintain the security of Orange sea communications—to prevent attacks on Orange shipping. Orange would fight a maximum-intensity defensive war intended to keep Blue from strengthening its position in the Far East. This was not a bad description of the Pacific War as actually fought. Orange would have to find and attack the Blue fleet before it got to the Philippines; if it arrived, its mobility would have to be impaired. Blue lines of communication would have to be attacked. The Japanese might occupy enemy territory. In that case their navy would have to escort shipping and protect it during and after a landing. The Orange navy would also have to protect Japanese shipping to insure sufficient imports to carry on the war. Even though Japan could receive massive amounts of material via protected sea routes from China, Manchuria, Asiatic Russia, Korea, and Sakhalin, in a war lasting perhaps more than two years she would also need southern trade routes for commodities such as oil and cotton. Japan was noticeably weak in tanker tonnage, so these ships would particularly need protection.

Even though Japan's war would be defensive, it would open with considerable offensive action, for which the 12 Japanese heavy cruisers already built or building would be suitable. The eight largest Japanese heavy cruisers were superior to all but the latest U.S. heavy cruisers. The other four (*Kako* class) were inferior to U.S. heavy cruisers, but compared well with the only U.S. cruisers normally attached to the U.S. fleet, the *Omaha* class, due to their greater radius of action. At shorter gun ranges, the *Omahas* were probably superior due to the higher rate of fire of their 6-inch guns. Combined with the three Orange battlecruisers, the Orange heavy cruisers would make a powerful advanced force. Six of the smaller Orange cruisers were already employed as destroyer leaders. At this time, the Japanese were building what they described as 1,315-ton destroyers (they were actually considerably larger), and within the available destroyer tonnage they could fill out the same six-squadron destroyer organization.[23] With the larger cruisers assigned to distant operations and seven smaller ones needed as flotilla leaders, the governing consideration for Orange would be numbers both for the fleet and for convoy duty. The sea lanes involved were short, and nearby bases and fortifications would make it unnecessary for the Japanese to build large cruisers. Orange had few ships suitable for conversion to cruisers. The Japanese quota of heavy cruisers had already been exhausted.[24]

By 31 December 1936, based on ships that would become over-age, Japan could lay down 50,955 tons of new cruisers. The War College had a variety of alternative cruiser designs from which the Japanese (like the Americans) might choose.[25] The treaty allowed Japan to build flight-deck cruisers, but that seemed unlikely given the numerous Japanese-held islands suited to seaplane operation.[26]

All of this logic implied that the Japanese would build 6-inch gun cruisers roughly equivalent to the (nominally) 7,100 ton-*Kakos*, which could mount as many as ten 6-inch guns. That would immediately give them five ships. The Japanese later announced that they were building four 8,500-ton *Mogami*-class cruisers, armed with 15 6-inch guns each. Thus, the War College prediction was not too far off the mark, except that the Japanese claimed that they could squeeze much more into the announced tonnage than any Western navy could. No one seems to have suspected the reality, namely that the *Mogami*s proved unacceptably flimsy and had to be rebuilt with much greater tonnage (which was never announced by the Japanese).

Given the assumed Orange program, what should Blue do? Ultimately Blue would want to gain control of vital Orange trade routes—the end-game blockade—while protecting its own. Possible Blue courses of action would be the "through ticket to Manila," or a fleet advance to an improvised base elsewhere in the Philippines, or seizure of an intermediate base short of the Philippines. The fleet would have to undertake very extensive scouting, and it would have to protect its lines of supply once it was established in the Western Pacific. In any case, Blue had to be ready at all times to face a maximum attack by Orange. War games showed that Orange generally tried to wear down Blue's battle line superiority (as provided by the Washington Treaty) using submarines, aircraft, and light torpedo craft (possibly including small torpedo boats not limited by treaty) based in the Mandates. They would have to be supported by Orange scouts. Blue would therefore need scouts to locate and defeat the Orange raiding/scouting parties. Enough Blue scouts would have to survive to ensure victory in a fleet action.

In wartime, when tonnage limits would no longer apply, Blue could build destroyers in three months, as U.S. shipyards had shown they could during World War I. However, cruisers would take at least a year. The Naval War College therefore argued that the United States should take advantage of the 10 percent of allowed destroyer tonnage that could be used instead for cruisers; every effort should be made to provide more cruisers. What type should they be?

During the U.S. fleet's advance to the west, it would need scouts. At the least, they would cover the large areas immediately ahead of the moving fleet. They would also screen the fleet so that the enemy's scouts would be unable to determine the U.S. formation or even its precise course. With the advent of aircraft, the area that had to be scouted and screened expanded enormously. To the college, the 10,000-ton flight-deck cruiser seemed an ideal scouting platform, because its aircraft could cover such large areas.[27] New technology offered it a considerable punch. It could operate dive bombers, which already seemed to be extremely effective against any ship short of a battleship. As long as aircraft from a flight-deck cruiser found the enemy first, the ship was more than a match for any enemy cruiser. If the aircraft did not find the enemy first, the flight-deck cruiser could be well protected. The college argued that flight-deck cruisers would provide aircraft during the early phases of the war. They would free the big carriers to concentrate on gaining and maintaining an initial advantage in the expected climactic fleet battle.

The Naval War College view was that U.S. 8-inch gun cruisers and flight-deck cruisers would be used mainly for offensive screening, scouting, convoying, and commerce raiding. Therefore, 6-inch cruisers would be used mainly on the flanks of the fleet to beat off torpedo attacks by enemy destroyers and also to back up torpedo attacks by U.S. destroyer on the Japanese battleships. Their high rate and volume of fire would suit them to this role. Screening and convoying would be secondary roles. The existing and projected 8-inch gun cruisers already offered a margin over their Japanese counterparts.[28]

No matter which type of cruiser the U.S. Navy chose to build, the available tonnage allowed for as many as eight flight-deck cruisers, which the War College's analysis suggested would be superior to 8-inch gun cruisers. Thus, if all 18 heavy cruisers were built, it would be possible to justify all eight flight-deck cruisers, each superior to any 8-inch cruiser for most cruiser duties—a remarkable achievement, as it would give the Navy 26 cruisers instead of the 24 the General Board had struggled for a year earlier.

The Naval War College report denigrated 8-inch cruisers as an artifact of the Washington Naval Treaty rather than a response to any particular tactical need; no one reading it would imagine that the General Board had insisted on such ships in 1920, before the treaty had been imagined. According to the report, these ships did not meet any particular tactical need, and were wanted only to offset similar ships in other navies. It seemed unlikely that more would be built in Europe.[29] The report argued further that heavy cruisers, particularly the faster French and Italian ones, had turned out to be poor sea boats. Too much had been sacrificed to cram 8-inch guns and protection into 10,000 tons. If such ships were given good protection, they would be slower than cruisers armed with 6-inch guns. In any case, the United States could not get enough 8-inch cruisers.[30]

The Naval War College had already demonstrated that small 6-inch gun cruisers were not worth building. It pointed out that by abandoning construction of 8-inch gun cruisers at 15 (rather than the allowable 18) and by transferring tonnage from destroyers, "we not only get better ships but more of them." College studies showed that although it was impossible to protect a cruiser against 8-inch fire at short ranges, a heavily gunned 6-inch gun cruiser had superior fighting strength to an 8-inch gun cruiser at ranges under 8,000 yards owing to its higher volume of fire.[31] Heavy cruisers might well encounter 6-inch gun cruisers at such ranges owing to low visibility or to tactical disposition. The 6-inch cruiser should have as many guns as possible, so that it could exploit their higher rate of fire, twice that of an 8-inch gun. Some of those guns might be disabled before the 6-inch cruiser got within effective range. To make certain it was superior once it was within range, the 6-inch gun ship should have at least three times the initial volume of fire as the 8-inch ship at the outset. That meant 12 6-inch guns against the usual eight 8-inchers in a British cruiser, for example, or 15 against Japanese cruisers with ten 8-inch guns. "As the life of any cruiser will be short after penetration starts, volume of gun fire will count heavily in determining which ship remains afloat." Firepower was particularly attractive in ships that might attack enemy trade, as the enemy would be forced to devote more of his limited cruiser force to convoy protection. The War College pointed out that protection could increase the life of a ship under fire and thus preserve her speed and increase the total amount of fire she could deliver. This comment reflected the fire-effect rules and hence the sense of how a ship would behave in combat. The trouble was that in a fight with an 8-inch gun cruiser, the 6-inch gun cruiser had to close the range quickly enough to retain firepower. A really fast 6-inch gun cruiser could not be protected at all, so by the time she got within range she might have no firepower left.

The War College view, then, based on gaming experience, was that the U.S. Navy should go as far as possible to build flight-deck cruisers and the largest 6-inch gun cruisers it could get by exploiting all treaty allowances. The latter was the 9,600-ton Scheme 2. Available tonnage would amount to eight flight-deck cruisers and six Scheme 2s. The resulting cruiser force would be considerably more powerful than Orange's—as long as all U.S. heavy cruisers were devoted to cruiser functions. In 1931, four heavy cruisers were used as flagships. One was flagship of the Asiatic Fleet. A second was flagship of the U.S. Fleet (the Navy was experimenting with the idea that the fleet commander should not be onboard a battleship engaging enemy battleships). Another was flagship of the Scouting Force; a fourth was flagship of the Scouting Force cruisers. All of them could be replaced by large light cruisers, but that would leave only two such ships for fleet functions such as beating off Japanese destroyer attacks and supporting U.S. destroyers. The War College therefore punted: Instead of those two large light cruisers, it might pay to invest the tonnage instead in six 6,500-ton cruisers, a type CNO had suggested for use as flagships and for other fleet duty.

In marked contrast to the tactical evaluation of Orange cruisers in the context of a Pacific War, the evaluation of the Red (British) cruiser force was political: The United States had to maintain parity in tonnage and in fighting strength.[32] It appeared that the British force as of 31 December 1936 (when the London Treaty expired) could be predicted much more accurately than that of the Japanese.[33] Alternative U.S. cruiser fleets could be rated as to total numbers, volume of fire, and types.

The War College staff spent its Christmas 1930–31 break working on the evaluation of the flight-deck cruiser. Laning himself wrote that he had long believed in the possibilities of this type of ship, particularly given the obvious power of dive-bombing—and the impossibility of stopping a dive-bombing attack. The problem, which the War College was unable to answer, was how to trade off the air power exercised by a flight-deck cruiser against gun power. That mattered, because a flight-deck cruiser was relatively expensive in terms of guns per ton. However, even building eight rather than four flight-deck cruisers would carry only a limited cost in total volume of fire.

Ultimately, then, the War College advocated building no further 8-inch cruisers and instead building a mix of flight-deck and conventional 6-inch cruisers. In its view, a flight-deck cruiser was likely to be more than a match for an 8 inch–gun cruiser. However, the college did not advocate simply building the largest possible 6 inch–gun cruisers. Instead it envisaged a trade-off between size and numbers. The college proposed a program of four flight-deck cruisers, two 10,000-ton 6-inch cruisers (convertible to flight-deck cruisers), and 18 8,250 ton-cruisers. This program would exploit the bonus tonnage available if 6 inch– rather than 8 inch–gun cruisers were built as well as the transfer of 18,300 tons of destroyer tonnage. Given the severe shortage of cruisers the study showed, the United States should immediately build as many as possible of a new unlimited class of sloops (up to 2,000 tons with 6-inch guns) to release real cruisers from convoy duty.[34] It should also prepare for rapid conversion of merchant ships to cruisers and carriers, with special emphasis on carriers, because the flight-deck cruiser aircraft would need landing decks. They would launch their aircraft before opening fire, but once they were firing they would be unable to recover their aircraft.[35]

A War College report submitted on 2 April 1931 emphasized the role of torpedo-firing destroyers in any future fleet battle. The Japanese would use their destroyers to overcome the treaty-mandated numerical superiority of the U.S. battleships. Conversely, in a fight against the British, who had treaty-approved parity in numbers, U.S. destroyers would attempt to tip the balance. The War College added that "unless such successful [destroyer torpedo] attacks can be prevented, they may prove the deciding factor in an engagement."[36] Torpedoes mattered because each torpedo hit would inflict more damage than several shell hits. Torpedoes could be delivered by destroyers or by torpedo bombers or by submarines. The college emphasized destroyer torpedo attacks because the other two means of delivery were unlikely to be as effective in a major fleet engagement. Torpedo-bomber performance was poor (torpedoes were very heavy, and aero engines not yet very powerful), to the point that in 1931 the U.S. Navy was seriously considering abandoning them. Submarines were not fast enough to keep up with the battleships in combat.

Further consideration of exactly what the battle line mission required further affected the desired characteristics of any new cruisers. If, as the War College argued, an enemy destroyer attack might decide the battle, any cruiser assigned to the battle line needed enough fire power to break it up. Since any of the 6-inch cruisers being considered could break up destroyer attacks, the deciding factor should be the ability to stop enemy cruisers backing up the attacking destroyers. That strongly favored a 12-gun cruiser protected against 8-inch fire, since the enemy might use heavy cruisers to back his destroyers. Comparison techniques developed for gaming showed that the 12-gun cruiser had greater fighting strength than the British *Sussex* class below 19,000 yards or the Japanese

Nachi below 13,500 yards.[37] It offered considerably more fighting power, ton for ton, than any Japanese or British cruiser that might be built for some years to come, barring radical changes in either country's building program.

In June 1931, CNO Pratt reversed himself and recommended the heavily protected 12-gun cruiser. The Secretary of the Navy approved this recommendation several days later. It is difficult to say just how much the War College and its game-based analysis were responsible for the change, although Pratt was certainly sensitive to War College arguments.[38] Pratt certainly did not accept the War College argument that the last three 8-inch cruisers should not be built as such, or that tonnage be transferred from the destroyer to the cruiser categories. Given the onset of the Great Depression, any such ideas may have been dropped because they could not be realized. However, the War College's arguments favoring large over small 6-inch cruisers—arguments based on war-game experience and on a war-gaming style of analysis—did carry the day once the U.S. Navy began building such ships. The U.S. Navy built seven large cruisers each armed with 15 6-inch guns—the *Brooklyn* class—plus two ships of a somewhat modified type (the *St. Louis* class).

The War College was asked again in 1932 to analyze U.S. cruiser requirements. That March, C&R claimed that there was "a strong and increasing sentiment" for a change in the characteristics of future 8-inch gun cruisers. Pressure for further cuts in spending and for further arms control would reduce capital ships to the point where large cruisers effectively replaced them.[39] C&R suggested that cutting cruiser speed by five knots and replacing the 5-inch antiaircraft gun with the new (developmental) 1.1-inch machine cannon would make it possible to add considerable armor.[40] The Secretary of the Navy asked the War College for an analysis.[41] The War College rejected the C&R idea: Speed was expensive, but cutting it back to 27.5 knots went too far.[42] It also rejected the idea of replacing the 5-inch secondary guns.[43] Above all, abolition of battleships was so bad a policy that the United States should reject it, and the War College should go on record again (as in the past) as strenuously opposed to it.

The War College also affected cruiser design in another way. Up to this time, all cruisers, including large ones, had been armed with torpedoes as well as with guns. Should that be the case this time? Given the treaty limit on overall cruiser size, the weight usually invested in torpedo tubes could be invested instead in antiaircraft guns or in armor, or both. The issue came up in General Board hearings. While the War College was working on overall cruiser characteristics, the board asked it to settle the torpedo question based on game experience.[44] In the games, cruisers in a general action had only about a 50-50 chance of firing torpedoes. The chance of firing torpedoes in a one-on-one encounter, which might often happen in wartime, would be even less. There was also the question of whether a shell hitting torpedo tubes might detonate the warheads.[45]

This study may have been the source of the War College perception that it was so important to deal with torpedo attacks by enemy destroyers. In games, destroyers succeeded in fewer than 2 percent of attempted attacks, but it turned out that many of the misses were due to errors by players and to lack of experience in conducting torpedo attacks. Thus, destroyers would probably do much better in reality.[46]

The War College reviewed experience in the major tactical game in each of the classes of 1923 to 1928; the major tactical game and the tactical phase of the major strategic game of the classes of 1929 and 1930; and of the two major tactical games played to date (January 1931) by the class of 1931, a total of 12 games. Not all had been carried through to a finish in which one side was completely routed, but all reached a decisive stage. The study illustrates the kind of data war gaming could provide.

A total of 311 cruisers were involved (108 Blue). Of these, 156 fired torpedoes (61 Blue, 95 enemy): 56.5 percent of Blue cruisers and 46.8 percent of enemy cruisers fired, an average of 50 percent. At the moment of firing, the average percentage damage already imposed on the cruisers was 29 percent, almost all of it by gunfire. Cruisers hit by torpedoes were so badly damaged that they could not fire their own torpedoes. A total of 99 cruisers (23 Blue) were sunk without firing torpedoes at all, an average of 32 percent. Of these, 53 percent were lost to gunfire and 30 percent to torpedoes. In many cases, gunfire was a contributing factor in the ship's loss, but the tabulation showed only the primary cause. At the end of the games, 102 cruisers still had torpedoes onboard (42 Blue), one third of all engaged. The total number of light cruisers lost was 190 (57 Blue), an average of about 61 percent. All of this understated the damage to cruisers, since it was now known that gunfire could detonate torpedo warheads. Moreover, as gun fire control improved, it seemed that effective gun range would increase to the point where torpedoes would be less and less attractive.

Until that year, cruisers had rarely been the object of air attack; they were damaged mainly by gunfire. The difference must have been the rise of dive bombing, which offered a good chance of hitting a fast-maneuvering cruiser. In the two tactical games of the Class of 1931, Blue launched early air attacks against the enemy cruisers, while the opponents concentrated on Blue capital ships and carriers. Out of 27 hostile light cruisers in the two games, seven were sunk immediately and three more sank later due to heavy bomb damage. None of the ten sunk had the chance to use their torpedoes. This amounted to about 40 percent of hostile cruisers lost to air attack. In the larger tabulation, the percentage of damage sustained by light cruisers up to the instant of firing torpedoes had to be estimated. However, limitations of torpedo fire due to damage received made it possible to come close.

The General Board took this experience seriously: It was more useful to protect a cruiser against air attack than to provide her with torpedoes for rare offensive use. On that basis, the board approved removal of torpedoes from existing U.S. treaty cruisers to provide space and weight to double their 5-inch antiaircraft batteries. Torpedoes were eliminated from the designs of the new better-armored cruisers. U.S. tacticians developed night tactics in which cruisers relied on their long-range guns and left torpedo attacks to destroyers.

Other navies thought otherwise. The Japanese gave their cruisers heavier torpedo batteries and spent considerable effort on torpedo tactics and on long-range torpedoes. These efforts paid off during the night actions in the Solomons during World War II. U.S. destroyers did poorly with their own torpedoes until late in the war, at Leyte Gulf in October 1944. The record of the Naval War College suggests strongly that this failure cannot be attributed to a lack of interest in torpedoes, but rather to poor execution of torpedo tactics. Evidence of sustained U.S. interest in torpedo attack includes the continued construction of destroyers with unusually heavy torpedo batteries.[47]

Cruisers at War: Three Years of Red-Blue Warfare

Gaming showed that the U.S. Navy could probably never have enough cruisers. That was graphically illustrated by a series of large-scale games simulating a Red-Blue war.[48] The first game in the series (Operations III, Class of 1932) was specifically designed to evaluate the flight-deck cruiser, and more generally the variety of cruisers then planned. BuAer chief Rear Admiral Moffett had recently testified before the House Naval Committee, and there was a real possibility that one or more would be laid down. After the game, Captain Adolphus Andrews, the War College chief of staff, considered the tactical employment of CLVs its most interesting aspect. The CLV had been introduced in the 1931 Red-Blue game in a conventional scenario. It had been very valuable in strategic scouting; three such cruisers formed and held a scouting line across the Caribbean despite strong opposition by Red heavy cruisers. Falling naturally into the dual roles of cruiser and carrier, and not being as efficient in either case as the pure type, the CLV was not as useful in ship-to-ship battle. In the 1932 game, Red had CLVs, conducting protective rather than offensive scouting. One question was how such a ship would function in low visibility, when she could not operate her aircraft. In this case, the CLV was placed behind the scouting line of cruisers for protection, but an enemy cruiser found her anyway. Because aircraft could not fly under some weather conditions, it was impossible simply to substitute CLVs and their aircraft for scouting cruisers ahead of a fleet. However, CLVs could back up heavy cruisers in a scouting line, providing the needed punch if more powerful enemy scouting forces appeared.

Using CLVs as scouts was controversial. When first proposed, they had been envisaged as cruisers with an extra punch offered by their aircraft. The idea of armed scouting had been rejected. Instead, it was assumed that bombers should be held back until the objective was found, then launched. In any case, a CLV did not have enough aircraft onboard to substitute for a carrier; if it was so regarded, it would be better simply to build carriers (however, carriers were limited by treaty, whereas CLVs could still be built in numbers). In the post-game wash-up, Captain S. C. Rowan, head of the Operations Department, pointed out that the value of the extra punch depended on the efficiency of bombing (presumably dive bombing) "which we at present don't know." Tests against the old armored cruiser *Pittsburgh* were ambiguous. In good weather, the attackers made 60 percent hits, demonstrating their skill with the Assistant Secretary of the Navy watching. When the Secretary of the Navy came out to watch in bad weather, they made no hits at all.

To compare a CLV with a 10,000 ton all-gun cruiser, three separate engagements were tried. Although a change in the fire-effect rules had just devalued the 1,000-pound bomb the CLV aircraft could deliver, in good visibility the CLV sank one light cruiser, two armed merchant cruisers, and a converted seaplane tender. In a second case, a mixed cruiser force (four CLVs, two heavy cruisers, six armed merchant cruisers) faced four Blue heavy cruisers and one large Blue light cruiser. In this case, one Red CLV sank two Blue heavy cruisers using 1,000-pound bombs, but Blue cruiser aircraft using light bombs were able to knock out the flight deck of one Red CLV (with all its aircraft). A third case was a low-visibility engagement in which CLVs could not fly aircraft. A Red CLV made contact with a Blue light cruiser at 12,000 yards and was sunk. A Red heavy cruiser then arrived and sank the Blue light cruiser. It was clear that the sacrifices in armament and protection required in order to provide air facilities made the CLV inferior as a *gun cruiser* to a conventional large cruiser armed with 6-inch guns.

The game was fought far from Blue's bases. That emphasized the need for long cruising radius, seakeeping qualities, protection, and high sustained speed—as differentiated from, and far more important than, high maximum speed. That, in turn, meant large ships. Although the Blue cruisers were individually more powerful than Red's, they suffered badly during raiding (anti-shipping) operations. Four of Blue's 15 cruisers were sunk. The Naval War College Research Department pointed out that, considering how few cruisers Blue had compared to the number Red had plus the size of the Red merchant fleet (some of which could be converted into armed merchant cruisers), the loss of cruisers would mean far more to Blue than to Red.

The Department of Research (which at the time meant Captain Wilbur van Auken) reached what must have been a shocking conclusion:

> This brings up the point of the desirability of *planning* for the *combined* use of cruisers and auxiliary cruisers by *Blue and Red*—if a *war* can be *visualized where Blue and Red* have such paramount interests together as to be allied. If such might be the case, it is to the *advantage* of Blue and Red to maintain a preponderance in cruiser tonnage. . . . In view of the present world's unsettled political situation and the common interests of Blue and Red, it seems worthwhile to propose for serious consideration and planning the question of 'pooling' *all* vessels of Blue and Red of *all* types to be *ready* to *meet any* emergency against *one* or *more* powers allied and prepare strategic and tactical plans. With Blue's *known* weakness in cruisers, and the delays in laying down more, it is vital to *plan* all *phases* of how Blue's Navy might be increased n power by aid from Red. It is a *fact* that

our combatant ships cannot be increased after a declaration of war to any great extent. During the World War, less than *ten* destroyers *authorized* for construction under the "250" program were completed. It also a fact that one battleship requires four years to construct *after* authorization at the minimum; an aircraft carrier, *three* years; one *new* cruiser (with *plans* completed), *three* years; one submarine of 1,700 tons, two and one-half years; and destroyers, about three months (*if* and *when* yards are organized and prepared to go into an emergency program). And now, ship yards are not as numerous or prepared as in *1917*.

If war should come within three years, it is manifest that the United States Fleet cannot be materially increased and would be *about* as it *exists* in combatant ships *today*, except all types would be *three* years *older*. It is therefore seen *how* the Blue types stand relative to Orange and Red. With no change to add *trained* fighting ships (for example against *Orange* and *her* Allies) it would *seem* that strategic and tactical studies *should* be made, games played and *plans* evolved, for the best *possible* use of *all vessels* of *all* types, bases, and the logistics for *Blue's* navy in a coalition with *Red*. *This* is the *only practicable* method of Blue's increasing her battleship, carrier, cruiser, or destroyer tonnage *besides* making *available* vessels for the convoy of our overseas force. An even though such may never come, the problem is one which merits the serious consideration of all officers of our Navy and Army [emphasis in the original].

This must be one of the earliest references to such a possibility in official U.S. Navy documents. Van Auken's analysis is in the OPNAV War Plans Division set of papers for Operations III, not in the War College set; it got to the Washington planners. Van Auken's coalition games were apparently not played; at least no records of such coalition games prior to 1941 appear to have survived. Van Auken's pessimism explains why rapid U.S. fleet expansion, a key World War II reality, did not figure in inter-war gaming. He repeated his call for coalition planning in a commentary on a later Blue-Orange game. There is no indication that it had any impact; it was absolutely the opposite of the U.S. official position. It says a great deal for independence of thought at Newport that van Auken did not suffer for his comments.

Van Auken did not say so, but an important question was whether cruiser operations against enemy trade—an important part of any campaign against Japan—were still viable, or whether such operations would have to be turned over to submarines. Several later war games examined the submarine option.

Unusually, the three games involved portrayed a war in which the U.S. strategy was to choke British trade, rather than the typical defensive war in the Western Hemisphere.[49] The basic U.S. strategy was to eject Red from the hemisphere while imposing economic pressure as a way of depriving Red of food and raw materials. The war had to be fought in such a way that Blue would not be so badly damaged that other countries, such as Orange, could take advantage of the post-war situation. Naval and military action would secure Blue's vital sea areas, dominate Red territory in the Western Hemisphere, prevent the passage of Red trade through the south Atlantic, and draw the Red fleet into a general engagement on disadvantageous terms. Blue could certainly block traffic through the south Atlantic, but the British could shift their traffic to the Suez-Mediterranean-Gibraltar route, outbound traffic passing down the West African coast and around the cape. Both navies threw out cruisers across the Atlantic narrows, between the Azores and the coast of Brazil. A clash between the British cruiser screen (intended to protect trade further north) and Blue raiding groups led to the destruction of two Blue heavy cruisers and the destruction of the flight deck of a Red CLV.

Tactical IV (played in December 1931 for the Class of 1932) was a continuation of Operations III.[50] It was a partial fleet engagement testing the effect of carrier superiority: Red had four fleet carriers, Blue one. Blue was stronger in battleships (12 to four), but Red was stronger in destroyers (60 in six flotillas compared to two destroyer squadrons and one destroyer division for Blue). Red still had one flight-deck cruiser, but she had no aircraft onboard. Blue's fleet was slower than Red's, and Red had reinforcements on the way. The forces were so placed that Red could open with a night destroyer torpedo attack against Blue.[51] The Red destroyers never managed to hit the Blue battleships, although they did sink one heavy and three light cruisers and 12 destroyers (Red lost 25 destroyers and two leaders). One Red half-flotilla got within six miles of the Blue battleships, but did not know it.

Before dawn, both sides launched big air attacks. Red's carriers were on the flanks and in the rear of his battle line. They sent out scouts, and then launched bombers and torpedo planes that hit the Blue battleships. Red met considerable opposition from Blue fighters and anti-aircraft guns. A later attack destroyed the flight deck of the Blue carrier using 116-pound bombs. Meanwhile, four Blue heavy cruisers caught one of Red's carriers and ultimately sank her at the cost of one heavy cruiser. Another Red carrier had her flight deck damaged. Both sides lost many aircraft. The carrier engagements once more demonstrated the vulnerability of carrier flight decks and also the danger of encountering enemy heavy cruisers.[52] The game also showed how important carrier speed was: Blue's 32.5-knot heavy cruisers easily ran down a 24-knot Red carrier (presumably either *Hermes* or *Eagle*).

That night, both fleets launched night destroyer attacks, the two striking forces meeting quickly after dark. In the melee, the side with superior numbers at the point of contact won. Each broke through to the other's battle line, albeit at a considerable cost. The game was called soon afterward.

The scenario continued with Operations III (Class of 1933), which was paired with Tactics III.[53] Like the 1932 game, this one was concerned with the attack on Red seaborne trade to exert economic pressure. Red might be able to occupy Portuguese possessions in the Atlantic (the Azores and the Cape Verde Islands); Blue might gain the use of Liberian ports. Blue would have liked to seize a base in the Azores, but could not do so as long as Red was of roughly equal strength. Overall, Blue suffered because it lacked an advanced base in the Eastern Atlantic. The issue of Red access to the Portuguese islands led to a discussion of the status of the Mandated Islands in a Blue-Orange war and the status of Canada in a Blue-Red war. The opinion was expressed that it would be Blue's policy to try to bring Canada into the war on the Blue side, despite her connection with Red.

A "special situation" to be considered envisioned a collision blocking the Suez Canal, causing much British shipping to be rerouted through the Cape Verde focal area, where it would be subject to attack by Blue forces and to defense by Red forces in West Africa. The Research Department considered the game useful as a reminder of how dependent the British were on overseas trade, but did not add that the only country more dependent on overseas trade than Great Britain was Japan, the likeliest future enemy. The game demonstrated that a huge number of cruisers and aircraft would be needed to prosecute a trade war against Red, and that it was futile simply to sweep enemy areas with large forces. The game also brought out the need for detailed coordinated search plans.

Both sides made extensive use of air power. In this game, Blue had six fleet carriers (three carrier divisions), although as yet the U.S. Navy had only two carriers, with a third under construction and two more authorized. A fourth carrier division consisted of a seaplane tender, an airship tender, and the large airship *Akron*.[54] There were also seven small converted carriers (two of which were initially troop transports), of which three would be available on the West and four on the East Coast 120 days after mobilization; and 14 converted seaplane tenders.

Both sides had flight-deck cruisers. The game again showed how vulnerable their flight decks were once they were found by enemy air forces, presumably because they could not put up extensive fighter opposition. Of three Red flight-deck cruisers sunk, one was lost to submarine torpedoes and two to enemy cruiser gunfire. However, all three Blue heavy cruisers and all three Blue flight-deck cruisers lost were sunk by bombing, presumably dive bombing. Even so, flight-deck cruisers seemed to be justified by their value in sudden encounters with enemy cruisers, and particularly in their ability to quickly develop contacts by air scouting.

In the corresponding Tactical III game (January 1933), each side sought a fleet action to relieve domestic pressure for decisive action. The Red commander was ordered to fight if the situation was favorable, but he knew that defeat would destroy the entire Red Empire. He told his government that conditions were unfavorable, but did not know whether it understood his fleet was inferior.[55] Both naval commanders knew that the Red fleet was the weaker of the two. In this game Blue did have a base in the Azores, but its hold was precarious. Each commander wanted to avoid fighting within the radius of land-based aircraft. For Blue that meant avoiding the area near Portugal; for Red it meant staying out of the area around the Azores.

Blue estimated that he was superior in capital ships and probably in the air, and about equal otherwise. He therefore chose to make an early air reconnaissance followed by an early maximum-strength air attack to slow the Red battle line, followed by a decisive general action. Blue's air plan set priorities: the enemy carriers first, then cruisers in the van (i.e., the enemy scouts), and finally the enemy battle line. In its critique, the Research Department argued that Blue should have concentrated everything on the Red battle line, "the foundation and most of the vital framing of the enemy's structure. The other arms form some of the substance, but all of the trimmings. In the give and take of a naval battle I believe the structure which loses some of its substance and all of its trimmings is much more able to stand up than one which has its vital framing and foundation weakened." The critique conceded that bombs and torpedoes would not give such visible effect on battleships as they would on cruisers and carries, but they would weaken the enemy battleships and thus make gunfire more effective. Aerial torpedoes could slow the enemy down and make it difficult for him to evade action.

Red had to weaken Blue's battle line. He hoped to do that using submarines, but the critique pointed out just how much they had to achieve; it seemed unlikely that Red had enough submarines to do that. The idea of using submarines against enemy battleships was of current interest at the War College. Red enjoyed an advantage in submarine speed, so that its submarines could maneuver to remain in position relative to the Blue battle line, whereas Blue's submarines could not maintain their positions relative to Red's battle line. By the end of the game, submarines on both sides had made successful attacks. The Research Department considered their hitting rate, 8.5 percent, far lower than that to be expected in war. The game showed that if the enemy sought action, the U.S. battle line could be maneuvered to draw the enemy's battle line into submarine-attack range. The threat of such attacks had been an important factor in British tactics at Jutland in 1916. During the war, the British had built fleet submarines specifically to exploit opportunities created by the maneuvers by their own fleet. The Research Department concluded that submarines would likely play an important role in future fleet actions. Tactics had to be adjusted to exploit their capabilities.

The Research Department considered air operations one of the most interesting parts of the game. "The air did in one hour what the destroyers had failed to do after working all night." Both Red and Blue air commanders were student officers who had had recent fleet aircraft experience in command of *Lexington* and *Saratoga*. Their tactics were what the fleet currently planned to use. The Blue air force attacked the Red carriers 140 miles away. It sank one Red carrier and destroyed the flight decks of three more. This was the sort of attack, staged when ships were far out of sight of each other, that seemed so revolutionary in the air-sea battles fought a decade later, for example at Coral Sea and Midway.

Either deliberately or by chance, Blue attacked when the Red planes were being serviced aboard. This attack was quite expensive: none of the Blue planes got back to the Blue carriers. Even after receiving tremendous damage from Blue, Red's surviving carriers managed to make three torpedo hits on the Blue battle line. Red later wrecked the flight decks of both Blue carriers. When the game ended, each side had four carriers, but their flight decks had been so badly damaged and aircraft losses so heavy that they were unlikely to play any further role in the battle that day. Blue lost a good proportion of its spotting aircraft. Without enough spotters, its battleships would have had to fight at reduced gun range, but the game was called before the battle lines engaged. Dive bombing sank three Red cruisers (including one flight-deck cruiser) and five Blue cruisers during the tactical game. Such attacks also cost two Blue cruisers 50 percent damage.

The Research Department asked whether the game was realistic. Would the U.S. population tolerate removal of the whole fleet from the Atlantic coast? Would it tolerate staking the safety of the country on a single decisive battle? The question recalled the experience of the Spanish-American War. As the Spanish squadron crossed the Atlantic, the politicians representing cities on the Atlantic coast demanded warships to protect their cities. Naval strategists had concentrated all the modern U.S. warships so that they could fight the Spanish. Fortunately, the U.S. Navy had managed to assign obsolete ships, even Civil War monitors, to the cities. Fortunately, too, the Spanish had not used the concentration of the U.S. fleet as an opportunity to strike at one of the rich coastal cities. The U.S. Navy had had no way of knowing where the Spanish squadron was headed. It took time to realize that it was in Cuba. After that the fleet concentrated off its Cuban port for a decisive battle. Would such concentration, particularly in distant waters, ever be possible in a future emergency? Without it, the U.S. fleet could not be sufficiently powerful to win the decisive battle.

As it turned out, no real public-relations problem developed when the Pacific Fleet was concentrated far from the Pacific coast in 1940. It deployed from its usual base at San Diego to its war base at Pearl Harbor. To some extent, the public may have been reassured by claims by the U.S. Army that its aircraft and coastal batteries were sufficient direct protection of the U.S. coast. Despite previous joint exercises with the Army War College, the Research Department did not raise the possibility that the Army's efforts would make just the sort of concentration the Navy required a practicable proposition.

The last game in the series was Operations IV of the Class of 1934, paired with Tactics IV.[56] This time, Blue sought a base in Portuguese Guinea, in Africa (the Bissagos Islands off the coast), from which its fleet could cover the entire western north Atlantic, the south Atlantic, and the Caribbean, and also disrupt Red trade along the African coast. Portugal was Red's ally, which justified seizure of the base. This was considered part of the second phase of the war, the attack on British commerce. Should the seizure of the base succeed, the next step would be to seek control of British trade routes north of Gibraltar. It was assumed that Red could not tolerate more than three weeks' interruption of the trade between Gibraltar and the Channel.

Geography favored Red, which had many bases in the theater of operations. Red's line of communications from Gibraltar would be short and well-protected, but Blue had a long line of communications stretching back from the African coast. Blue was not only far from home, but also far from the target area: 1,800 miles from Gibraltar, 2,200 from Trinidad, 3,400 from New York. The situation was not too different from that in an Orange war, with Blue penetrating deeper and deeper toward Orange through areas full of Orange bases. Red's only logistical problem was that it had been cut off from New World oil supplies, hence had to rely on European and East Indian sources (the Middle East was not yet a source of oil). However, Red had at least six months and probably a year of oil stored at home.

To fight so far from home would be a tremendous undertaking. The writer of the critique, Captain George B. Wright, pointed out that either Red or Blue would need some compelling reason to undertake so vast a task as extend its naval power into the other's sphere of control in a war between them. He did not rule that out, and he pointed out that governments might undertake such efforts "particularly if the risks of such an undertaking are not clearly foreseen by responsible naval authorities." It would be vital to show what the risks were, "and if there is any reasonable way in which they may be avoided or minimized." The main lesson of the games was to show how difficult either side would find it to project its power across the Atlantic. Red projection was tested in games in which Red tried to seize or defend a base in the Caribbean or in Canada. The current game tested Blue projection in the opposite direction. Operations became prohibitive, or nearly so, when the power on the offensive lacked both allies and a secure base capable of both receiving and maintaining the fleet. Even so, Wright could imagine several possible forces that would justify a transatlantic war: public opinion, economic pressure, social conditions at home, international relations, or political factors.

Red's fleet covered Red trade well enough that in order to block that trade, and win the war, Blue had to destroy that fleet. Thus, Blue's problem was really to bring the Red fleet to battle. Equating the destruction of Red's fleet to strangulation of Red's vital commerce was equivalent to the assumption in the Orange war plans and games: The Orange fleet had to be destroyed as a prerequisite. The other question in the Red-Blue game was how to compel Red to fight even though both sides understood that Red's fleet was inferior, particularly in long-range gunnery.

It was possible that Blue's seizure of a base off of the African coast would affect Red public or political opinion in such a way as to force the Red fleet to fight, but that could not be guaranteed. The Research Department pointed out that the previous year's Blue "Estimate of the Situation" had missed the key point of Blue's battle line superiority, concentrating instead on operations near Cape Verde to control Red trade.

Blue estimated that Red would see little point in accepting battle anywhere south of the Azores-Madeira line, since in that case it would be accepting the same disadvantage of a long line of communications that Blue had accepted. Red would most likely attempt to wear Blue down as it moved north in preparation for a fleet engagement. This was much the attrition war the U.S. Navy assumed the Japanese planned against a U.S. fleet heading west.

Red formed both a main body and a separate raiding force built around its three battlecruisers and its carriers. The raiding force was intended to operate against Blue carriers and other heavy units using its aircraft and night destroyer attacks. Not only did the main Red raiding force attack on Blue develop into an air attack as intended; Blue's main counterattack was also by air.

Once the raiding force was heavily engaged with the Blue covering force, the Red main body turned away. Only at the end of the game did it approach the Azores to fight a decisive day battle against the main Blue fleet. The threat of the Red battleships forced Blue to expose its battleships to harassing attacks while accompanying his base force and train to the Azores. The Red battle line remained intact, but the Blue battle line was exposed to air and submarine attack. This was much what the Japanese managed in Blue-Orange campaigns: attrition, possibly without any decisive battle.

The key strategic point was that each commander had to understand what the mission really was. Was it simply to attack trade, or was it to gain supremacy by destroying the enemy's fleet? Students tended to concentrate on the immediate task of attacking enemy trade. Few of them realized how much superior Blue was, due to a combination of superior battle line and air strength. Several student plans might have been different had they realized the actual relative strength of the two sides. That was a very serious criticism since the whole point of much of the analysis done by the War College was to make such relationships clear. No one mentioned dispersing shipping as a means of protecting trade, yet, Wright wrote in the critique, "it is the cheapest and in war the most commonly used form of protection."

Like many other games, this one involved very large air forces. Red had eight carriers to Blue's seven, but Blue had many more aircraft: 798 to 479.[57] Each side had three flight-deck cruisers, each with 18 heavy dive bombers onboard. Blue also had six seaplane tenders supporting 54 floatplane torpedo bombers and 54 patrol planes, compared to four Red tenders supporting 48 patrol planes. Blue carriers were generally much faster than Red's (except for *Langley*, still a carrier at this time) and much longer-legged. How much did the difference in carrier numbers count? The main point developed by the critique was that massive numbers of aircraft were needed. When the enemy had adequate air forces, only aircraft could develop contacts with screening ships into useful information about the enemy's position and disposition. There was little point in mounting air attacks against an enemy with an approximately equal air force unless some particular effort was made to catch his aircraft on deck. It seemed that long-range shore-based aircraft would be very useful for this purpose, because their attacks would not reveal the position of the carriers. To the Research Department, outstanding features of the exercise included large-scale air attacks by both Red and Blue.[58]

Neither side could prevent the other from scouting by air. Generally each carrier division (two carriers) located its objective with its own bombers and then sent out strikes. That seemed necessary given the rapidly changing situation and the great dispersal of enemy forces. Blue lost more aircraft when one of its carriers was torpedoed than by all previous air operations. At the end of the problem, Blue had one intact carrier, one with some underwater damage, and two seriously damaged with their flight decks out of action. Three more had been sunk. Research attributed the relatively poor showing of the Blue carriers to a fundamental mistake: Blue had used carriers to overwhelm the modest Red shore defenses. In doing that, Blue had left the Red carriers full freedom of action. Blue had forgotten the developing rule that there could be no security as long as the enemy carriers remained.

Red found that a scouting flight out to 150 miles was not nearly as effective as had been imagined, as it found only one of the Blue carriers. Red had been standing by with carrier bombers and patrol planes to attack based on this one flight. On the other hand, Blue needed three carriers to support his landing operation, because its fighters were distributed among them. That brought the carriers close to the beach, hence vulnerable to discovery and attack. One of the players thought that Blue should have used fewer of his carrier aircraft to support the landing. Had he used some of them as scouts, he would have found a Red force to the east.[59] Air scouting was done mainly by carrier aircraft and patrol seaplanes, not by cruiser or battleship aircraft.

As in previous games, it was assumed that whoever attacked an enemy carrier first had an excellent chance of destroying all or part of her flight deck, rendering her ineffective. In this particular game, one Blue carrier was sunk by initial air attacks and one had her flight deck ruined. One Blue flight-deck cruiser was sunk by air attack and the other two had their flight decks ruined. Two Red carriers were sunk and a third had her flight deck ruined; all three Red flight-deck cruisers were sunk. Initial air attacks also sank four Blue heavy cruisers, two Blue light cruisers, and two Blue destroyers. Red lost five heavy cruisers and six destroyers, including a flotilla leader. Capital ships on both sides were hit, but not sunk. In addition, Red submarines sank the damaged Blue carrier that had retained her flight deck, nearly sank another (she was later sunk in a night destroyer attack), and badly damaged a third. Blue battleships and cruisers were hit. One of the two damaged flight-deck cruisers was sunk, together with three heavy cruisers (two of them damaged and hence slowed). The last Red air attacks sank a Blue carrier and ruined the flight decks of two more.[60]

The action was somewhat unrealistic in that at least one large air battle involved aircraft of such disparate types that they would not have worked together. For example, there was an air battle between 30 Red bombers and 42 Blue observation aircraft from Blue battleships. Normally the bombers would have so outperformed the observation aircraft that they would never have fought each other at all.

Both sides used flight-deck cruisers as carriers rather than as cruisers (for screening). The Research Department pointed out that flight-deck cruisers had been conceived as improved scouting cruisers that could use their aircraft to make and develop initial contacts better than normal cruisers because they had more (and higher-performance) aircraft. This idea was not tried out. In the game, these small carriers of "a distinctly offensive type" proved useful. Their early loss on both sides could be attributed to the way they were used—in easily located positions, which the enemy did not have to disperse to find—rather than to their assigned tasks.

When Red aircraft struck the Blue main body, the Blue flight-deck cruisers (with bombers onboard) tried unsuccessfully to help repel the attack; one was sunk and another lost her flight deck, so that flight-deck cruiser aircraft in the air had only one ship left on which to land. At this point, 40 of the original 54 bombers were in the air, and they put the flight decks of two Red carriers out of action. Nine of them recovered onboard the remaining flight-deck cruiser. They were launched again to sink the Red carrier previously crippled. Another 12 took off but missed the Red fleet; ten more were re-serviced and took off. These last ten sank a Red flight-deck cruiser. Surviving bombers from the Red flight-deck cruisers landed on Red carriers, reinforcing their air groups.

During Red air attacks on Blue and Blue air counterattacks, Red's three flight-deck cruisers were kept near each other and near the Red carriers; all were lost to air attack. Their 54 bombers were launched against the Blue carriers, but this order was then cancelled in favor of attacking a Blue heavy cruiser. Before they arrived, this ship had already been sunk, and they were ordered back. En route they found and attacked Blue cruisers, 21 bombers being shot down in the process. The survivors helped defend the Red carrier on which they were to land. Survivors eventually landed on a Red flight-deck cruiser.

Postscript: The Fate of the Flight-Deck Cruiser
The flight-deck cruiser was undoubtedly the most radical element of the War College's vision of a future balanced U.S. fleet. In August 1931, the War College declared the cruiser of paramount value in the "service of security and information" (scouting and screening) and in sea control. The latter meant finding and destroying individual enemy ships. Their aircraft should be first scouts, second dive bombers, and third (if practical) defenders. In Blue-Orange games in 1931–32, these ships gained early information of the enemy upon which major operations were based. The Naval War College evaluation was that even if the flight-deck cruisers were sunk after carrying out their scouting mission, they had already done their share in making the overall operation a success. Generally, they were better able to hide than the big carriers due to their small size and their high speed. They showed valuable flexibility. Against all that, their flight decks were just as vulnerable as those of the big carriers, and without those decks, or in weather that prohibited flying, they lost their special scouting value. Without her aircraft, the flight-deck cruiser could not stand up against other cruisers with more powerful gun armament; she had to be supported tactically by other cruisers.

CNO Admiral Pratt, who took credit for the clause of the London Naval Treaty that allowed the United States to build flight-deck cruisers, included one in the projected 1933 program.[61] The General Board submitted characteristics in January 1931 and that May it included a flight-deck cruiser in the projected 1932 program. This recommendation was repeated in November and December 1931. This proposal was serious enough that the Bureau of Construction and Repair produced a set of contract plans for the ship, tentatively designated Cruiser No. 39 (this hull number was actually applied to the heavy cruiser *Quincy*).[62]

Support for the flight-deck cruiser was by no means unanimous.[63] For example, in September 1930, CinC Battle Fleet argued that to combine a gun cruiser with a flight deck was "unsound in every way." He pointed to the severe fire hazard presented by both airplanes and their fuel, "so great that we shall have even under the most advantageous conditions, a timid cruiser, one fearful of being brought into position where its guns are to be used. It is my experience that airplane carriers demand for their protection, even in problem work [i.e., fleet exercises] more potential offensive power than they themselves can exercise....The more the Navy is reduced [by treaty], the more important it becomes that every unit of the Fleet shall be offensive in its character, and the offense to be contained within itself."[64] Moreover, it seemed that a single 4- or 5-inch shell could put the ship out of action by setting her fuel or airplanes on fire. However, Admiral Pratt completely rejected these arguments.

The FY33 program was intended as the first in a ten-year program submitted by the General Board in December 1931. Even after the first year's program was cut to conform to President Hoover's proposed budget, it included one flight deck cruiser. This program was abandoned due to the Depression. In September 1932, the General Board recommended two 20,000-ton carriers for FY34. It omitted the flight-deck cruiser, which was clearly less important. The two carriers were built as *Yorktown* and *Enterprise* after the FY34 program veered back and forth in content. The flight-deck cruiser vanished from the board's formal construction plans after that.

The main sponsors of the flight deck cruiser disappeared in 1933. Admiral Pratt retired as CNO. Rear Admiral Moffett was killed when the airship *Akron* crashed. Admiral Laning left the War College, and flight-deck cruisers disappeared from its games. The death of Admiral Moffett was probably by far the most important. He enjoyed enormous political influence; he can presumably be credited with House interest in flight-deck cruisers and also with the perception, in 1931, that Congress would authorize such a ship, but not pure 6-inch gun cruisers. For its part, from 1932 on the General Board was more interested in the two large carriers (the *Yorktown* class) then planned. It saw them, and a further smaller carrier that could be built on the rest of the available tonnage, as a better way to provide more aircraft to the fleet. There was no one of Moffett's stature to press the argument that even with all of the carriers, the fleet would not have nearly enough aircraft, or that the fleet badly needed dispersed as well as concentrated air power. This argument, moreover, was closely associated with simulations of an Orange war. By 1934, War College influence was waning (see the next chapter).

In 1934, the War Plans Division, which received reports of war games, asked for at least one flight-deck cruiser; the General Board sought fleet opinion. The ship, if any, would be included in the FY36 program. In October 1934, Fleet CinC Admiral Reeves wrote a letter strongly opposing the idea on the ground that carriers and cruisers "simply do not mix"; the proposed cruiser did not have a sufficiently long flight deck. "From fleet experience, a flight deck 200 ft long will restrict flight operations to an intolerable extent." That such a deck would be extremely dangerous was shown by the numerous barrier crashes onboard existing carriers with much longer flight decks leading up to their barriers. Reeves was certainly pro-aviation; he was "completely cognizant of the vital need for more and more airplanes in the fleet particularly for scouting." His solution was "vigorous promotion of the patrol plane program including the preparation of additional tenders and patrol plane bases."

The question the General Board asked was whether it was worth building a flight-deck cruiser for tests, or whether the concept could be tested using suitably modified real carriers—but the need for such a ship declined as new carriers were completed. Not only would five carriers be ready before a single flight-deck cruiser could be completed, but a new conference due in 1935 might allow the United States more tonnage. Since the Japanese had already announced their withdrawal from the treaty system, there was reason to imagine that the 1935 conference might fail altogether, ending all limitation. In fact, the new conference removed total tonnage limitations, but U.S. legislation continued to key construction to the earlier limits, something no one could have expected. Any new flight-deck cruiser would hardly be begun before the conference changed the rules. On the other hand, the fleet had a pressing need for more conventional cruisers.

Moffett's successor at BuAer, Rear Admiral Ernest J. King, proposed a new version of the flight-deck cruiser in March 1935. He cited new high-lift devices as a rejoinder to Reeves's critique. King had nothing like Moffett's clout. The idea was formally rejected by the Secretary of the Navy on 29 March 1935 on the grounds the General Board had stated: All cruiser tonnage should go into conventional types. CNO recommended that no further steps be taken until "it is practicable and allowable to build types other than those now authorized." BuAer did not give up. It revived the flight-deck cruiser idea in February 1936, when proposed disposal of the ten *Omaha*-class light cruisers to Brazil would free up tonnage.[65]

The flight-deck cruiser idea survived as late as 1940, when the Navy developed a large new cruiser program, but no such ship was ever built. It might be argued that the World War II conversion of light cruisers to light carriers (CVL) may have been inspired by the flight-deck cruiser idea, but these ships were never used the way flight-deck cruisers were played in war games.

6. Downfall

The role of the Naval War College as the U.S. Navy's primary tactical laboratory did not last. In October 1934, the Naval War College was transferred from OPNAV to the U.S. Navy school system administered by the Bureau of Navigation (BuNav, later renamed the Bureau of Naval Personnel).[1] In August 1932, its chief, Rear Admiral F. B. Upham, argued that the college was no different in principle from the schools he already administered—the Naval Academy and the Naval Postgraduate School.[2] The argument made here is that, at least up to 1934, that was not at all true: The Naval War College was at least as valuable as a laboratory, thanks to its unique war-gaming capacity. The BuNav argument had no chance of succeeding as long as the CNO was a former Naval War College president who had relied on war-gaming arguments to support policy, particularly air policy. It also seems likely that, as Naval War College president, Admiral Laning would have argued forcefully for the special capability his school represented.

It would seem either that Upham was ignorant of the special role of the Naval War College or that he was aware that it was becoming much less popular within the upper reaches of the Navy. In August 1932, Admiral Pratt's term as CNO still had nearly a year to run, as did Admiral Laning's term as Naval War College president. The president succeeding Admiral Laning, Admiral Luke McNamee, clearly considered the laboratory role vital. Pratt's successor as CNO, Admiral William H. Standley, took the result of the 1933 operations game seriously—he ordered a dramatic change in the war plan. That makes the shift to BuNav under Standley's tenure somewhat puzzling. The difference may have been connected with the departure of Admiral McNamee from the college. He requested early retirement to become president of the Mackay Radio and Telegraphy and Federal Telegraph companies, which had important strategic roles. It is, of course, also quite possible that Admiral McNamee decided to leave the college halfway through his term because he saw the change of status coming.

It is also possible that McNamee's successor, Rear Admiral Edward C. Kalbfus, took the change as an indication that the role of the college was due to shift in favor of its educational mission. He seems to have emphasized that mission, not just inside the college but also outside it, in the form of the correspondence course. In 1934, the fleet was just beginning to expand, and it could be argued that war was coming. In that case, it was may have seemed much more important to teach as many officers as possible to make sound decisions (using the methods developed by the college), than to provide the Navy with detailed advice as to the characteristics of its ships or the nature of its war plans. After all, in the latter mission the college was in effect competing with other naval organizations.

The run-up to the London Naval Treaty illustrates one of Admiral McNamee's last contributions, which was based on war gaming. In January 1934, the Secretary of the Navy asked the General Board for guidance as to the U.S. position at the forthcoming conference, which would frame a treaty to replace the existing Washington and London naval arms-control treaties in 1936. Advice was urgent because a U.S.-British preparatory meeting was scheduled for 18 June–19 July 1934. Late in February, McNamee wrote a memo for the board.[3] Although the memo nowhere refers to war gaming, it is evidently based on recent game experience. McNamee began with the "present disturbed state of the world." The "one clear outstanding fact is that the safety and security of our nation will depend to a large extent upon an efficient Navy...." That was no surprise, but what followed was: McNamee pointed out that without an adequate base in the Philippines, the fleet could not in itself exercise decisive influence in the Far East—the Japanese were just too strong. That echoed the conclusions of the 1933 operations game and its 1934 sequel, in which the fleet could not hold Dumanquilas, its temporary base, while protecting a vital convoy bringing it supplies. "A war between Japan and the United States alone under present conditions would involve us in losses entirely out of proportion to any possible gain."

McNamee was also apparently aware of Captain van Auken's call for cooperation or even alliance with the British. Although the U.S. Navy was not strong enough to restrain the Japanese, "the British and U.S. navies combined could enforce any reasonable restrictions on Japanese policies." Since British interests in the Far East were far greater than those of the United States, McNamee thought "it would appear wiser for us to support her where it is to our interest to do so, than to be placed in the position of fighting her battles for her. The more clearly Great Britain understands that this will be our policy, the surer we will be of obtaining joint action if and when it may become necessary."

Without adequate defense, the Philippines were a liability, and any attempt to fortify them now would create grave tensions with Japan. McNamee suggested some possible offers to Japan in return for more favorable naval ratios. Overall, he called for decisions prior to the conference, for example as to whether the Navy should be able to control the western Pacific or should be limited to control of the eastern Pacific and the western Atlantic. This was very much a consequence of campaign gaming, which showed just how much naval power a western Pacific offensive would require.

After 1934, the War College stopped providing advice to the U.S. naval leadership based on war gaming. It is not at all clear what happened, but it does appear that the War College was moved from OpNav to the navy educational system, and its educational role superseded its think tank role. At the time, it was led by Rear Admiral Edward C. Kalbfus, shown here as an admiral at the change of command ceremony when he relinquished command of the Battle Force. It is possible that the driving consideration was that the fleet was beginning to expand and that the War College had to reach a wider audience. Kalbfus certainly thought that way; he was very much interested in the college's correspondence course. (NHHC NH 120063)

McNamee also pointed out that U.S. policy of encouraging general reductions in naval power by U.S. example had been a total failure. The United States could out-build both Britain and Japan. Doing so would improve the U.S. strategic position and also provide shipyard employment. He was far from alone in pointing this out, and his vision was realized in the Vinson-Trammell Act passed later in the year. This was not an inference from gaming, but games did show just how heavy wartime losses might be.

Overall, McNamee recommended extending the London Naval Treaty with only minor modification because it provided equality with the British, "thereby insuring to us the maximum naval security attainable." It defined a "treaty" navy "which Congress and the people now understand as the meaning of 'Adequate' Navy, which they have always been willing to support." It put the United States in a stronger political position not to ask for anything more; considering the political situation, anything else would be worse. If proposed changes came from another power, the United States would not be caught up in justifying unacceptable ones. Finally, McNamee thought that the British wanted and needed a better ratio with Japan than the United States did, but it wanted the United States to press for that ratio.

McNamee's belief that Congress and the American people would accept a treaty navy was certainly in line with the Vinson-Trammell Act, and it may have helped inspire it. The act replaced separate authorizations (not appropriations) with a blanket authorization for ships to fill the tonnages allowed under the Washington and London (1930) Treaties as an effort to maintain a "modern Treaty Navy." The Naval War College had already studied the characteristics of a "balanced Treaty Navy," but it is not clear whether McNamee had that work in mind at the time.

For McNamee, desirable modifications to the London Treaty of 1930 included clarification of the meaning of "flight-deck cruisers" to insure that planes could fly off as well as on. He argued that the United States should accept the proposed abolition of submarines if all countries agreed to scrap fully and at once, and to accept inspection. The destroyer quota should not be changed "unless every submarine in the world is destroyed. They might otherwise be supplied by an allied minor power." McNamee absolutely rejected any change in cruiser ratios.[4]

For the British, the most important goal in a new treaty would be a dramatic reduction in the size—hence the cost—of new battleships, to 25,000 tons with 12-inch guns. To McNamee that was by far the worst danger in a new treaty. War-game experience (not cited) showed how much bomb and torpedo damage battleships could be expected to receive *before* they faced other battleships in the western Pacific. It took considerable tonnage to provide the requisite protection as well as protection against existing battleship guns. McNamee argued that the battleship tonnage limit should be increased to 40,000 tons so that ships could be armed with the 16-inch/50-caliber gun, which was known to be far superior to the 16-inch/45-caliber model arming existing ships, not to mention the 14-inch gun. "If the tonnage of a battleship is reduced it cannot retain the characteristics

The decline of the War College (or, rather, that of its influence) might be considered a reaction to the air-mindedness shown by Admiral Laning in 1930–33. It took a long time to build the remaining carriers that could be acquired within treaty limits, and the U.S. Navy did not accelerate carrier construction once treaty limits (on total tonnage) had been removed by the London Naval Treaty of 1936. However, any very visible naval program had to survive attacks by a very powerful pacifist lobby. Naval rearmament, under the Vinson Act of 1934, was justified as maintenance of a modern treaty Navy, and there was widespread sentiment within the Navy that any further expansion had to be tied in some way to the earlier naval arms treaties and the force ratios between countries they embodied. Any such policy had to reckon with the ratio between battleships and carriers embodied in the 1921 Washington Naval Treaty, at a time when carriers were very much experimental ships. New carrier construction, on a very modest scale, was embodied in the 1938 Vinson-Trammel Act, which in turn was a reaction to Japanese aggression in China. Congress provided only enough tonnage for two new carriers of *Yorktown* size, of which only one, *Hornet*, was ordered before war broke out in 1940. The question is whether the naval leadership was interested in any more than that. *Hornet* is shown newly completed, late in 1941. (NHHC NH 81313)

which are the chief arguments for the retention of the type." McNamee also wanted to keep the 10,000-ton cruiser—the fruit of the War College's analysis a few years before. "The keenness of our rivals to abolish it should alone be enough to arouse our suspicions."

McNamee warned that the British and Japanese might collude; some British statesmen had written that in the event of American withdrawal from the Philippines (which Congress was considering), the old Anglo-Japanese Treaty might be renewed. This was probably the only unrealistic point he made.

McNamee's arguments seem to have appealed to the General Board. What actually happened was complicated enough so that many of McNamee's points were ultimately lost. President Franklin D. Roosevelt was in a delicate position. There was enormous public support for continued reduction in naval spending through a new treaty. However, at the same time it was clear that (as McNamee wrote) the process of arms reduction by example was failing, and the Japanese were building and modernizing as rapidly as they could, to an extent U.S. naval intelligence could not know. Roosevelt's solution was simple. He sent Norman Davis, who had been negotiating arms-control treaties for some time, to London with a proposal for another 20 percent cut in fleets. The President was probably aware that no such proposal had the slightest chance of success. At the same time, CNO Admiral Standley went to London as the U.S. representative who had the best chance of securing agreement from the Royal Navy. It is not clear to what extent the President was aware that the British were already moving toward rearmament in the face of Japanese aggression in China, where they had important interests.[5]

When the Americans met the British for preliminary discussions, it became clear that the British desperately needed more cruisers. In order to reduce overall tonnage, they wanted to reduce the size of battleships and carriers, although they soon understood that they would have to accept the larger battleships the U.S. Navy wanted. Their new orientation showed in comments that the Admiralty was now in the ascendant (as it had not been during the 1930 negotiations) and that although the British wanted to continue the treaty system, they would let it die if they could not get a satisfactory one.[6]

From 1931 onward, a dominant "fleet faction" within the Japanese navy increasingly demanded that the "insulting" 5:5:3 ratio favoring the two Western powers be scrapped. When representatives of the three powers met in London in October 1934 to prepare for the coming conference, the Japanese offered what they knew would be unacceptable proposals, such as the abolition of "offensive" warships like carriers—the warships the Western powers would need to defend their interests in the Far East. On 29 December 1934, the Japanese abrogated the Washington Treaty, as was their right. It was probably clear that the treaty system was finished; to the Western powers the 5:5:3 ratio was an essential guarantee that they could defend their interests in East Asia in the face of the Japanese. Japanese delegates did attend the 1935 London Conference, but they withdrew on 15 January 1936. The United States rejected parity, leaving the other Washington Treaty powers to negotiate a new treaty. Italy withdrew to protest denunciation over her aggression in Abyssinia (Ethiopia), leaving the United States, Britain, and France.[7]

Without any agreement by Japan to abide by fixed ratios of total tonnage, the London Treaty of 1936 eliminated treaty ratios altogether, concentrating on limits on individual ship characteristics. The United States managed to fight off the British attempt to cut battleship size, but the British did manage to impose a new upper limit of 8,000 (rather than 10,000) tons on cruisers. That was very much against Naval War College advice, but it was probably a necessary concession.[8]

The new treaty reopened the question of exactly what sort of navy the United States needed. In 1930–31, the college's answer was based on war gaming and on the sort of reasoning it taught. In 1935, the General Board asked senior officers, including the new president of the Naval War College, Rear Admiral Kalbfus, for proposals. Kalbfus was certainly no stranger to the laboratory role of his college, having served as dDirector of War Plans.

The shift in acceptable Naval War College advice seems evident in what Kalbfus offered. He and CinC U.S. Fleet were asked to propose a post-treaty building program.[9] Kalbfus took his time to answer; CNO's questionnaire was issued on 1 April 1935 (with a reminder on 6 July), but Kalbus' reply was dated 1 October 1935. Given previous Naval War College memos, it is striking that Kalbfus made no reference whatever to gaming experience. He preferred to reason from basic political principles. The program had to accord with national policy and it had to be shaped to gain public support. Neither could be quantified. Kalbfus thought the public would probably be willing to extend the 5:5:3 ratio embodied in the earlier treaties. He emphasized his fear that anything more aggressive would place the onus for starting an arms race upon the United States. This was rather different from Admiral McNamee's advice a year earlier. Kalbfus saw the ratio as a guide to what was politically possible. The Navy had not even reached the limits set by the earlier treaties. Once that was done, it could continue building to maintain the 5:5:3 ratio as other countries built beyond their treaty fleets. The main change that Kalbfus envisaged was to use the number of ships rather than their tonnage as the standard, as that would enable the U.S. Navy to build ships of the types it needed.

The closest Kalbfus came to using gaming material was a comment that the existing fleet was satisfactory to fight a defensive war against Orange, but it was inadequate for a transpacific offensive campaign. In that case, Blue would needed at least a 2:1 ratio. Kalbfus considered it unwise, at least at the outset, to strive for that on a numerical basis. He suggested instead that the United States should seek effective superiority on a unit-by-unit basis. "Later on, should conditions arise to create a national sentiment in favor of increasing the navy, the opportunity should be seized to build up to a point at which war could, with reasonable assurance of success, be carried to Orange and Orange vital trade."

Kalbfus recommended building three battlecruisers instead of three battleships. In effect, he cited gaming experience, pointing to the long period of cruiser and commerce warfare that would probably precede a decisive battle-line engagement. Without battlecruisers, the U.S. Navy had to depend on air strength to deal with the four Japanese *Kongo*-class battlecruisers. "If for any reason our air forces were unable to operate in an area into which enemy battlecruisers had projected themselves, the latter would be a formidable menace to such light forces, convoys, and shipping of ours as might be there." To Kalbfus, sacrificing three new battleships was done to exchange "a tactical advantage during a possible general engagement for strategical and tactical advantages on perhaps numerous occasions during the pre-battle period." If the new Blue battlecruisers were

It certainly seems arguable that the heavy emphasis on battleships in the late pre-war building programs came at the expense of carriers—but it is also arguable that the existing U.S. Navy battle force was obsolescent and badly needed replacement, whereas the existing carriers were relatively young. There does not seem to have been significant sentiment in favor of changing the balance between battleships and carriers. The Royal Navy had a much larger carrier-building program (six ships), but all but one of its existing carriers were clearly obsolete. The main pre-war effort began with the *South Dakota*–class battleships, of which *Massachusetts* (BB-59) is shown in January 1946. Gaming did not figure in their design, which initially called for slow ships (about 24 knots) unsuited to carrier warfare. The ships' speed was drastically increased only when the U.S. Navy became aware (through radio interception of a report on trials of the battleship *Mutsu*) that the Japanese battle line was significantly faster. The crash redesign worked, but its nature testifies to the lack of earlier analysis. (NHHC NH 46430)

powerful enough, they might work effectively with the battle line. The new battlecruisers would have a speed of 30 knots; new battleships would have a speed of 21 knots, like existing U.S. battleships. If no battlecruisers were built, an increase in U.S. air forces would be the main way to counter the enemy's ships. That would mean more patrol planes.

A study of U.S. cruiser requirements, which has not survived in the Naval War College document, showed a need for 53 to 60 ships— on the basis of a numerical ratio, not a study of a particular kind of war through gaming. As of 31 December 1936, the United States would have 37 cruisers. Orange would have the same number; the 5:3 ratio would require 61 U.S. cruisers. If Orange scrapped over-age cruisers (an unlikely possibility), 28 would be left, so the U.S. figure would be 47. Kalbfus recommended 57 as a conservative figure. If the U.S. Navy also built battlecruisers, all 20 new cruisers could be 6-inch gun ships. Beside the *Brooklyn* class (10,000 tons) the Navy was then building, there was also interest at this time in 7,000-ton light cruisers—a type the college had explicitly ruled out in its deliberations a few years earlier.

In calling for ten squadrons of destroyers, Kalbfus noted that two of them would be used to screen carriers, which implies that he envisioned the carriers operating in groups. This was certainly not the lesson of war games, in which carriers survived by operating singly.

In the questionnaire (and in Kalbfus's answer), carriers came after battleships/battlecruisers, cruisers, and destroyers, suggesting that they now enjoyed a low priority. Kalbfus pointed out that Orange had a total of six carriers built and building, so in order to maintain a 5:3 ratio, the United States needed ten. He pointed to the desirability of having carriers re-service planes launched by battleships and cruisers, hence the potential value of two small carriers to work with the battleships. This was certainly contrary to previous Naval War College advice. Kalbfus tempered his advice: "[I]n view of the need for striking power in our air force…it is considered that we cannot afford to build vessels for servicing purpose alone within the 5:3 carrier ratio. In time of war, possibly converted merchant vessels could be utilized for this purpose." Kalbfus thus ended up calling for three more big carriers like *Yorktown* and *Enterprise*. If the smaller *Wasp* was not laid down until after the treaties lapsed, she should be re-ordered as a large carrier.

Kalbfus estimated that on the outbreak of war the Navy should have 12 patrol plane squadrons (18 aircraft each) based on the needs of a step-by-step advance through the Mandates, using these aircraft for patrol and screening. More would be needed as the fleet advanced more deeply into the western Pacific. This might be read as based on wargame experience, but not explicitly so.

Kalbfus's list of desired new ships in priority order placed battlecruisers first, then battleships, then submarines, then destroyers, light cruisers, and—finally— heavy carriers. His proposed building program was two battlecruisers, two battleships, ten submarines, ten destroyers, five light cruisers, one carrier, one minelaying cruiser, three tenders, one repair ship, and one net layer (to protect an advanced base). Even this list does not make it clear whether naval aviation was no longer considered very important. All U.S. carriers were under-age and entirely modern. All U.S. battleships were over-age and obsolescent at best. Nearly all U.S. destroyers were obsolescent at best, and nearly all of the submarines were obsolete.

The pressure to build as much carrier capacity as possible certainly evaporated after 1934. Whether that can be traced to the decline of the Naval War College is not certain. Surviving records do not show intense interest in building carriers. In commenting on a proposed ten-year building program in March 1938, the General Board mentioned that "in view of the arising questioned value of aircraft carriers relative to increasing improvement in characteristics of land planes, it is believed that if more carriers are desired they should be built as soon as possible."[10] The sense may have been that a gap was opening between the performance of land-based and carrier-based aircraft. The longer the Navy waited for new carriers, the worse the gap and the lower the value of new carriers. However, it cannot have been clear what was going to replace sea-based aircraft. Moreover, in 1938 BuAer was actively trying to develop very high-performance carrier-based fighters that would more than close the gap. It succeeded with the Vought F4U Corsair, which won the 1938 fighter competition and flew in 1940.

Kalbfus's call for battlecruisers resonated with the General Board. In 1935–36, the U.S. Navy's designers produced a variety of possible designs for the first two U.S. post-treaty battleships. The General Board favored a 30-knot ship with nine 14-inch guns (as signed, the London Treaty limited ships to that caliber). This was Kalbfus's battlecruiser. In addition to its ability to deal with its Japanese counterparts, it was fast enough to work with carriers. The alternative was a slower but better-armed and -armored ship. CNO Admiral Standley, acting as Secretary of the Navy in the Secretary's absence, rejected the fast design, adopting instead a 27-knot ship armed with twelve 14-inch guns. It was built as the *North Carolina* (the 14-inch guns were replaced by 16-inchers when the Japanese refused to agree to limit themselves to 14-inch guns). The really fast battleship returned only in 1938 when the *Iowa*s were designed, and they combined speed, firepower, and armor in a package that was impossible on the more restricted displacement available in 1936.[11]

As the world situation worsened dramatically from 1937 onward, U.S. Navy new construction policy seems to have been less and less connected to the lessons of campaign games. Comparisons of overall strength, generally in numerical rather than tonnage terms, seem to have dominated. In 1938, faced with Japanese aggression in the Far East and German aggression in Europe, Congress passed the second Vinson Act. The 1938 act simply increased the size of the Navy by 20 percent. The increase in carrier tonnage was slightly greater, from 135,000 to 175,000 tons.[12]

The character of the Vinson Act and the absence of detailed reasoning leading up to it suggest how badly the influence of the Naval War College had declined. To continue to use the tonnages set at Washington in 1921 to set relative battleship and carrier strengths was to imagine that almost nothing had happened in the interim to naval warfare, despite rather dramatic improvements in carrier air capability. Neither of the other major navies was following suit; both were building new carriers in considerable numbers.[13] It might be argued that the absence of the Naval War College laboratory showed in U.S. reluctance to order more carriers.

To complicate interpretation, the shift in strategy (due to Naval War College analysis based on gaming) changed the basis for air requirements. Until 1933, it was assumed that the fleet would fight its way through the Mandates to reach the Philippines as soon as possible. Once it reached the Philippines, it would fight the decisive engagement that would open Japan to blockade. En route it would face large numbers of Japanese aircraft, which the treaties did not limit. Campaign games included enormous air battles, with a high rate of wastage. It made perfect sense for Admiral Pratt to say that the fleet needed vast numbers of aircraft in order to survive long enough to fight the fleet battle at the end. Gaming data also showed that the direct contribution of aircraft to that decisive battle would be limited, because it was difficult for aircraft to damage enemy battleships. The aircraft could, however, strip the Japanese fleet of its other ships, adding to the U.S. advantage in a more powerful battle line (air spotting would also make U.S. gunnery far more effective).

It was one thing to defend the fleet as it passed through the Mandates en route to the western Pacific. It was quite another to move into the Mandates island by island. The Japanese would certainly defend their islands. Their aircraft would be able to hop from base to base to concentrate aircraft: That was exactly how they tried to defend Saipan in June 1944. However, once an atoll had been seized it could support both an air defense force and long-range seaplanes capable of scouting and also—importantly—of bombing. War games incorporated such aircraft (on both sides) as long-range bombers. Whether these games reflected BuAer's vision of seaplane power or were trying out a new tactic is not clear. There was apparently no direct attempt to determine how effective a seaplane force could be compared to the carrier force, or even how roles should be apportioned between them.

The new step-by-step strategy also made scouting more important. A fleet thrusting through the Mandates needed air and surface scouts to ascertain whether an attack on it was imminent. A fleet seizing one island after another needed to know much more about the position of enemy forces. Among other things, it had to decide which atolls were most vulnerable to attack, and which the enemy might find it easier to reinforce. To the extent that an advance through the Mandates might draw out the enemy's fleet—again, as it did at Saipan in 1944—it was necessary to know where that fleet was, even when it was nowhere near the advancing U.S. forces. Step-by-step also emphasized the ability of the seaplanes to relocate as the fleet seized successive island bases, since seaplanes could operate from any sheltered body of water, such as the lagoon of an atoll. To do that, the fleet needed far more numerous seaplane tenders.

In a world of treaties, the greatest virtue of a seaplane force was that it was entirely unrestricted by treaty. There were several attempts to limit the total number of military aircraft, but they all foundered on the reality that civilian aircraft, which could not possibly be limited, could be converted for military use. As was previously noted, when the war gamers at Newport wanted to test the idea of a long-range seaplane scout, they provided commercial Dornier Do X flying boats, the largest of their time, to one fleet or the other.[14]

The major change in the early 1930s was that seaplane performance improved to the point where, for an interval, it seemed that seaplanes might be effective long-range bombers. By that time, long-range seaplanes were already fitted to carry 1,000-pound bombs, but performance was less than impressive.[15] In 1933, however, BuAer decided to take advantage of evolving aircraft technology to seek a combination of longer range and higher performance.[16] What BuAer got was so impressive that the resulting airplane was designated a patrol bomber (VPB) rather than simply a patrol plane (VP): It was the Consolidated Catalina (PBY).[17] At least on paper, it was nearly as good as a land-based bomber. In reality, big seaplanes could not compete with land-based fighters, but they were effective in war games.

At least in theory, using seaplanes in a step-by-step advance would much reduce the need for carrier aircraft. Instead of having to weather continuous attacks from the Japanese-held atolls, the fleet would seize them one by one. Heavy seaplane bombers based on atolls could neutralize air bases on other islands. Bases not seized outright could be gassed repeatedly to make their use impossible.[18] As noted previously, such neutralization was a staple of war games.

The step-by-step approach changed the role of Marine Corps aviation. In 1938, the General Board was asked to estimate what the Marines would need to seize island bases. Marine aviators were expected to specialize in the support of Marines on the ground, so it was assumed that an assault on an atoll would be led by Marine aircraft. That opened the question of exactly how these aircraft were to reach the target island. The three possibilities were to place them onboard fleet carriers, or to carry them onboard freighters or possibly merchant ships converted into auxiliary carriers, or to use aircraft with sufficient range to reach the target area from land bases. The third was dismissed out of hand. Since no converted merchant ships would be immediately available when war broke out, the second entailed assembling knocked-down aircraft one by one and launching them from the water, far too cumbersome a possibility. That left the fleet carriers. Marine aircraft could go onboard each of the existing four large carriers as a fifth squadron. Nothing came of this project; Marines found themselves supported by Navy aircraft in the initial Pacific landings. Not until 1944–45 was there serious planning for carriers embarking Marine aircraft to provide close-air support.[19] It is striking that the step-by-step strategy did not give rise to any pre-war plans for dedicated Marine Corps aviation carriers.

What does this say about war gaming and its influence? The step-by-step strategy certainly reflected the effect of war gaming. It is possible that estimates of the level of air support the landing Marines needed resulted from gaming. However, they were expressed in terms of yardage to be covered, which suggests that they came from map exercises and staff estimates. Certainly no discussion of the Marine aviation issue mentioned gaming evidence.

All of this confuses any interpretation of the decline of the War College as laboratory. The radical changes in overall strategy and in the role of seaplanes make it impossible to say that it was connected with distaste for naval aviation compared to, say, conventional capital ships. On the other hand, there seems to have been considerable ambiguity about

Anyone arguing that the War College declined because it was too air-minded must deal with the reality that so much was invested in seaplanes and their tenders. Seaplane tenders had the significant political advantage in that they were not limited by treaty or in any other way, yet they could support very significant offensive air capability (which was, however, probably much overrated at the time). Many of the tenders, moreover, were sophisticated ships with significant protective armor. This is *Albemarle* (AV-5), photographed on 30 December 1943 in the Atlantic, with a PBY Catalina on her deck aft. (National Archives 80-G-450247)

the role of the seaplanes. When the General Board held hearings on their characteristics, all its witnesses pointed to scouting as their primary role.[20] Yet BuAer often pointed to the value of sea-based bombers. In 1935, it issued a specification for a four-engine flying boat with a substantially larger bomb capacity.[21] In 1938, it asked for much larger flying boats, ultimately selecting the Martin PB2M Mars. They, too, were described as bombers as much as scouts.[22]

The changing role of the seaplanes—and the step-by-step strategy—made it important for the fleet to move seaplanes forward as it advanced through the Mandates. In 1934, the seaplanes were part of the base force at Pearl Harbor. They had mobile tenders, but mainly to support them once the fleet reached the Philippines. There would hardly be time for the seaplanes to deploy and act as the fleet pursued the "through ticket to Manila." In September 1937, however, the patrol planes were transferred to the Scouting Force.[23]

If indeed the seaplanes could deal with Japanese land-based air forces in the Mandates, then the role of the carriers would change dramatically. Instead of having to beat off nearly constant enemy air attacks, they would deal mainly with the enemy's carrier force, either in the Mandates or in the climactic fleet action. The sort of unlimited carrier air arm envisaged in the 1920s might not be needed. This argument seems not to have been made explicitly, at least in surviving documents, but it may help to explain the shift in policy that roughly coincided with the downfall of gaming as laboratory. This explanation is an alternative to a simpler one: that with the death of Admiral Moffett, naval aviation no longer had a powerful enough advocate forcing the whole fleet to accept the somewhat futuristic view of naval aviation associated with the Naval War College.

If the carrier story is open to alternative interpretations, pre-war U.S. cruiser development seems to show the collapse of Naval War College gaming input much more clearly. In the fall of 1934, with the 1930 London Treaty still in force, the question of replacements for the first two of the ten-year-old 7,500-ton *Omaha*-class cruisers came up. About 20,000 tons was available, counting unobligated tonnage, the tonnage of the two old cruisers, and savings on various ships. The Bureau of Construction and Repair produced a variety of sketch designs for ships displacing 7,000 to 10,000 tons. The only agency that could compare their fighting power was the Naval War College, using the techniques it had developed to rate various ships. Its data showed roughly what it had argued a few years earlier: The best bargain was a pair of 10,000-ton ships (the largest possible). The result was two modified *Brooklyn*-class cruisers, *Helena* and *St. Louis*.[24] Any decision as to replacement of the eight remaining *Omaha*-class cruisers was deferred pending the outcome of the London Naval Conference.

Despite the General Board's endorsement of the college's recommendation that any new cruisers be the largest possible, there was fleet sentiment favoring a few smaller cruisers for battle-line work, presumably meaning to beat off enemy destroyer attacks and to stiffen U.S. destroyer attacks. Thus, in March 1935, OPNAV asked the designers for studies of smaller cruisers. In June 1935, they submitted sketches of 5,000- and 7,000-tonners.

At about the same time Commander, Cruisers Scouting Force (Rear Admiral Thomas C. Hart) argued that the major Naval War College proposal that torpedoes be omitted from cruisers be overturned.[25] He pointed out that all other navies armed their cruisers with this weapon; "if this is a case wherein we claim all the other navies to be wrong and we only to be correct, it would seem that for our own purposes our argument needs to be very strong." He pointed to the weakness of the Naval War College argument. It was based on experience of major fleet actions (as gamed). It was well understood by this time (also from gaming) that a major action would be the culmination of a campaign, in which cruisers might well find themselves individually—or in small numbers—fighting enemy ships. Torpedoes might then prove quite valuable. The college had not taken such combat into account.

The U.S. negotiating position at London in 1935 followed the recommendation to concentrate on large cruisers, but in the end the British managed to cut maximum cruiser size to 8,000 tons. What should the United States build after 1936? In the past, the Naval War College would have offered a proposal based on gaming experience, or perhaps a variety of smaller cruisers would have been tested in games. Nothing of the sort happened. The only U.S. experience with modern cruisers in the allowed size range was with the ten 7,500-ton *Omaha*-class light cruisers built under the 1916 program. In 1936, they were over-age and obsolete. It might be imagined that any new cruisers would fill their existing roles. Some were flagships of destroyer and submarine flotillas. Others were intended to help the battle fleet beat off enemy destroyer attacks, a role being taken over by the larger *Brooklyn* class, whose more numerous guns fired considerably faster.[26] The *Brooklyns* had been conceived specifically because the Naval War College argued that smaller cruisers were not worth building for that purpose.

The Bureau of Construction and Repair was asked to lay out a spectrum of cruiser designs within the 8,000-ton limit.[27] All of them incorporated torpedo tubes. Initially, the General Board preferred the largest possible ship, with the usual combination of 6-inch and 5-inch secondary guns. Its proposed FY38 program was three such ships. It turned out that no satisfactory two-caliber design was possible within the specified displacement. The General Board chose the next largest possibility, a 6,000-tonner armed with 5-inch guns. It could be justified as a fleet cruiser capable of working with or against destroyers. Four could be built on about the same displacement as the three larger ships, hence at a similar cost. None of this was based on Naval War College analysis; it does not even appear that the college was consulted. The FY38 cruiser was very much the best that could be achieved on an unsatisfactory tonnage. For FY39, the Bureau of Ordnance promised a solution employing an embryonic dual-purpose 6-inch gun. Using that weapon, an apparently satisfactory 8,000-ton cruiser could be designed. The FY39 program included two such ships. Once war broke out in September 1939, the London Naval Treaty could be considered dead. The two 8,000-ton ships were replaced by modified *Brooklyns* (*Cleveland* class). That must have been a considerable relief to BuOrd. Its twin dual-purpose 6-inch mounting was not ready until after World War II began. It was incorporated in the very large *Worcester*-class cruiser ordered in 1943.

Another cruiser development almost certainly did not involve the college or gaming. In March 1938, the Secretary of the Navy (presumably meaning the CNO) proposed a new type of large cruiser armed with 10- or 12-inch guns, hence capable of overcoming all existing cruisers.[28] Initial design studies showed that it would displace about 18,000 tons or more (26,000 tons with three triple-gun 12-inch turrets), hence would be well outside the cruiser category as defined by the London Naval Treaty. It would be defined, then, as a capital ship. One question was whether the U.S. Navy would willingly sacrifice battleship tonnage (as provided under the Vinson Acts) to build such ships.

There is no indication that the idea came out of gaming. The General Board cruiser file (420-8) includes a 4 April 1938 paper strongly advocating the new type of cruiser, written by Captain A. J. Chantry, who was then head of Preliminary Design in the

Bureau of Construction and Repair.[29] Chantry had been a member of the Naval War College Senior Class of 1936. His 1938 paper is in the form of the college-advocated "Estimate of the Situation," but it is far less analytical than the paper the college produced in 1931 to guide future cruiser policy. Given its timing, it may have been mainly an attempt to back the Secretary's request rather than an independent proposal that the United States should build a super-cruiser.[30] Chantry based his estimate of what the Japanese were likely to build on

> the national character of her people; the aspirations and aims encompassed in her national policy; geographic factors; and in particular, her present and future economic position.... In considering Japanese character it is perhaps sufficient for present purposes to note that her people are zealots, almost fanatical in type, and only removed in moderate degree from the influences of feudalism. Nationalism is highly developed backed by determination of the strongest character. The Japanese believe that they are the children of Heaven and nothing can divert them from their national idealisms. Life is held cheaply in support of their country. They are the most resolute of peoples, the most fearless and determined adversaries of those who seek to oppose them. They are secretive.... It is safe to assume that [Japan] now seeks the hegemony of all China and areas to the north and desires to extend her influence to the East Indies and South Seas as opportunity presents...there is almost no limit to Japanese conceptions of Empire. Coupled with the fanatical determination of her people it makes Japan a "problem child" in the family of nations; one whose every move deserves the most serious attention and consideration of major powers.

Some of Chantry's tone may be traceable to the recent shock of the Japanese attack on China in 1937, in the course of which the U.S. Yangtze River gunboat *Panay* had been attacked and sunk, apparently deliberately. The invasion and such incidents were then leading to the beginning of U.S. mobilization, in the form of the second Vinson Act and less public debate on whether to increase production of naval weapons and equipment. Japan was also in the process of rejecting attempts to convince her to abide by the limits stated in the new London Naval Treaty. However, Chantry was going far beyond the usual cool Naval War College analysis. War College papers did not dwell on national characteristics the way Chantry did.

Chantry went on to point to Japanese dependence on imported raw materials; Japan had amassed sufficient supplies to last for the early stages of a war, but would have to rely on imports anyway. None of this can have been unfamiliar. Earlier analyses of likely Japanese naval construction would not even have mentioned it. A list of courses of action open to Japan repeated that Japanese secretiveness would be a major factor in her grand strategy. Japan would counter opponents' greater strength with "cunning," meaning that Japan would develop "weapons of attack for which her opponents have no counterpart readily available." This last was an accurate prediction of Japanese policy, which emphasized the development of individually superior weapons such as the huge *Yamato*-class battleship, the Zero fighter, and the Type 93 "Long Lance" oxygen-propelled torpedo.

Chantry correctly predicted that the Japanese would build warships individually superior to their foreign counterparts. He parted company with the Naval War College in not considering overall Japanese requirements, hence looking at likely Japanese priorities. Instead, he focused on one part of a possible Pacific War, a war against trade that the *United States* would prosecute. The question was how the *Japanese* would defend their own trade. This was very different from the issues typically raised in gaming. The Naval War College certainly agreed that Japan would live or die based on its ability to import what it needed, but that led to a U.S. strategy of blockade. Games explored raiding, not as a means of crippling Japan, but rather as a way to draw off Japanese forces to allow the U.S. fleet greater freedom of action.[31]

Chantry pointed out that a Japanese convoy strategy would fail if the convoy escorts could not deal with individually powerful raiders (presumably cruisers). However, the raiders would be drawn into focal areas (where shipping routes crossed) as they ran down their victims. By placing individually powerful ships in the focal areas, the Japanese would gradually annihilate the raiders. It would not matter if the United States built large numbers of cruisers, as long as they were individually inferior to those the Japanese had. It happened that the focal-area defense had been planned by the Royal Navy during the latter part of the 19th century. It was the alternative to convoys that the British adopted at the time as a means of protecting their trade. It is not clear that Chantry or anyone outside the Admiralty knew this in 1938; the focal-area approach may merely have been the obvious one.

Chantry cited persistent rumors that Japan was already building cruisers somewhat larger than the 10,000-ton limit embodied in the treaties. Other countries might do the same; Chantry cited the German pocket battleships (essentially cruisers armed with 11-inch guns) as well as very fast light Italian cruisers. He suggested that the Japanese would use their extra tonnage to mount heavier guns—9-, 10-, or 11-inch calibers. In fact, the Japanese cruisers displaced considerably more than 10,000 tons. No one knew that until the end of the Pacific War. The Japanese used the extra tonnage to gain conventional advantages like greater speed, but the ships did not mount heavier guns.

Chantry pointed out that having freed themselves of treaty restrictions, the Japanese might choose to build somewhat larger and better-protected ships armed with more 8-inch guns . Alternately, they might jump to a super-cruiser type armed with much more powerful guns. Japanese secrecy made it impossible to know whether super-cruisers were planned. Chantry argued that the possibility could not be dismissed. Any of these ships might indeed be extremely expensive, but the Japanese might well build them anyway. Moreover, the Japanese had a national tendency to spring surprises on their opponents.

Chantry argued that the United States should anticipate the possible construction of Japanese or other super-cruisers by building competing ships, the type the General Board was considering. The U.S. Navy needed something superior to either the super 8-inch gun cruiser or the semi-capital ship, since it could not be sure what the Japanese were building.

All of this was a very *non*–Naval War College argument dressed in Naval War College clothing. Chantry never asked the usual Naval War College questions. What were overall Japanese requirements? What were their priorities likely to be, given likely U.S. action in wartime? There is no evidence that the college was ever asked for its views, or for that matter whether it ever volunteered any.[32] For the moment, enthusiasm for big over-armed cruisers waned, because it seemed more important to use available tonnage for real battleships (the *Iowa*s were being designed). However, the super-cruiser idea returned in 1940, when the Two-Ocean Navy Act of July 1940 dramatically increased available tonnage and all restrictions associated with the London Naval Treaty had died. The big new cruiser program included the six *Alaska*s, which certainly filled Chantry's requirements.[33] As for the Japanese, they seem to have decided to build their own super-cruisers only in 1941, and then as part of their main fleet (they were to replace the *Kongo*-class battlecruisers).[34] These ships were never laid down.

In 1938, the General Board was considering a larger cruiser program. The individual tonnage limits embodied in the London Naval Treaty could be modified by invoking the escalation clause: If one of the signatories of the Washington Treaty (i.e., Japan) refused to accept the limitations set at London, the others could modify the London Naval Treaty. That actually happened with battleships (the upper limit was changed from 35,000 to 45,000 tons). A 26 May 1938 General Board memo suggested that it should also be applied to cruisers.[35] "The employment of cruisers in war, particularly against Orange, leads to the assumption that we require a considerable number for assignment to trade routes, convoys, and independent duties in general." That implied a need for large rather than small cruisers, and probably for 8-inch gun cruisers, the construction of additional units having been banned by the London Naval Treaty. Moreover, the new triple 6-inch gun in the *Brooklyn*s had not yet been tested. The U.S. government did not raise the cruiser issue when the escalation clause was invoked to permit construction of the 45,000-ton battleships. Surviving documentation gives no indication of why, but it seems likely that escalation was limited because there was still such strong anti-military sentiment.

Again, this possibility invited the sort of analysis that in the past had been done by the Naval War College based on gaming experience. The 1938 General Board memo is concerned with a simple numerical comparison of U.S. and foreign cruiser fleets, and with possibilities within the tonnage available under the Vinson Act following the four 6,000-ton ships of the 1939–40 (FY38) program. Size therefore had to be traded off against numbers, which was exactly the sort of consideration the college had taken into account in 1930–31. This time, it was handled as a simple list of alternative possibilities, without even the college's comparison of fighting value.

7. Conclusion: Games Versus Reality in the Pacific

Gaming had at least three possible functions. One was to explore possible wartime situations in ways full-scale exercises could not. Military judgment based on experience could often foresee outcomes, but not when entirely new technology was involved. This was the laboratory function performed by the Naval War College. The one entirely new technology of the inter-war period was naval aviation. Gaming provided the U.S. Navy with an invaluable sense of how that technology would affect war not only on the tactical, but also on the operational and strategic levels. For students at the college, the games offered a glimpse of what a future naval war would be like. No full-scale exercise could have done the same.

A second function was to teach students how to fight. The Naval War College emphasized its careful rational way of deciding what to do. Not long ago, the Naval War College placed Admiral Nimitz's "Gray Book" (in effect, his command diary of the Pacific War) online. The "Gray Book" can be considered an extended exercise in the "Estimate of the Situation," the decision-making basis that the college taught Admiral Nimitz when he was a student.

In January 1942, a British officer criticized American naval officers for tending to estimate the odds before making decisions.[1] In effect, he was comparing what Newport taught with what his own Tactical School taught. The lesson the British derived from their World War I experience was that they should have shown far more aggressiveness and initiative. It can be argued that this was an illusory lesson; the massive World War I Grand Fleet was like an army, in which excessive initiative could have had devastating consequences. In any case, the Royal Navy's tactical school awarded points for aggressiveness, even when the decisions taken proved unfortunate.[2] In effect, that was a judgment on Britain's enemies, that they could be bluffed because they were not nearly as professional as the Royal Navy. That judgment proved quite correct when the Royal Navy fought the Germans and the Italians. The U.S. Naval War College based its course on the idea that the enemy would be as professional as its own students. The students had to be taught to out-think a fully witting enemy. In comparing the U.S. and British approaches to gaming and to decision making, it would be interesting to ask whether the British approach would have been as successful against the Japanese.

Perhaps the greatest gap in Naval War College simulations of combat was the tempo of an actual battle. The complexity of the rules made it impossible for move to follow move fast enough to give students a sense of the confusion generated in an actual battle. There certainly was an attempt to simulate the "fog of war," but instructors pointed out again and again that in games students had far more time to make decisions than they would have in reality. In theory, full-scale exercises would have filled some of this gap, but in reality they proved insufficient. That seems to have been particularly the case in the night battles in the Solomons in 1942–43.

A third gaming function was to understand or even predict the behavior of foreign powers. That was the most difficult of all, because it required that some players simulate alien ways of thinking. Could the U.S. government have guessed that, under intense pressure in 1941, the Japanese would have chosen to fight even though their own analysis showed that they would lose a Pacific War? Much the same might be asked about the Japanese reaction to disaster. After the war, many Japanese said that they knew the war was lost when the Americans took Saipan in June 1944. Yet it took Japan more than another year to accept defeat. Would gaming have provided insight into what happened? Such questions were outside the purview of the Naval War College. However, since 1945, there have been attempts to game foreign governments' thinking, generally at the level of the State Department.

How well did the Naval War College envision the Pacific War? After the war, Admiral Nimitz famously said that in the course of the games, at one time or another, someone or other at the college experienced everything that happened in the Pacific, other than the kamikaze warfare at the end of the war. The Naval War College archives of the time included material on all the games, and at least some students must have taken advantage of them. Moreover, the OPNAV War Plans Division, which actually planned the war, obtained copies of much of the gaming material. Some of the game material survives only in the War Plans Division files.

Did the Naval War College predict the course and character of the Pacific War in detail? Of course not. On the technical level, it never predicted that carriers would displace battleships as capital ships. Its estimates of weapon effectiveness showed that it would still take either battleship firepower or torpedoes (typically launched by destroyers or submarines and much less often by aircraft) to destroy an enemy's battleships. That was not entirely misleading. The number of carrier aircraft required to sink the huge

War gaming prepared officers to wield carrier airpower, even if they were not aviators. Both commanders at Midway, for example, were surface officers, and one of them, Admiral Raymond F. Spruance, went on to command the fast carrier task force. Spruance had been a student and then an instructor at the War College. He later testified to the value of war gaming in his own professional development. He is shown here as Commander, Central Pacific Force, U.S. Pacific Fleet, 23 April 1944. (National Archives 80-G-225341)

Japanese *Musashi* in October 1944 were entirely unimaginable in the pre-war period. Even this attack did not sink or even seriously damage other Japanese battleships in the same formation.³ They survived to attack the U.S. escort carriers covering the Leyte Gulf landing. However, the college did make it clear that carrier air power would be extremely important, and it did show that it would be essential in assaulting Japanese-held islands during the fleet's step-by-step advance to the West.

On the strategic level, the college did not envisage the type of coalition war that would lead the United States to defend Australia and thus to fight on Guadalcanal and in the Solomons. These places hardly figured in pre-war thinking. The coalition games envisaged by Captain van Auken were never played. The college did play coalition games in 1941, when U.S. ships were already working with the Royal Navy to help protect convoys and also to help deal with German and other Axis fleets. However, these games did not touch on the real problems of such operations, such as differences in operating practices and problems of potentially split command.

Similarly, no one envisioned the dramatic German success in Europe that opened the Far East colonies to Japanese attack in 1941. No Naval War College game foresaw a situation in which the United States would have to divide the fleet between Atlantic and Pacific to deal with simultaneous threats in both oceans. However, there was at least one War Plans Division study of a U.S. strategy to fight a Red-Orange coalition, which would have had exactly that effect. In such a war, the Atlantic threat (Red) would constrain what the United States could do against Orange, and vice versa. Moreover, the prefaces of the series of major Red-Blue games mentioned the possibility that Orange would exploit U.S. weakness due to the fight against Red. That might be considered broadly analogous to Japanese exploitation of the defeat of the European colonial powers in 1940, and to the consequent weakening of the British. In 1938, the War Plans Division studied a situation in which the U.S. fleet would move to Singapore and help protect the British Far East empire against the Japanese while the British faced the Germans in Europe, but that operation was not, as far as it seems, gamed. The War Plans Division study seems to have been carried out in the context of secret staff talks between U.S. and British officers in 1938–39.⁴

The various sorts of battles that the U.S. Navy fought in the Pacific certainly did figure in the mass of games. They included not only the battle-line engagements that might be associated with the pre-war Navy, but also many carrier battles, and even night cruiser battles like the ones fought in the Solomons. The run of war games does read like an encyclopedia of possible kinds of future warfare in the Pacific.

Most important, throughout the inter-war period the games educated the U.S. Navy in the problems of a war projected across the vastness of the Pacific. No navy had ever fought successfully so far from home. The only precedent, in the inter-war period, was the disastrous Russian performance in the Russo-Japanese War of 1904–1905, which culminated in the destruction of the Russian Baltic fleet at Tsushima. In that war, the

Probably the greatest contribution the Naval War College made to victory in World War II was its demonstration that the strategy of steaming directly to the Philippines was bankrupt, even before the destruction of the battle line at Pearl Harbor. This, in turn, led to the development of the amphibious force that was so successful in both the Atlantic and the Pacific during World War II. Here, amphibious tractors (Amtracs) pass the battleship *Tennessee* (BB-43), which is shelling Okinawa, 1 April 1945. (NHHC NH 42390)

problem for the Russians had been logistical—merely getting to the Far East in condition to fight. After World War I, the U.S. Navy would face both the logistics and whatever attacks the Japanese could mount as it passed through the island chains they had gained as a result of the war. Even the logistics would be daunting. In 1919, when the U.S. Navy first seriously considered a Far East war, most of the U.S. fleet burned coal. Ton for ton, coal did not provide enough energy to move the fleet across the Pacific. Replenishment was difficult at best. A U.S. naval war could not really be fought in the Far East until the fleet had converted to oil fuel. That process began just before World War I, but it was very much a post–World War I development. Oil fuel made it possible to project naval forces and whatever they might convoy across the Pacific.

The war's possibilities and consequences had to be envisaged in the context of new weapons: large numbers of aircraft and also long-range submarines. Gaming was a good way to do so. Looking back, we find it difficult to appreciate just how novel a Pacific naval war would be. After World War II, the Royal Navy circulated an analysis of U.S. performance in the Pacific that gives a sense of what was achieved. The writer likened the sustained U.S. fleet operation in the western Pacific, which was based at Ulithi, to a war fought off the U.S. coast using a base on the other side of the ocean at Gibraltar.

The U.S. Navy used war gaming to explore this kind of warfare and to make it familiar to its officers. Admiral Nimitz's remark cited above may seem to refer to islands around and on which successive generations of War College students fought. Some students certainly did become familiar with places they would fight years later, but that was a minor advantage. Much more important, gaming familiarized students with the impact of the new aviation technology, new kinds of command and control, fleet submarine operations, and also (from 1935 onward) with concepts such as amphibious assault against defended islands. Like aviation, amphibious warfare was a new and nearly unexplored branch of naval warfare. For the U.S. Navy and the Marine Corps, it had little significance until war gaming revealed that it would be essential as part of a step-by-step advance across the Pacific.

Gaming forced officers to define and then to confront the most likely Japanese strategy, which was to impose attrition as the U.S. fleet moved west. Exploration meant working out what the Japanese would have to do to execute various alternative strategies. To reach that conclusion, the Naval War College used games that explored a wide variety of possible Japanese strategies. The extent of that exploration made it reasonable to assume that the Japanese would follow the expected attrition strategy. This type of exploration was necessary because other kinds of evidence of Japanese thinking were so limited, thanks to tight Japanese security. The U.S. Navy monitored Japanese fleet exercises—which turned out to reflect Japanese war planning—using broken Japanese codes and also traffic analysis. However, the fruit of such communications intelligence was very closely held, and it could not be used to shape war games at Newport.

The inter-war games embodied a particular approach to naval strategy, which the gamers correctly understood applied to both the U.S. and Imperial Japanese Navies. It emphasized the role of the main or battle fleet as a shield behind which other naval activities could be carried out. That is why both navies looked toward a decisive fleet engagement, which would determine whether either could freely use the sea. For example, a convoy approach to shipping protection was viable only under the cover of a dominant fleet. Freed of that dominance, the enemy would be able to destroy escorts and convoys.[5]

Through the inter-war period, the Naval War College—and indeed the U.S. Navy—and the Japanese naval leadership considered battleships the deciding factor in battle. The disparity in numbers of battleships enforced by the inter-war arms control treaties made it likely that the Japanese would adopt an attrition strategy.[6] They spent heavily on improving their ships, but their prospects were bleak. Until 1936, nine Japanese battleships faced 15 U.S. ships. If the battleships were of roughly equal combat value, the relative power of the two battle fleets would be proportional to the square of their numbers: a ratio of 81:225 against Japan. Even if the three oldest U.S. battleships were

The straight-through strategy was impossible because the fleet had no forward repair facilities; if it suffered underwater damage (as was likely), it had nowhere to go beyond Hawaii. Once war began, a variety of portable base units was created, most spectacularly floating dry docks that were towed to combat zones. Here, the cruiser *Canberra* (CA-70) and the destroyers *Claxton* (DD-571) and *Killen* (DD-593) share repair space in the floating dry dock ABSD-2 (Advanced Base Sectional Dock) at Manus in the Admiralty Islands, 2 December 1944. (National Archives 80-G-304096)

not counted, the Japanese would face an 81:144 ratio. The war college understood that their strategy would concentrate on changing those odds as the U.S. fleet headed west through the Mandates. Inevitably, that meant using weapons not limited by treaty, such as land-based aircraft. That, in turn, inevitably made U.S. naval aircraft more important than they might otherwise have been.

The Japanese attack on Pearl Harbor can be seen as the ultimate attrition operation.[7] In 1941, the U.S. Navy had nine battleships in the Pacific, three of the best modernized ships having been transferred to the Atlantic. It also had two new battleships, which were not yet entirely ready (*North Carolina* and *Washington*). Once war broke out, the United States would presumably have transferred all of its best ships to the Pacific, giving the U.S. fleet there a total of 14 battleships compared to ten in the Imperial Japanese Navy, with two more nearly ready. When Admiral Yamamoto Isoroku planned the attack on

Pearl Harbor, he wrote that his primary objective was to destroy at least four U.S. battleships. That would have eliminated any U.S. superiority, at least in numerical terms. Yamamoto also said that he could "run wild for six months," which would be roughly the period before additional U.S. battleships already under construction could enter service. Yamamoto was willing to sacrifice the carriers in the attack force to gain this level of attrition. Even more than the Americans, the Japanese naval high command did not consider carrier aircraft an essential arm in a battleship-on-battleship fight.[8]

The Pearl Harbor attack itself did not figure in any surviving War College game. The gaming system based on estimates of the situation was in effect rigged against any operation as risky as the Japanese attack on Pearl Harbor. Under the game rules, had the defenses been alerted, the Japanese might well have suffered heavy losses. The Japanese themselves thought it entirely possible that they would lose several carriers. The technique taught by the War College forced players to balance possible gains against probable losses. It did not take national psychology into effect. There is some evidence, for example, that Yamamoto thought that the U.S. public would be so shocked by the loss of battleships at Pearl Harbor that it would shrink from fighting a protracted war. Others in the Japanese leadership took the opposite position: that the attack would enrage the U.S. public exactly as it did. It is not clear why no one, Japanese or American, considered the sudden attack on U.S. forces in the Philippines, which the Japanese considered an inescapable requirement, would not equally enrage the U.S. public. The gap in Japanese perception of U.S. national psychology might be matched by the inability of U.S. gamers to estimate the effects on the Japanese of a U.S. carrier air attack on Tokyo, a feature of one of the games. One Naval War College game did include a surprise submarine attack on the battle fleet at San Diego, which reduced U.S. superiority over the Japanese battle fleet.

The games tended to focus participants' attention on the way each side was likely to shape its strategy to use its inherent advantages to offset its inherent disadvantages. For the United States, the advantages were greater capital ship strength at the outset, thanks to the ratios embodied in the inter-war treaties, plus a much larger industrial base. To some extent, the advantages inherent in the treaties were offset by U.S. unwillingness to build up to the allowed limit. In most types of ships in 1941 the Japanese enjoyed much better ratios of strength than they might have. This disadvantage was alleviated after 1942. Since 1938, the United States had been building up under the second Vinson Act. The great U.S. disadvantage was a combination of distance and the lack of secure bases—and, even more important, repair facilities—in the Far East.

For Japan, the most important inherent advantages were distance from the United States and the geography of the Mandates in the Pacific. The disadvantages were industrial. Game designers might doubt that the United States could instantly build large numbers of new ships, but notes for participants often pointed to the much greater strength of the U.S. aircraft industry. We now know that the Japanese were well aware that they could not match U.S. numbers, and that they hoped to develop special weapons of various sorts as equalizers: To this extent, Chantry was right. The Naval War College never seems to have envisioned this approach, even during its analysis of the cruiser issue in 1930–31. Note that Chantry used Naval War College methodology to reach his conclusion.

The Naval War College fought its games using existing technology, although it was sometimes criticized on the ground that its view of air operations was on the optimistic side. This argument was made particularly strongly once champions of air power like Moffett and Pratt were gone. Thus, nothing in gaming reflected the vast jump in aircraft performance after 1941, which was achieved largely due to the advent of a new generation of engines. It can be argued that improvements in aircraft on both sides more or less cancelled out. Air weapons used against ships and ground installations did not change nearly as much. Thus, game rules involving dive bombing and air-launched torpedoes seem to have been good predictors of wartime weapon performance.

The Naval War College did not predict the impact of electronics on the coming war. The U.S. Navy began experimenting with radar in 1939, and became aware of British experience using radar to control fighters in 1940. Neither development seems to have affected gaming. Nor does either seem to have penetrated through to many U.S. Navy decision makers like, for example, those on the General Board concerned with future building programs. For carriers, the central fact of World War II was that efficient fighter control cancelled the pre-war assumption that, once found by enemy aircraft, a carrier was probably doomed, or at least destined to be put out of action. A vital secondary reality was the advent of navigational beacons that made it practicable for carrier aircraft to operate much further from their ships. Thus, gaming as practiced at the Naval War College between wars could not have predicted an air-to-air victory like that won in the "Turkey Shoot" of the Battle of the Philippine Sea. Against that, at least one game ("Siargo"/Surigao) did include successful defense of a fleet against enemy carrier aircraft.

Some games carried important strategic lessons. For example, in several games played in the 1930s, U.S. forces raided Japanese commerce, not so much because they could inflict decisive damage, as because they could compel the limited Japanese fleet to divert forces to shipping protection and thus not to deal with a simultaneous U.S. fleet operation. This logic applied in 1942–43 when a U.S. submarine offensive against Japanese trade began. The Japanese lacked the surplus naval strength to set up a convoy system while maintaining their fleets. Eventually they were able to build escorts, but to have done so before the war would have been to accept what they considered excessive weakness in their main, shielding, fleet. The failure of the Japanese to protect shipping is often attributed to cultural factors and an inability to see the wider naval picture, but the pre-war games show that it was inherent in the logic of a Pacific war fought against an adversary with limited industrial strength.

Overall, the games provided a correct assessment of a naval strategy the Japanese would feel compelled to adopt—not because of some quirk in Japanese culture, but be-

The pre-war War College did not analyze a submarine campaign against Japan in any detail, but it did show, in a series of games, that anything except unrestricted submarine warfare (which at the time was considered entirely objectionable) would not be successful. Once war broke out, unrestricted submarine warfare against Japan was authorized. Initially hampered by torpedo problems, it proved extremely successful. *Barb* (SS-220), which made celebrated attacks in Japanese home waters, is shown in San Francisco Bay near Mare Island, 3 May 1945. (National Archives 19-N-83952)

cause of the logic of sea power that applied to both sides. The strength of the gaming system, as it was used at Newport, was that it avoided any reference to cultural peculiarities on either side. It was a clear-eyed means of understanding the logic of the situation. The rationale was that any additional cultural factor would reduce effectiveness on one side or the other. For the historian, what is remarkable is that so much that generally considered culturally determined in wartime Japanese naval thinking can be attributed instead to the logic of the situation. That is evident in the record of the war games; without this record it would probably be missed. Conversely, understanding the logic of the situation was a great advantage for a U.S. Navy, which might otherwise have misunderstood what the Japanese were doing.

For example, Japanese submarine commanders generally avoided attacking merchant ships and naval auxiliaries. They considered that their quarry was U.S. capital ships. The post-war evaluation was that, infused by the spirit of the samurai, these commanders considered attacking anything short of a battleship or carrier beneath them. However, the logic of the situation was compelling. Japan could afford considerable losses to merchant ships as long as the shield represented by the main Japanese fleet survived. When the decisive battle was fought, the degree of attrition imposed on U.S. capital ships would determine the outcome. Conversely, unless the U.S. main fleet was worn down, the smaller Japanese fleet was doomed according to accepted theories of combat power. It made good sense to limit risks to valuable submarines (which could not easily be replaced, given the weakness of Japanese industry) to attacks that were likely to be worth the trouble.

The U.S. Navy had no such priority, because it already enjoyed such crushing superiority in capital ship numbers that it did not have to attrite Japanese capital ships before the decisive battle. Sinking those capital ships would be helpful, but it did not carry nearly the same significance as it did for the Japanese. That was true even after Pearl Harbor, because by 1941 the United States was building so many new battleships that it would soon be able to overwhelm any Japanese force. It turned out that battleships were not decisive to anything remotely approaching the extent that anyone imagined in 1941—but the war games certainly did emphasize the considerations that actually affected the Japanese. Moreover, pre-war games did show that a campaign against Japanese trade could bleed the main Japanese fleet by forcing it to detach ships and valuable personnel and, it turned out, scarce aircraft to protect merchant ships. That campaign might or might not be decisive, but its diversionary impact would be considerable. Pre-war gaming had suggested as much. As it happened, the U.S. submarine campaign was devastating, but it would have been worthwhile even had that not been the case.

The correctly understood Japanese strategy in turn defined a particular kind of air-sea warfare, which shaped the development of U.S. naval aviation in the inter-war period. If that seems grandiose, it is apparent when U.S. naval aviation is compared to that developed by the Royal Navy. The British invented carrier aviation, and during the inter-war period they invested heavily in it. Unlike the U.S. Navy, they associated it mainly with a decisive naval battle. They therefore concentrated on finding their enemy first and attacking first. There was no sense that a fleet would have to fight a lengthy series of air-sea battles before it could even get to fight an enemy battle fleet. Gaming showed Americans that this would hardly be the case. Simply to get to the Far East and a hoped-for decisive battle, the U.S. fleet would have to beat off repeated heavy air attacks, many of them mounted by shore-based aircraft. This latter point mattered. The inter-war Royal Navy was convinced, incorrectly, that aircraft developed to operate from carriers would necessarily be inferior to their land-based counterparts. That was acceptable as long as carriers fought only other carriers. It was entirely unacceptable if carriers had to beat off repeated attacks mounted from land bases. Gaming forced U.S. officers to confront exactly that situation, and therefore to demand high performance of their fleet aircraft. That had enormous wartime consequences.

More generally, gaming forced U.S. officers to take into account many operations that would have to be undertaken before any hoped-for decisive fleet action. They learned, for example, that carriers could be subject to surprise attack by surface ships. Not only did these valuable carriers have to be screened, but their commanders had to mount constant scouting flights to insure against the possibility of surface attack. That may seem obvious, but it was not. In June 1940, the British lost the fleet carrier HMS *Glorious* to two German battlecruisers because she had not put up scouts, hence could not simply evade the attack. U.S. escort carriers were surprised off Samar in October 1944, but their commander knew how to react. He had thought through situations that included this one. The absolute need to screen fleet carriers against surface attack was involved. The escort carriers were uncovered largely because Admiral Halsey kept his battleships with his fleet carriers when he steamed north to attack what turned out to be a Japanese decoy force. Provision had been made to detach the U.S. battleships in the event of a Japanese surface threat, but Halsey understood that his battleships were integral with his carrier force.[9]

The comparison with the British is natural because the Royal Navy and the U.S. Navy were considered roughly equal as the two leading navies of their time, and because for most of the inter-war period both developed their strategies and their tactics specifically to deal with Japan. Both shared the idea that the ultimate stage of a war would be a strangling blockade, and that the prerequisite for such a blockade would be a decisive battle in which the main Japanese fleet would be destroyed. The U.S. Navy seems to have gone considerably further in laying out this rationale and thinking through the whole campaign.

That was not the war that either navy actually fought. The U.S. Navy was fortunate in that the broad scenario for its strategy forced it to contemplate the sorts of combat it encountered. This was because it had to plan on a passage through chains of islands that Japan controlled. The Japanese attack on Pearl Harbor eliminated the battleships that might have sought an early decisive battle, but in 1941 the United States was building a new fleet of fast battleships, which would enter service in 1942–43. Until they arrived, the air-sea war that preceded any decisive fleet battle in war games was a good model for the war the Navy actually fought.

By way of contrast, the Royal Navy envisaged a relatively unopposed passage to its war base at Singapore. Operating in that area, its fleet would threaten vital Japanese sea lines of communication, so the Japanese would feel compelled to fight. To some extent, this was a version of British World War I naval strategy, the difference being that the Japanese, unlike the World War I Germans, had a vital reason to challenge the British fleet.[10] The consequence of the British strategy was that the key part of the war plan was logistical rather than tactical; the fleet needed sufficient supplies to get to the Far East. En route it might be threatened by Japanese submarines, but not by constant air attack. Unfortunately for the British, however well they analyzed the problems involved, this analysis or gaming did not prepare them for the sort of constant air-sea attack they encountered in European waters during World War II. There was no way that a British

Nimitz could say that thinking or gaming the central war plan could prepare the navy for what it would later experience.

It seems, moreover, that the British did not undertake the sort of cross-analysis of possibilities the U.S. Naval War College taught. Their Far East war plan seems to have assumed that the Japanese would choose to steam south to meet the British fleet somewhere near Singapore, far from the support offered by land-based aircraft. It is not at all clear from later accounts that the British ever asked whether the Japanese would be so cooperative![11] The key assumption seems to have been that the Japanese would find intolerable the presence of a British fleet athwart the trade routes to the south, just as the U.S. Navy assumed that the Japanese would find American possession of the Philippines intolerable.

However, there was a vast difference. In a war between only Britain and Japan (or Britain plus European partners against Japan), the Japanese had an alternative source of supply across the Pacific in the Western Hemisphere. The loss of the southern routes would be a serious blow, but it might not be all that fatal. The Japanese might choose to keep their fleet closer to home and to other Japanese resources. They could not do the same if they fought the Americans, because in that case their supplies really would come from the south. The Naval War College did consider the one alternative possibility, namely that the Japanese could rely instead on supplies from the Soviet Union. War games did include cases in which the Japanese chose not to attack the Philippines. British accounts of the Singapore strategy do not mention any alternative plan in the event the Japanese chose not to take the bait represented by a British fleet based at Singapore. The main problem they considered seems to have been how to protect Singapore until the British fleet arrived there.

The issue of bait and timing became crucial during the last pre-war years. About 1936, the First Sea Lord, Ernle Chatfield, wrote that his greatest nightmare was a war against both the European fascist powers and Japan. His only solution would be to send a fleet east, destroy the Japanese fleet, and then move what was left back west to confront the enemy European powers. This strategy was barely practicable because the French would be allied to the British, and they would more than balance the Italians. The Germans were only beginning to build up. Before 1939, the British discussed the need to build a two-ocean navy, but it was never affordable. Moreover, the Japanese did not cooperate. They waited until the Germans had overrun Europe, destroying a large part of the Royal Navy in the process, and putting the French out of action. By 1941, the British no longer had a big fleet that could be sent east to deal with Japan before returning west to beat down the Germans. Even had the Japanese entered the war at the same time as the Germans, the British still faced the question of how long they could maintain a fleet at Singapore, waiting for the Japanese to take the bait it represented.

These are exactly the sorts of issues that campaign games at Newport explored, albeit for the U.S. Navy rather than the Royal Navy. That they have never been raised in the 70 years since World War II seems to be a clear indication that the British never had any equivalent to Newport and its gamers—and that they suffered for that lack. The apparent failure of British planning illuminates the success of the Naval War College gaming system in preparing the U.S. naval leadership for the war it would fight.

The Pacific War opened with a stunning surprise attack at Pearl Harbor and with an expected sudden attack on the Philippines. Many pre-war officers credited the Japanese with a penchant for surprise attacks. The question of how to keep such a surprise from being decisive was a theme of many war games. The problem was how to force the Japanese to open the war with an overt act that would not be devastating. This act in turn would justify U.S. counterattacks that might blunt the Japanese strike. In December 1941, the U.S. Navy sought exactly such a solution by commissioning some small craft to patrol around the Philippines. The idea was that the war would open with a Japanese attack there, and that the Japanese would feel compelled to sink the patrol craft before they could land troops. In the event, none of the small craft was sunk, and the main surprise attack included the Pearl Harbor raid. Some games included surprise attacks (such as a submarine attack on the battleships at San Pedro), but they were not nearly as devastating as the one against Pearl Harbor.

That omission made less difference than might be imagined, because the games gave a sense of how a protracted Pacific campaign would be fought. The battleships sunk or disabled at Pearl Harbor would have been essential in the fleet action that the Naval War College (and U.S. war planners) assumed would be the culmination of a trans-Pacific campaign. Before that, the Japanese would concentrate on wearing down the U.S. fleet so that their own less powerful battle line could win. The wearing-down campaign would be fought mainly by naval aircraft and submarines. That is why war games, particularly those depicting a full campaign or a large part of one, so often involved huge numbers of aircraft.

The game also involved large numbers of submarines on both sides. In 1932, when he compiled lessons learned, Naval War College President Rear Admiral Harris Laning wrote that Orange submarines had often exerted decisive power—meaning that the U.S. Navy should invest heavily in anti-submarine warfare. The U.S. Navy installed sonars onboard its new destroyers. That did not seem remarkable when World War II broke out, but it seems notable that other navies with access to this technology seem not to have done so on anything like the same scale. The British did install sonars on a large scale, but mainly because they were concerned with protecting trade against a future U-boat campaign, something of much less concern to the inter-war U.S. Navy. One indication of relative U.S. unconcern with a submarine campaign against shipping was that, unlike the Royal Navy, it made no attempt to produce convoy escorts after the 1930 London Naval Treaty drastically reduced its stock of older destroyers. Thus, the U.S. sonar program was designed mainly to protect the battle fleet against the threat of Japanese submarines based in the Mandates—against a threat evident in gaming experience. The sonar-equipped destroyers turned out to be very useful during the Battle of the Atlantic, a problem not foreseen until 1941.[12]

Conclusion: Games Versus Reality in the Pacific

The pre-war Naval War College did not envision a new Battle of the Atlantic, but it did foresee large-scale anti-submarine warfare because the Japanese submarine force was expected to be a substantial threat. On this basis, the U.S. Navy spent considerable efforts developing sonar and anti-submarine tactics, although once war came it was not nearly as advanced as the Royal Navy. U.S. industrial resources did provide large numbers of escorts and escort carriers, as well as much-improved sonars. The destroyer escort *Harmon* (DE-678) and another destroyer escort (unidentified) are shown alongside the destroyer tender *Dixie* (AD-14) in Hawthorne Sound, New Georgia, in May 1944. (National Archives 80-G-238416)

From the early 1920s, the message was that the fleet would have to steam through areas in which it faced constant attack by aircraft, submarines, and surface torpedo craft. There would be no well-defined threat axis, because the fleet would find itself in the midst of Japanese-held mandated islands. Prior practice, both in the U.S. Navy and abroad, was to cruise in columns, in a rectangular formation. Such formations were easy to maintain, but difficult to maneuver. Typically they could be converted into battle formation once an enemy fleet was sighted. They were poorly adapted to a situation in which the fleet faced constant threats from unknown directions. The alternative, invented by Commander Roscoe C. MacFall as a War College student, was a circular formation. It may have been tried for the first time during Tactical Problem IV (Red-Blue, played in November 1922 for the Class of 1923).[13] This was also one of the first games critiqued by Captain Harris Laning in his new role as head of the Tactics Department. Laning described the proposed formation in considerable detail "because it seems to have possibilities. It is the only one so far produced at the War College that permits a change of front of a huge fleet without either causing confusion among the many ships when they rush to a new position, or sacrificing security and ability to deploy." A more conventional formation was difficult to turn. In the long run, the most important virtue of the circular formation was that it provided an all-round defense. The circular formation was used again in a Blue-Orange game, and again Laning described it in great detail. One of MacFall's fellow students, Commander Chester W. Nimitz, was assigned as aide to U.S. Fleet commander Admiral Samuel S. Robison when he left the War College. He sold MacFall's idea, and by 1930 it had been adopted as a standard formation. Meanwhile, it invariably featured in the staff solutions to the estimates of the situation.

MacFall thought of the circular formation as a cruising formation rather than as a battle formation, because battle meant a fight between two battle lines plus other warships. However, the battles in the Mandates would generally pit the U.S. fleet against Japanese aircraft and submarines. The circular formation *was* the battle formation in these cases.

Air-sea warfare in the Pacific was generally much more like the attrition war expected in the Mandates than like the culminating fleet battle envisioned as their sequel—hence, the importance of the circular formation. As it happened, the circular formation was also very well adapted to carrier operations, and even to multi-carrier ones. It reflected the Naval War College perception that a Pacific War would necessarily be a protracted campaign.

The Naval War College envisaged three Japanese means of attrition. One would be air attack, using either carriers or land-based aircraft, or a combination. A second would be submarine attacks. A third would be mass surface torpedo attack, particularly at night. Although the Pacific War did not turn out quite as expected, the Japanese took a similar view: The college successfully predicted the methods they would emphasize. In doing so it also predicted what many, looking back, see as peculiarly Japanese emphases.

Thus, the war games reflected an expectation that the Japanese would mount large night torpedo attacks on the U.S. fleet. The night battles in the Solomons should not have been nearly the surprise that they were. There seem to have been two reasons this was not the case in reality. First, the pre-war U.S. Navy did not conduct extensive night exercises because they were rather dangerous, hence politically unacceptable. As a consequence, it did not gain the night expertise achieved by either the Imperial Japanese Navy or the Royal Navy.[14] That meant it did not come to appreciate the special requirements of night combat. Worse, radar seemed by itself sufficient to solve the problem of night combat, in effect to turn night into day. That was not quite what happened. Radar plus the plot in a combat information center really did change night combat for the U.S. Navy, but that took time to develop. Before that the Navy suffered badly in the Solomons. However, it did not do so out of deep ignorance as to what night torpedo combat could mean—the pre-war games made that fairly obvious.

Pearl Harbor clearly changed the war from the expected immediate step-by-step advance through the Mandates to something much more protracted, and far more reactive to what the Japanese were doing. Thus, there was no direct war game equivalent to the desperate fight around Guadalcanal or, for that matter, to Midway. However, there were plenty of carrier-versus-carrier battles. There were also cruiser battles fought as part of campaigns to attack Japanese trade as a distraction. There was considerable game (and full-scale) experience showing that the U.S. battle line would have little or no part in a carrier-versus-carrier struggle. In that sense, Midway was fought as though Pearl Harbor had not happened—except for the perception that further major losses would be disastrous.

Once the U.S. fleet had been rebuilt, the pre-war concept of a step-by-step advance through the Mandates was revived. As expected, the Japanese held back their battle fleet while contesting the U.S. advance mainly with aircraft. There was, however, a major surprise. The war games showed that there would be huge wastage of aircraft on both sides. For the U.S. Navy, the lesson was simple. As war approached, production not only of aircraft, but also of pilots was stepped up. It was assumed that the Japanese would do the same, although it was known that they could not remotely match U.S. aircraft production. That was not true; until 1944, the Japanese persisted in training only small numbers of new pilots. They also failed to rotate exhausted pilots out of combat. The effect on the Japanese of large-scale pilot and airplane losses were therefore underestimated. It seems to have been assumed that Japanese naval aviation could be eliminated only if the Japanese carriers were sunk.

The effect of such misperception was evident after the Battle of the Philippine Sea in June 1944. During the battle, U.S. commander Admiral Raymond Spruance considered protection of the invasion of Saipan his primary responsibility.[15] He was unwilling to pull his fast carriers away from the island in order to pursue the Japanese carriers. The Japanese found the U.S. carriers before they themselves had been detected. In the past, this would have led to serious U.S. losses. By this time, U.S. fighters and fighter control were so effective that the Japanese strike force was massacred. The air portion of the battle was called the "Turkey Shoot." Because of the perception that no such massacre could be decisive, Spruance was widely condemned. As Pacific commander, Admiral Nimitz told the fleet commander for the next major operation, Admiral William F. Halsey, that the Japanese carriers were his primary target, even if that meant sacrificing the landing at Leyte Gulf.[16] It seems not even to have been suspected that at the time the Japanese carriers were effectively toothless, and that they were almost certain to remain so.

At Leyte Gulf, the Japanese used their carriers, nearly all of which had survived the Philippine Sea battle, as bait to draw Halsey away from Leyte Gulf and the invasion shipping, leaving it open to their powerful surface fleet. When the Japanese surface force actually attacked, Halsey was roundly criticized. His pilots had attacked this force the previous day, and had reported routing it. Halsey was later criticized for naively believing the pilots, despite considerable experience that they often exaggerated. On the basis of

The one great failure of War College analysis was the kamikaze. No one seems to have realized that the Japanese would have to resort to special tactics as they ran out of experienced pilots, probably because no one aware of the huge U.S. pilot replacement program could imagine that any serious navy lacked an equivalent. Once the Japanese had too few effective pilots, suicide tactics were the only ones that had much chance of success; they were also consistent with Japanese culture, which glorified individual sacrifice through suicidal attack. The War College tended to treat potential enemies as Americans who spoke other languages, but shared U.S. thinking processes. The carrier *Randolph* (CV-15) is shown at Ulithi alongside a repair ship after a kamikaze damaged the after part of her flight deck. (National Archives 80-G-344531)

this belief, Halsey felt justified heading north to attack the Japanese carrier bait. The Japanese surface force turned around and struck Leyte Gulf. The invasion shipping survived due to a heroic stand by U.S. destroyers and escort carriers off Samar. War gaming did not describe anything like the complex attack mounted by the Japanese.[17] However, it did raise exactly the question Spruance and Nimitz faced: What was the mission (or the priority) of the commander of a force covering an invasion?

The kamikaze surprise is another aspect of the same failure of perception. The Japanese adopted kamikaze tactics because by late 1944 they no longer had sufficient numbers of pilots who had any chance of attacking effectively using conventional tactics. The choice they made certainly did reflect Japanese culture; no one in the West would have used suicide tactics on a large scale. However, Western accounts of Japanese culture certainly did discuss their willingness to sacrifice lives freely "for the Emperor." There were even examples of deliberate suicide attacks earlier in the war. No one included mass suicide attacks in war games, but mass air attacks were common. Given the difficulty of estimating how well defenses would do, it is not clear that including the kamikazes would have made an enormous difference at the gaming level. Moreover, by the spring of 1945 it seems to have been clear that kamikaze attacks were a rational response to the success of U.S. fighters and anti-aircraft fire. It would actually cost fewer pilots' lives (in their function as kamikazes) to inflict a given level of damage on the U.S. fleet.[18]

What does all this say about gaming at Newport as preparation for World War II? Above all, gaming was an excellent preparation for an entirely new kind of war. Gaming teased out the ways in which aircraft could or would fight at sea in large numbers. Full-scale exercises could and did test carrier-versus-carrier and carrier-versus-battleship tactics, but they could not possibly simulate a protracted grinding campaign. Moreover, gaming could show what would happen as temporary limitations on aircraft capability were overcome. Looking back, we see officers who really did operate aircraft complaining before the war that what the War College allowed was often unrealistic, but aircraft were developing so rapidly that this was more virtue than vice.

The entirely new type of war could not be simulated on a full scale: No one was going to see what happened if massed torpedo bombers encountered massed fighters, or if they attacked a battle line at sea. There was some simulation, and there were dummy attacks with practice weapons, but they were clearly unrealistic (among other things, no one was shooting back). In all other cases, actual combat experience gave some sense of what future warfare would be like, although all types of naval technology had moved on considerably since 1918.

Gaming portrayed not just short battles but whole campaigns. Nothing but gaming could have simulated this, and the virtues of gaming explain why so much game material ended up in the files of the War Plans Division of OPNAV. An individual student would experience a year of games and perhaps, as a staff member, another year or two. In the 1930s, the Research Department sometimes explored the lessons of multiple games, and it produced a summary of highlights.[19] However, anyone exploring the War Plans Division files saw much more, and got a much better sense of how alternative approaches to the fundamental Pacific problem played out.

Moreover, in the course of the games players tried many ploys that could feed into future plans. For example, in one game the Blue commander withdrew one of his carriers to make a surprise strike on Tokyo. Because the game did not take Japanese policy makers into account, it was impossible to decide what the effect of the ploy would be. It was, however, interesting enough that it was included in a series of notes on the games produced by Naval War College's research director Captain van Auken. It is impossible to say whether it had any later consequences.

Gaming, at least as it was practiced at the college, offered no insight into the way the Japanese thought. The Japanese as they were simulated at the college were Americans provided with the Japanese fleet and Japanese bases. That is why it is so impressive that the sort of tactics the Japanese actually used during the war were evident in the games. On the other hand, gaming at this level could not provide answers to questions like what it would take to end the Pacific War. There is no evidence of much insight of that sort anywhere in the U.S. government of the time.

Pre-war gaming only lightly explored the problems and potential of coalition warfare, because the pre-war United States was neutral, with a foreign policy that emphasized neutrality. As fought, the Pacific War was very much a coalition effort, but the most important impact of the coalition was in the geography of the war. Defense of Australia was a vital theme in the Pacific, and it largely accounted for the devastating fight around Guadalcanal. It could not and did not figure in pre-war gaming (Australia was generally a friendly neutral).

Gaming had far less impact on U.S. performance in the Atlantic War. No such war was contemplated for most of the inter-war period (with the exception of various Red-Blue exercises). It amounted to a replay of World War I, hence anathema to many Americans. Moreover, the Atlantic War was far more a coalition affair, and also far more affected by strategy for the land and air components of the war. Convoy warfare did not entail major new operating concepts like those featured in the Pacific War. War games played in 1941 were clearly intended mainly to familiarize officers with the conditions of the undeclared naval war in the Atlantic then being fought. All of this having been said, it seems unlikely that the U.S. Navy would have been nearly as effective in the amphibious phases of the Atlantic War had it not developed a form of amphibious warfare to fight in the Pacific.

What does all of this mean for war gaming? War gaming was best at exploring a new kind of warfare, for which prior experience was not very relevant. That meant above all air-sea warfare in the Pacific, with amphibious warfare a close second. War gaming forced practitioners to think in campaign terms—not in terms of a single battle. It was consideration of a Pacific campaign that led the U.S. Navy to understand what air-sea battle would be like, and what it would mean. The great contrast with the Japanese and the British is that neither thought through a protracted Pacific campaign. It is difficult not to see war gaming as the difference. Nothing short of war gaming could have provided a campaign perspective.

War gaming was also valuable because it allowed free exploration of possibilities, away from any publicity or leakage. War gamers could envisage options that were entirely unacceptable politically during the 1930s—and that were sometimes actually realized during World War II, such as coalition warfare. Secrecy mattered, and it proved possible to enforce.

Conversely, war gaming was a poor way to investigate the problem of how to end the Pacific War—that depended very much on how the Japanese thought, which was just what the War College could not simulate. The assumption through all the games was that the Japanese government would behave rationally; it would come to terms (presumably short of unconditional surrender) when it knew that it was beaten. No war game envisioned the bloodbath at Okinawa or the much worse one the Japanese planned for the invasion of the Home Islands. Nor, incidentally, did war planners who had learned their trade largely by gaming (simulation).

Appendixes

A: Playing the Games

There were two types of game, the large-scale (chart) maneuver and the smaller-scale tactical game (board maneuver); all of the games were called maneuvers rather than games. Notes in the account of Chart Maneuver I (Class of 1928) show how a large-scale game was played. Typically each student turned in an "Estimate of the Situation." All were evaluated, but solutions (estimates and basic decisions) were produced by two staff members as a basis for further play. No student solutions were collected for the 1928 game, which began with a fixed set of orders for the two CinCs.

Normally solutions included radio frequency plans; in this particular case, Blue and Orange frequency plans, indicating which ships talked to which commands, were provided.

Normally each CinC held a conference to explain to his subordinates his estimate (commander's intent), his decision, orders, and plans, and also such doctrines as he might choose. He gave copies to all the subordinate commanders and to the game director. The game director chose the length of each move. When ready for a move, the director would post an announcement giving the serial number of the move and the date and hour when it began and ended, and also the weather and visibility. He would set a clock dial to the game time at the beginning of the move.

For the first move, each player turned in a tracing showing what his force would do. A plotter transferred these movements onto the master chart in the game room. Once all the first-move tracings had been submitted, a player could see whether or not there would be contact between the forces. The assistant director for plotting would transfer students' tracings onto his own chart showing what had happened during the move.

If there was no contact, the next move would be called. Even without contact, players had to work up fuel and similar accounts, and they might send dispatches. In some cases this work-up delayed play; the game director advised players to get their tracings ("flimsies") in first, then dispatches, and only then fuel account and plans for the next move. Players also had to turn in their "record of move" so that a history could be kept. The flimsy showed what the player intended to do if nothing happened; the record showed what the player actually did. There was provision for changing intentions, but only if the player did not try to go back in game time.

Commanders of forces not in sight of each other were placed in separate rooms or otherwise separated to keep each from knowing what the other's force was doing. All communication between individuals not assumed to be in the same ship was by message. The rules included transmission and decoding delays.

If ships came into contact, either in sight of each other or in sight of smoke, the director or his assistants informed the commanding officers of the ships, and asked what their action would be. The resulting movements were plotted. Players could ask for more information, but were reminded that "every commander always suffers from lack of information." The game continued once contact was cleared up. Play was often delayed by the mass of messages sent upon contact; players were reminded to write up and hand in messages as quickly as possible, because in reality there would be little time to waste once in contact with the enemy. In some cases, players intended to send radio messages but delayed until later moves, even though the messages affected action during earlier moves. Players also sometimes took too long to decide what to do.

Once contact became general, action might be moved to the game board or the game might be called. If the maneuver was to continue as a chart maneuver, contacts would be handled and arbitrarily adjudicated by the director so that the maneuver could proceed. The director would decide whether to play a situation out on the game board.

A record of each move was kept, and at the end of the game the records were all turned in so that a narrative could be constructed. Accounts of games in this book are based on such narratives, which in some cases included commentary on students' performance.

The tactical game was conducted in much the same way, each move equivalent to three minutes of real time unless otherwise specified.[1] As in the strategic game, it was important that players see no more than they would in reality. Typically, screens were used to cut off visibility; alternatively some players could be placed out of sight of the board and of any other player handling forces not in sight from his force. A player would make his moves on tactical plotting sheets. They might be transferred either to the game board or to the master plot, or to both. Commanders of submarines (individual players) were permitted to see the board only when their craft were surfaced, and when permitted by the director.

Players turned in special blanks at the end of each move. If a move blank or other form called for a ship doing more than the rules allowed, the umpire would return the form for correction. In the event of undue delay, the director might simply require a force to continue what it had been doing on the previous move. No ships would be moved on the board until the director ordered them moved. All would then be moved promptly.

Once turned in, moves could not be changed except for ships whose action was based upon the action of other ships, or in case a compelling but unforeseeable factor intervened, such as a torpedo hit. Should a new move be ordered before damage had been determined, the value of the ship's own action would be appropriately revised when damage was known, but the director would decide when to apply damage by sinking or losing speed. Damage would be classed as above or underwater, each with its own effects. Damage was expressed as percentage of original life in 14-inch hits. It was counted in units of 10 percent, rounded down (e.g., 0 to 9 percent was counted as 0). As soon as it was determined, damage was reported to the commander under whose personal command the ship came. All damage during a move became effective at the end of the move.

B: War Game Rules—Aircraft

The war game rules ("Maneuver Rules") were key to the simulation the Naval War College sought to create. They reflected the fleet's understanding of current naval technology. In addition to their use at the college, the rules were the basis for the umpire rules used during full-scale fleet exercises. Thus, they affected the fleet's experience of simulated combat far beyond the War College. The fleet's experience fed back into the rules, from formal reports of the Fleet Problems (which survive in Naval War College files), from replies to correspondence sent as the rules were revised each year, and from discussions during game wash-ups, particularly of the annual "big game." Members of the Research Department were present at the wash-ups, and surviving copies of discussion mention possible changes to make the rules more realistic. For aircraft, contentious issues included the effect of fighter defense and the damage effect of dive bombing.

Most, but not all, of the annual editions of rules have survived in NWC files.[2] It is possible to follow changes because each set of rules was a compilation of sheets, each of which, from 1927 on, was dated. The first of the new-type sets of rules was issued in 1919. However, the earliest full set that survives is dated June 1923 (new sets were issued each June). Notes to some of the games show that there were initial difficulties over the effectiveness of air attack. To some extent, experiments against real warships in 1920–21 may have resolved them, but important questions could not be resolved in this way. It might be clear, for example, how much damage a 1,000-pound bomb would inflict *if* it hit, but it was much more difficult to say what chance it had of hitting a moving, maneuvering ship compared with, say, a 14-inch shell. Much the same could be said of air-launched torpedoes. It was also difficult to evaluate anti-aircraft and air-to-air fire.

In most cases, the analysts at the college could point back to combat experience and exercises as a basis for rules. For example, they knew the hitting rates of heavy guns in gunnery practice, and they also knew hitting rates experienced in World War I combat. They could use combat experience to estimate how much worse crews would do in reality, compared to how well they performed in peacetime. Aircraft were a very different proposition. No one could say, for example, how anti-aircraft fire would affect the accuracy of a dive-bombing pilot. The analysts did try to use gun-camera footage to translate practice dog-fights into estimates of combat performance, but they knew that this was far from adequate, and many in the fleet told them so. Air-to-air tactics were still so primitive that it was impossible to say how formations of fighters would fare against bombers. Aircraft performance was difficult to factor in.

It might be argued that World War I experience was relevant. It appears that little data had been collected during the war, for example on air-to-air combat. Wartime exigencies had precluded systematic training of fighter pilots, so that it might be argued that the much more highly trained future pilots would perform very differently than their predecessors. There had been, moreover, little experience of the sort of mass-on-mass air engagements that figured in the games. Evidence of the effect of anti-aircraft fire was anecdotal at best, and future guns would have much better forms of fire control—hence, were likely to be far more effective than those used during World War I. Naval War College files seem to reveal no analysis of World War I air warfare data, such as it was. Perhaps more remarkably, there also seems not to have been any attempt to base rules on experience in the local wars of the inter-war period, particularly the Japanese war against China and the Spanish Civil War. They were certainly studied, but it may be that hard data were lacking.[3]

Accounts of war games sometimes feature air battles which seem entirely unrealistic. For example, battleship and cruiser scout planes, with their limited armament and bomb capacity, sometimes successfully attacked enemy carriers, wrecking their flight decks, despite the presence of their fighters. In some cases, bombers and even low-performance observation aircraft shot down fighters, because the rules made that possible.

For a modern reader, the game rules give a sense of how the U.S. Navy saw the inter-war evolution of carrier aviation. This account, though considerably simplified, also gives a sense of just how detailed the rules were and of how much effort the War College had to expend to take them into account during a relatively fast-moving game. The rules could not simply be codified into slide rule settings or graphs. Many of them involved judgment, and players could seek advantages by using them creatively.

Until 1930, aircraft could be detected only by eye. The rules indicated at what range aircraft could be seen. For example, a VT above the observer could be seen from five miles away. In particularly high visibility, another mile was added. A surface observer could see an airplane on land two miles further away. In 1930, sound detectors, which armies were then using, were added to the game.[4] These detectors were employed on land to defend a base against air attack in one game (this one alerted defending fighters). It is not clear whether they were ever considered by the U.S. Navy for shipboard use. The June 1941 rules still included sound equipment, by then quite obsolete, but also included the statement that "a few installations exist of other (secret) means of detecting airplanes at a distance of several miles [i.e., radar]. The Director will rule as to their availability on request."

For simplification and comparison with game and wartime experience, this summary is limited to carrier aircraft. It therefore excludes extensive discussions of air spotting for heavy guns (an important consideration in inter-war battle fleet tactics, but irrelevant during World War II); seaplanes; and rigid airships. The rules were never intended as general guides to materiel and to tactics. The 1929 aircraft section, for example, warned that

> players should have a thorough knowledge of the characteristics, performance, and use of each class of aircraft, and a general knowledge of aircraft tactics.... These rules are based on the best information obtainable and for average conditions but should not be considered literally as results that will always obtain in actual practice.... For instance, the rules relating to visibility represent ranges at which objects are visible to the eye, but in order to permit aircraft to employ the principle of surprise with which they are endowed, it cannot be expected that such swiftly moving objects will always be seen at the same definite range. The Maneuver Rules concerning aircraft are laid down to give a fair basis for development of tactics and strategy, and also to emphasize certain characteristics of aircraft, such as their fragility of construction, liability to breakdown, and short life, as compared to other naval materiel. At the suggestion of naval aviators, breakdown penalties have been kept quite severe in an attempt to duplicate war time operation with average material, and to discourage excessive operation. It is well known that in peace time, airplanes can be made more reliable than these figures indicate...effectiveness is all relative and...the life assigned a battleship is probably as open to argument as is the effectiveness of the air, submarine, or destroyer arm.

In the account that follows, different parts of the aircraft rules are treated separately to show how they changed from 1923 onward. Strikingly, the 1939 rules opened with a discussion of what sort of airfields would be needed by modern heavy aircraft, a natural issue in games concerned with seizing and developing advanced bases, but one absent from previous sets of rules.

The aviation part of the rules was Section J, which in the 1923 edition came last, after submarines, and was 15 pages long. It included tables of aircraft performance. It referred to rules in other sections, such as those for visibility (Section D). When he answered the 1935 call for revisions, BuAer chief (and later CNO) Rear Admiral King pointed out that aircraft were the only naval weapon not yet really tested in war, so to some extent the rules reflected opinion as well as experience.[5] Questions of air-to-air combat and modern anti-aircraft effectiveness were particularly difficult to answer.

* * *

The Airplanes
In 1923, the only surface attack aircraft were torpedo bombers capable of carrying either a lightweight (1,650-pound) torpedo or one 1,500-pound bomb. Patrol planes (seaplanes) were credited with four 230-pound bombs. All aircraft came in both land and seaplane versions. Performance figures included maximum and cruising speeds; a VT cruised at 70 knots but could make 90 or 95 (landplane version); a fighter had a maximum speed of 130 or 135 knots. Data included the speed the airplane could make while climbing, and its ceiling, span, and gross weight. Endurance was also given: 6.75 hours for a land-based VT, 4 for a land-based fighter. All aircraft other than fighters had radios.

In 1926, for the first time, the game rules were based on the performance of actual U.S. Navy aircraft. From then on, standard practice was to assume that both sides used U.S. aircraft. In a very few cases it was accepted that a foreign navy had aircraft for which the U.S. Navy had no equivalent.[6] Overall, crediting foreign navies with U.S. types of aircraft was a way of avoiding the problem of Japanese secrecy. No one really knew how Japanese aircraft performed. In effect, the inter-war college assumed that U.S. aircraft represented the state of the art and, true to the conservative practice of avoiding giving the U.S. side any particular advantage, credited the Japanese with aircraft quite as good.

In the 1925 rules, the heavy carrier aircraft could function as torpedo bombers or level bombers or scouts. VTs could carry one 1,000-pound bomb or two 500-pounders or 30 25-pounders—in each case, enough to wreck a flight deck. Dive bombing did not yet exist for the Navy, although the Marines were developing it. There were also single- and two-seat fighters. Fighters could carry four 25-pound bombs.

The 1925 rules pointed out that aircraft should be organized into tactical groups before taking off, although they could be reorganized once in the air if the necessary signals were available. In daylight without fog, planes could rendezvous in accord with specific instructions, if they could see some recognizable object. The 1926 rules went further: Prior to the game, airplanes should be organized into groups, wings, squadrons, divisions, and sections as suitable, in accordance with standard procedure, and operated as units (this was simplified to squadrons in the 1933 rules). Again, after taking off, tactical units might be changed, however. Ideas about organization were linked with estimates of how many aircraft could attack together or could join to fight an enemy's air strike.

Carrier Air Operation
The 1923 rules specified how quickly aircraft could be flown off and landed on carriers: one fighter every 15 seconds, but one heavier VT or VS or VO every minute. Either two fighters or one VT could be brought up from the hangar in 2 minutes. To recover aircraft, a carrier had to steady herself into the wind for six minutes, after which airplanes could fly on at two-minute intervals (twice that at night). They could be struck below at the same rate. However, a carrier could no longer launch or recover aircraft after receiving the equivalent of a single 14-inch penetrating hit.

The rules took the dangers of flying into account. In the initial rules, an airplane had a 1 in 20 chance of being wrecked when landing on the carrier (1 in 10 at night). Umpires decided how many aircraft were wrecked. In some games, so many aircraft were wrecked that air reconnaissance became impossible. As airplanes improved, the wreck figures were reduced. In 1925, the chance of being wrecked while landing in daylight was given as 1 in 30 (1 in 20 in darkness), a 50 percent improvement. In 1927, rules were made more elaborate and the chance of being wrecked while landing on a carrier depending on the type of airplane and the weather. In smooth weather, a fighter had a 1 in 25 (0.04) chance of being wrecked every time it landed. However, in rough weather this rose to 1 in 10. As aircraft improved, the 1929 rules reduced the chance of being wrecked to 0.03 under all sea conditions (0.06 at night). Comparable figures for a landing field ashore were 0.001 and 0.03. In 1933, the chance of being wrecked while landing on a carrier was reduced to 0.01, which was still ten times what a pilot faced ashore under the earlier rules.

In the 1925 rules, aircraft could not land on a carrier with bombs or torpedoes onboard. This rule made for very high expenditure of weapons whether or not aircraft found targets, with serious logistical consequences. In one game, bombs were in such short supply that a badly damaged carrier transferred hers at night to undamaged carriers. In another game, one of three carriers soon had no torpedoes left. In the 1930s, the Bureau of Ordnance found ways to fuze bombs so that aircraft could land on carriers while still carrying them. This improvement was reflected in the 1939 rules. Even then, aircraft could not land on a carrier with their torpedoes onboard.

According to the 1925 rules, an airplane had to be serviced for at least half an hour between flights if it had been out to less than half its endurance. It would need more work after a longer flight, up to two hours if it had been out over three quarters of its endurance. Catapults could be fired once every two minutes. Scouts should fly in pairs, because if operating singly their chance of being unable to return was 1 in 20.

The time required to recover each airplane was crucial. A carrier could not fly aircraft if she could not take them onboard when they returned. After flying out and attacking, a returned airplane would have only limited remaining fuel, hence limited remaining time in the air. The shorter the time to land airplanes, the longer they could remain in the air. Reeves's innovations were reflected in the 1929 rules. Aircraft could fly on at one-minute intervals, and the rule that the flight deck had to be clear for an airplane to land was eliminated. The 1933 rules cut the daylight landing interval to half a minute. At the same time, launching intervals, which were already short, were further reduced. A fighter could be launched every 10 rather than 15 seconds. The change for torpedo bombers was much greater, from the earlier two minutes between takeoffs to 15 seconds. This change much reduced the advantage a group of small carriers had enjoyed. If torpedo bombers took so long to launch, several small carriers launching more or less simultaneously were needed to place a large strike force in the air. A single large carrier could do far better if she could launch the same aircraft very quickly. To the extent that grouping carriers was inherently dangerous, the change in rules favored dispersing large carriers as widely as possible. That had enormous impact. Note that it coincided with the appearance of the two large U.S. carriers *Lexington* and *Saratoga*.

Other altered rules made sense given barriers and arresting gear. Now it was assumed that

> at least a quarter of a carrier's planes would normally ride the flight deck. Half could be accommodated there "in such a manner that either flying off or flying on, but not both simultaneously, can be carried on." A maximum of nine aircraft could fly off as a group.

The rules changed in 1936 to allow for launch *and* recovery with aircraft on deck. All of a carrier's aircraft could be parked on deck, but in that case airplanes could take off but not land (i.e., a deck-load strike could be set up). With three quarters of the aircraft on deck, airplanes could take off (if the aircraft were all aft) or land (with all of them forward), but not both simultaneously. However, with half the airplanes on deck, presumably all amidships, aircraft could take off and land simultaneously. No more than half the aircraft could be in the hangar at any one time. Such practices differed completely from those in other carrier navies. Launch interval was given as ten seconds, and landing interval as 30 (intervals would double in darkness). A new feature was that unscheduled launching could occur five minutes after the decision to launch—as when enemy ships were suddenly found—but only out to a radius of 100 miles. Preparations for longer flights would take 15 minutes. The 1936 rules gave refueling time for different types of carrier aircraft, the least being 20 minutes for a fighter. Rearming times were also given for bombs and torpedoes.

Carriers had flight-deck catapults throughout the inter-war period. Although they were mentioned, they were not the subject of any rules. No catapult launch rate was given. The 1939 rules pointed out that they could be used even if the carrier did not turn into the wind, as long as the wind was forward of the beam. Otherwise the carrier had to turn into the wind (five minutes) and the apparent wind had to be within 10 degrees of the ship's centerline. That might be a real advantage. Presumably this new rule was framed because the new carriers had more powerful (and faster-firing) catapults, but carrier deck catapults were not heavily used until late in World War II.

The 1941 rules, the last written before war broke out, added that carriers had arresting gear fore and aft, hence could land planes going either ahead or astern. Such gear had been in place at least since 1939.

Bombing
Initial estimates of bombing accuracy were optimistic. For airplanes flying at up to 1,500 feet, it should take two bombs to make one hit on a large or intermediate ship, such as a battleship or cruiser; up to 4,000 feet, four; up to 8,000 feet, six; and up to 12,000 feet,

ten (twelve for an intermediate ship). Airplanes dropping bombs from below 500 feet would be destroyed by their blast. The 1924 rules recognized that the hitting figures were valid only for target practice. They might decline sharply in combat, for example when bombers met opposition. These modified rules added a table: When met by planes of fighting strength considerably superior to the bombers, the hitting rate would decrease by 40 percent. The umpire was to assign further decreases due to maneuvers by the target ship or speed or anti-aircraft fire. It was assumed that bombers would face either fighters *or* anti-aircraft fire, in the theory that ships would not fire with friendly aircraft overhead.

The next year these figures were changed again. Under target practice conditions, a bomb dropped from 1,500 feet should have a 40 percent chance of hitting (less than before); from 8,000 feet, 15 percent; and from 10,000 feet, 10 percent— all against a large ship. Soon it was accepted that bombers would generally attack in formation, their bombs forming salvoes equivalent, in theory, to the salvoes fired by a ship's guns firing together. As with gunnery, the chance of at least one hit depended on the spacing of the projectiles, which in turn depended both on the spacing of the airplanes and the precision with which their bombardiers released them. Much also depended on how the ship moved—maneuvered—between the moment the bombers committed to an attack and the moment the bombs arrived. Typically, a master bombardier had to track the ship to estimate its course and speed before releasing bombs.[7] According to the 1926 rules, a 15-knot ship attacked from 4,000 feet would present no problems. The faster the ship and the greater the altitude, the greater the chance would be that radical maneuvers would throw off a bombardier. Thus, in 1926, it was estimated that a bomber attacking a 35-knot ship from 10,000 feet would lose about a quarter of its effectiveness.[8]

In the 1925 and later rules, a single 500-pound bomb was credited with the ability to destroy half a carrier's flight deck, a figure retained in later sets of rules. The 1926 rules listed the weapons most suited to various targets, the 500-pound or 1,000-pound bomb being best against carrier flight decks (but 2,000-pound bombs and torpedoes were best to destroy a large carrier). Under the 1925 rules, a direct hit by a 2,000-pound bomb would cause above-water damage equivalent to four 14-inch hits. One hit by a 1,000-pound bomb would cause the equivalent of two such hits. The supposed impact of bomb hits was sharply reduced in 1931, in effect devaluing the new dive bombers with their 1,000-pound bombs.

As a way of better understanding air weapons, the 1931 rules pointed out that the different bombs were about the weight of different standard shells: 2,000-pounder for 16 inch, 1,000-pounder for 13 inch, 500-pounder for 10 inch. This was intended as a rough guide to which bombs should be used against which targets. Two 1,500-pound or four 1,000-pound or eight 500-pound hits on a carrier were equivalent to one 14-inch hit. However, it would take four 1,500-pound bombs to have the effect of one 14-inch hit on a battleship, whose life might be ten to twenty such hits.

Appendixes

Aircraft could also deliver chemical bombs. Under the 1931 rules, four such bombs hitting a carrier would inflict the effect of one 14-inch hit. However, the effect of gas was transitory, typically lasting only an hour. Six gas bombs hitting a battleship were credited with the effect of one 14-inch hit. As noted previously, gas bombs were a fixture of most inter-war rules, and in many games airfields ashore were kept out of action by repeated gassing. Rules included considerable information about the effect of gas weapons and there was much interest in ways of defeating gas attacks.

The 1929 rules introduced the key inter-war technique for air attack against small moving targets: dive bombing. At minimum altitude for horizontal bombing, 1,000 feet, a bomber was expected to have a 50 percent chance of hitting a large target. Normally, it would never attack from so low an altitude, because it would be so easy a target for anti-aircraft fire. If the level bomber attacked from 10,000 feet, just above maximum anti-aircraft altitude, it would have only a 10 percent chance of hitting a ship, even if the ship did not maneuver at all. Navies adopted strafing tactics to dislocate shipboard anti-aircraft guns so that bombers could attack from lower altitudes. It was up to the game director to decide how well such attacks would work.

A dive-bomber pilot pointed the nose of the airplane directly at the target and, to some extent, the pilot could adjust as the target maneuvered.[9] Anti-aircraft guns typically could not hit a rapidly diving airplane at all, so the dive bomber could survive to release its bombs at low altitude. Release altitude was typically set by the altitude at which the bomber pulled out after releasing its bomb; the bomb was released about 600 feet above that. Typically, the pull-out altitude was only 500 to 1,000 feet. Under the 1929 rules, a dive bomber pulling out at 500 feet had an 80 percent chance of hitting (reduced to 65 percent if the bomber pulled out at 1,000 feet).

The 1931 rules handled bombing in a more sophisticated way. Level bombing was given a slightly better chance of success (12 percent from 10,000 feet). However, anti-aircraft fire was also rated more effective, so that the lowest altitude for such bombing was 2,000 feet, with a hitting rate of only 26 percent against a large non-maneuvering ship. The rules now also took into account opposition by fighters during the bombing run. How effective that might be depended on whether there were more or fewer fighters than bombers. There was no attempt to go into greater detail as to the degree of superiority or inferiority. A superior number of fighters could reduce the bombers' effectiveness by 60 percent; an inferior number by 20 percent.

Dive-bombing accuracy was now estimated by reducing the level achieved in target practice by 25 percent to take wartime conditions and anti-aircraft fire into account. By this time, the figures given in 1929 were considered somewhat optimistic, so that in 1931 the chance of hitting a large ship after a 500-foot pull-out was given as 54 percent (the corresponding 1929 figure would have been 60 percent). The higher the pull-out, the worse the accuracy, but even a 2,000-foot pull-out was expected to give a 21 percent chance of hitting a target. These figures explain why dive bombing was so exciting a tactic.

It seemed that if even a few dive bombers managed to reach an enemy ship, they were virtually certain to score hits. By this time, too, U.S. dive bombers could deliver the heaviest available bombs: 1,000-pounders.

At this point, it was accepted that the only effective antidote to dive bombers was anti-aircraft fire, mainly by automatic weapons. Anything heavier required fire control that could not follow the bomber through its dive. The effectiveness of a dive bomber under heavy fire would be halved (light fire would cost only 10 percent). Comparable figures for a horizontal bomber were 43 percent under heavy fire and 18 percent under light fire. However, since the horizontal bomber was much less effective in the first place, it might be that a combination of maneuver and anti-aircraft fire would negate it altogether. By way of comparison, even halving that 54 percent hitting rate would make for many hits. Another table gave reductions due to target maneuvers.[10]

Bombing data were further simplified in the 1933 rules, with unopposed dive bombers attacking a large ship simply being given a 40 percent hitting rate. That compared to 27 percent for level bombers attacking a large ship from 2,000 feet, and 13 percent for the same target from 10,000 feet.

The value of bomb damage was significantly reduced. A 1,000-pound bomb hitting a battleship had 0.58 of the effect of a penetrating 14-inch hit, the common currency of War College damage tables. It had 0.75 of that effect against a carrier or a cruiser—which would have less overall lifetime.

The figures for accuracy were based on actual results in the air section of the annual gunnery exercises. A 1936 study of the rules tabulated results achieved by dive bombers in 1929–33: 37.5 percent against large targets.[11] On this basis, the rules gave dive bombers 40 percent hits. In 1936, the percentage of hits was reduced by a quarter—to take the "fog of war" into account—to give 28 percent hits against large ships.[12] In 1929–33, the comparable probability for horizontal bombing was 18.18 percent, which fell to 17.9 percent for 1929–35. The practice targets were smaller than enemy capital ships, which is why somewhat higher hitting percentages were assumed. For example, the 17.9 percent hitting rate of 1929–35 would be corrected up to 19.25 percent against real ships. Taking off a quarter for operational factors gave 15 percent.

When the rules were revised in 1936, bombing effectiveness was significantly reduced, because it was now evident that the explosion of its bomb would destroy an airplane as low as 1,500 feet. In the past, it had been assumed that the critical altitude was 400 to 500 feet. The minimum altitude for level bombing was increased to 3,000 feet, the hitting rate falling accordingly. These rules omitted the percentages of past rules, instead changing the percentages previously accepted. New bombing-effect diagrams separate from the rules themselves made it possible to convert hits and near-misses into equivalent above- and below-water 14-inch hits.

The 1937 rules largely reversed the relationship between dive and level bombing, perhaps because the U.S. Navy now had Norden precision bombsights for level bombing and also because it seemed that machine guns had bested dive bombers. Thus, a percentage of direct bomb hits under the full range of conditions showed 16 percent for dive bombing a large ship, compared to 8 percent for level bombing from 9,000 feet and 15 percent from 6,000 feet (30 percent from 3,000 feet). However, another table showed that dive bombing retained much more of its effectiveness at higher target ship speeds. Effectiveness was halved for a ship making over 25 knots, but the equivalent factor for level bombing from 9,000 feet was 0.3. War experience would show that all of these figures were rather optimistic, but that dive bombers could hit moving ships whereas horizontal bombers generally could not.[13]

At the same time, the reduction in assumed bomb effectiveness was reversed. This time a single above-water hit by a 500-pound bomb was considered equivalent to one above-water and 0.25 below-water 14-inch hits; a single 1,000-pound bomb was equivalent to two above-water and 0.75 below-water 14-inch hits.

Bombs Versus Carriers
Through the inter-war games, the single crucial fact of carrier warfare, highlighted by the game rules, was that a few bombs could wreck a flight deck and eliminate a ship as a carrier. Early figures for bomb damage showed, among other things, that a single 500-pound or 1,000-pound or 2,000-pound bomb—or even two 100-pound bombs or ten 25-pound bombs— would destroy a carrier flight deck. Later rules differentiated between hits on each end of the flight deck, leaving a ship with half her deck. The U.S. Navy placed arrester gear at both ends of its carrier flight decks partly to increase flexibility (the carrier could recover aircraft while steaming astern, her air group assembled at the after end of the flight deck ready to fly off), but also so that the carrier would retain some capability even if the other end of the deck was destroyed.

The 1928 rules included estimates of the time it would take to repair flight-deck hits. A navy yard or base would require 30 days for permanent repairs to a flight deck ruined by a 2,000-pound or 1,000-pound bomb, or ten days for temporary repairs. A tender could do a temporary repair in 15 days, a ship's company in 20 days. Corresponding figures for a hit by a 500-pound bomb (half a flight deck) were 20 days and seven days at a navy yard, 10 days by a tender, and 15 days by ship's company. Aircraft could fly off a temporarily repaired flight deck in twice the time normally required. With the forward end of the flight deck destroyed, aircraft could fly off in one and a half times normal time. With the after end destroyed, they could not fly on at all, but they could fly off under normal conditions.

The 1937 rules were written as the first of the new carriers, the *Yorktowns*, were completed. They had light wooden flight decks that would be far easier to repair than the steel decks of the past. These rules reduced the effect of bombing flight decks, so that both 1,000-b and 500-pound bombs would temporarily destroy only half a flight deck. That was reasonable: The new U.S. carriers had open hangars that would dissipate the effects of a hit. More important, the new rules allowed for quick temporary repairs onboard a ship. Now, a ship's company could repair anything less than two large bomb hits in 12 hours, and two to four in 48 hours. However, more than four hits could not be repaired onboard. Because a carrier could repair her own flight deck, it now mattered how many airplanes survived the damage. Thus, the 1937 rules indicated that each large bomb would destroy a fifth of the aircraft onboard the ship when it hit. This rule placed a premium on flying off the maximum possible strike.

Note that these rules did not make the point that foreign navies, with their steel-decked carriers, could not enjoy the new advantages. They would still be difficult to repair after battle damage.

None of the rules, moreover, took into account the profound difference between U.S. and foreign carriers. A foreign carrier whose hangar was integral with her hull, which meant all British and Japanese carriers, would suffer far worse damage if that hangar was penetrated. An open U.S. hangar would dissipate any explosion inside, but stout hangar walls would not. Had this difference been clearer in the rules, they might have come closer to predicting what actually happened to the four Japanese carriers at Midway.

The 1939 rules were even more optimistic from the carrier's point of view. Under these rules, a carrier should be able to repair fewer than two 500-pound bomb hits in four hours, and fewer than two 1,000-pound hits in six; two to four large bomb hits could be repaired in 36 hours (but, as in 1937, anything more than four bomb hits could not be repaired onboard).

Torpedo Bombing
Under the 1923 rules, aerial torpedo tracks would be plotted by the torpedo umpire. Half the torpedoes would run properly. Aircraft could lay smoke screens, an important factor in game tactics.[14] Not only could an aerial smoke screen hide ships, it could also hide aircraft about to emerge to launch torpedoes. In one game, defending aircraft laid a second smoke screen specifically to frustrate torpedo bombers emerging from their own (the bombers') smoke screen, preventing them from finding their targets in time to release their weapons.

Presumably as a reflection of developing tactics, the 1928 rules called for concentrated attacks by torpedo bombers, if possible, simultaneous with bombing. The earlier figure of half of torpedoes running normally was upped to three quarters, unless curved fire (angled fire) was used, in which case only half should be expected to run normally.[15] Of torpedoes running normally, under good conditions, at ranges of no more than 2,000 yards with favorable track angles and targets not radically maneuvering, 50 percent could be expected to hit a large target (35 percent could hit a small one like a destroyer). Umpires had to decide how much to reduce hits under unfavorable conditions. Attacks by hostile aircraft during the approach might "somewhat" reduce the efficiency with which the surviving torpedo bombers completed their attacks, "but it is assumed that defending planes discontinue their attack when attacking planes can come under effective gunfire." The anti-aircraft rules, which were not yet firm, would be applied to a torpedo bomber coming under fire before it could drop its torpedo. These 1928 rules also gave greater credit for radical target maneuvers.

In view of later experience, these rules much overstated the effectiveness of torpedo bombers. Presumably reflecting sobering fleet experience, the 1931 rules cut the percentage of success for aerial torpedoes to 60 percent (40 for curved fire), but allowed a maximum range of 4,000 yards.

Air-to-Air Combat

For air-to-air combat under the 1923 rules, the umpire would decide the outcome based on numbers, types, and the previously mentioned Lanchester's Square Law (fighting strength was proportional to the square of numbers). Land-based aircraft were given an advantage over seaplanes, formations were considered stronger than individual airplanes, and airplanes flying low over water were considered stronger in defense than at higher altitude. Superior altitude was an advantage in attack, particularly if the attack was a surprise. Estimated relative fighting strength equated fighters and torpedo bombers and scouts (relationships with patrol planes and spotters were more complicated). When many airplanes were engaged, ordinarily 10 percent would be shot down every three moves (nine minutes).

As an example, the outcome of a fight between 50 fighters and 40 torpedo bombers (total 90) lasting 18 minutes was initially determined by their relative fighting strength (by the square law) of 25:16. In the first nine minutes, 10 percent would mean nine aircraft in all shot down. The number each side lost would be proportional to its total numbers, meaning 5.5 and 3.5, respectively (rounded off as 6 and 4). On the next move, then, the strengths would be 46 and 34 (fighting strength 100:55). On this move another 10 percent would be shot down, and losses would be 5.2 and 2.8, respectively (5 and 3). On the third turn, that would leave 43 and 29 airplanes in the air on the two sides. This simple melee calculation assumed no coordination between airplanes—fighters at this time lacked radios, although pilots could use hand signals. Fighter control of any kind was not envisaged. Fighters and torpedo bombers were treated equally, presumably on the theory that the rear-firing guns of the latter were at least equivalent to the fighters' one or two forward-firing guns.[16]

The rules became more sophisticated as the post–World War I U.S. Navy gained experience with aircraft. The 1925 rules envisioned very short engagements: Airplanes would not remain in firing range and position for very long. However, a single bullet could bring down an airplane. This was very different from the cumulative damage envisioned in ship-to-ship combat. Given superior performance or even position, moreover, a pilot could break off (or avoid combat) at will. Planes of equal performance would engage only if both were willing or if one was in a favorable (higher-altitude) position. The square law was modified to take account of different performance by different types of aircraft: One single-seat fighter was considered equivalent to half a two-seat fighter (because the latter had a rear gun) and also to half a VT (again, because of the rear gun), despite its performance edge. Because it was so difficult for several airplanes to attack together, the square law was now applied to airplane-on-airplane combat, the maximum advantage being 4:1 (one two-seat fighter versus one single-seat fighter, for example). The rules pointed out that estimates of relative strength were more applicable to formations, where relative positions of aircraft gave a maximum volume of fire and thus mutual protection, than to engagements between single aircraft.

Rules for air-to-air combat became more complex as air tactics developed. According to the 1926 rules, the maximum number of flexible-gun aircraft that could form a single fighting unit was 24; the maximum number of fixed-gun aircraft was nine. This is why, in the wash-up after a game in the early 1930s, an Army Air Corps student spoke of 72 fighters, of which only 9 would dive to attack at any one time. This time, the cumulative-hit model of damage used for ships was also applied to aircraft, which were assigned lifetimes calibrated in machine gun hits: 20 for a single-seat fighter, 25 for a two-seater, and 35 for a single-engine VT. That was realistic—a single hit could destroy an airplane, but most single hits would not. Even during World War I, airplanes got home with bullet holes in them. The hitting power ratios first used in 1925 were slightly modified, strength being 2 for a single-seat fighter, 4 for a two-seat fighter, and 5 for a single-engine VT. It is not clear why the VT was considered more effective than any sort of fighter.

The aggregate hitting power on each side would be the total of airplane life multiplied by airplane hitting power. The square law was replaced by a ratio of relative fighting strengths. As in 1925, total casualties in a fight would be a proportion of the total number of aircraft: 15 percent for fewer than 36 aircraft, 10 percent for 37 to 90, and 5 percent for more than 90. Instead of proportioning losses based simply on numbers, they were based on relative total fighting strength. There were further rules for one-on-one combat between "stunters"—highly maneuverable aircraft such as fighters—and non-stunters such as a VT or VP. It would take four minutes for a stunter to finish a fight with a non-stunter, or eight minutes to end a fight with another stunter.

By 1933, the fleet had learned far more about air-to-air combat, and the rules were rewritten. Like other contemporary air forces, the U.S. Navy organized its fighters in threes, of which two would attack a formation and the third would cover them.[17] The revised 1933 rules for air-to-air combat reflected new tactical experience, but it was still assumed that a flexible gun used to defend a bomber was more effective than the two fixed guns of an attacking fighter. This logic justified the current tactical practice of using three-fighter groups attacking together, two of the fighters firing at the single target, and the third protecting them. It was assumed that the fighters would attack from greater altitude and that they would initially enjoy some surprise. The targeted bombers would maintain their formation when attacked.

If fighters attacked dive bombers and enjoyed a 2:1 advantage in numbers, the fighters would lose 15 percent of their numbers, the dive bombers 20 percent. With 3:1 odds, the situation would improve: the fighters would lose 10 percent, the bombers 25 percent. Against slower but more heavily armed torpedo bombers, at 2:1 odds, the fighters would sacrifice 20 percent to kill 25 percent of the bombers. In fighter-fighter combat, with the numbers roughly even, each side would lose 15 percent in each attack. These rather pessimistic figures (for the fighters) reflect the reality that they did not yet much outperform their targets—a reality that would change dramatically over the next few years.

When the rules were revised in 1938, it was still assumed that flexible rear guns onboard bombers would be very effective. U.S. naval fighters still had only two guns of their own, although that would soon change dramatically. The rules also allowed for attacks by aircraft such as observation planes and bombers, with single fixed forward guns. They were credited with half the effectiveness of fighters. No one seems to have pointed out that there was a world of difference between a fighter pilot with lengthy gunnery training and a bomber or observation pilot with little or no gunnery experience. Moreover, bomber performance was still very close to fighter performance. On this basis, it was now assumed that the defenders had a real edge: With 2:1 odds, the fighters would lose 30 percent to shoot down 20 percent of the bombers.[18] These figures were wildly unrealistic in that bombers were rarely able to shoot down fighters in any numbers. Experience in Spain and in China was showing that high-performance bombers often could penetrate to their targets, but also that in doing so they rarely destroyed large numbers of attacking fighters.

Finally, in 1941 it had to be admitted that situations would be so varied that the game director would have to assign permanent or temporary losses after considering the conditions of the fight, allowing for factors such as altitude, formations, and positions permitting mutual support. The previous table of relative losses was repeated unchanged. However, the comment was added that when a group of airplanes was attacked by three times their number, both sides would incur the expected losses and the attackers be driven off; they could not resume their mission for 30 minutes. A group making an aerial attack could not engage again for ten minutes.

There was no mention of fighter control. At this time, numerous U.S. official observers, mostly naval, were working in the United Kingdom and onboard British ships, and their reports had given full details of fleet fighter operations and also of the fighter control being exercised ashore. Presumably, the new rules reflected combat experience to some extent, as U.S. files of the period show extensive reports from the observers.

Anti-Aircraft Firepower
Anti-aircraft guns were considered effective only up to a 10,000-foot altitude or a 5,000-yard range. Effectiveness would vary with the state of the sea, light, and visibility, and with the type, number, position, course, and speed of the aircraft, and also with the amount of warning given. No figures were given in the 1923 rules. Later, the effectiveness

of a carrier was rated equal to that of two battleships or battlecruisers, or to that of six light cruisers or 24 destroyers. The 1933 rules differentiated between heavy, moderate, and light levels of anti-aircraft fire. Heavy fire could inflict 20 percent casualties on airplanes bombing horizontally from 5,000 to 8,000 feet. Light fire could inflict 5 percent casualties. Heavy fire could inflict 40 percent casualties on dive bombers (light fire could inflict 10 percent). Coordinating dive bombing with a horizontal bombing attack would halve the casualties suffered by the horizontal bombers. Experiments had not shown how difficult it was to hit dive bombers.

The levels of anti-aircraft fire were associated with types of ships and with whether the ships were also fighting a surface battle. When not under effective surface fire, a division of battleships or cruisers or a carrier could develop heavy anti-aircraft fire. The next level down (moderate) was associated with individual battleships and cruisers.

The 1935 rules went into greater detail about anti-aircraft batteries. It took one 1.1-inch gun or two .50-caliber machine guns as equivalent to a heavy anti-aircraft gun, an unrealistic rule because the machine guns and the heavier guns had such different characteristics.[19] Heavy, moderate, and light batteries were now defined in terms of numbers of guns: 20 for heavy, eight or more for moderate, and fewer than eight (but more than one) for light. At this time, the standard battery for a U.S. battleship or a recent cruiser was eight 5-inch guns plus .50-caliber machine guns (each of the two big carriers had 12 5-inch guns). New battleships then in the design stage had 20 5-inch guns. The rule stated that the guns of a division of ships, operating 1,000 yards apart or closer, could be considered one battery. That explains why, in some games, carriers operated close to battleships to share their anti-aircraft firepower.

In 1936, further conditions were added. Standard figures envisaged a ship under fire by a similar ship, but not firing her own guns. If the ship was not under effective fire, the effect of her anti-aircraft fire would be increased by 20 percent, but if she was firing her own secondary battery, it would decreased by 20 percent. If the main attack was preceded by at least nine light aircraft (e.g., dive bombers or fighters) bombing and strafing, the effect of a ship's anti-aircraft fire would be halved. On the other hand, now the anti-aircraft batteries of all ships within 3,000 yards could be grouped together. Half a ship's anti-aircraft guns were considered effective on each side. These rules stated that the machine guns were effective out to 5,000 yards, each machine gun being considered equivalent to half a larger-caliber (5-inch/25-caliber or 5-inch/38-caliber) anti-aircraft gun. This last was clearly excessive; in the 1937 rules, machine guns were effective only within a slant range of 3,000 feet. The 5,000-yard figure was the horizontal (not sloped) range to which heavy anti-aircraft guns could fire.

By this time, the fleet was exercising anti-aircraft guns systematically, so the 1937 rules included tables of expected performance using a battleship as the unit of effectiveness. Airplanes were assumed to attack in groups of no more than 60 (wartime experience would show that practical attack groups were considerably smaller). The situation would show how many units were firing at the group, and that in turn gave a figure for effectiveness, which might be diminished by further factors. No more than eight ships could fire at any one attacking group. Guns and machine guns were counted separately, so that a battleship had a unit of guns and one of machine guns, but a carrier had two units of machine guns. An old destroyer armed with a single 3-inch/23-caliber gun was credited with only 0.1 unit of anti-aircraft guns, which might be considered grossly optimistic. A loss of 40 percent would abort an attack. For example, a single firing unit could exact 60 percent losses in a five-airplane group under ideal conditions, or 40 percent of a 15-airplane group. Doubling up helped: Four units could wipe out a five- or ten- airplane group, or destroy 40 percent of a 40-airplane group. An additional table showed that by flying higher-level bombers could reduce their losses to only 20 percent of the basic figure at 9,000 feet.

By this time, it was beginning to be apparent that it was difficult to hit dive bombers. In their approach, guns could exact only a tenth as great losses as with level bombers, because the dive bombers did not have to make a long, straight and level approach. When diving, they would face only machine guns, which were credited with better results: The standard figure would only be halved.

The 1937 rules also went into some detail concerning the effects of strafing and light (116-pound) bombing by fighters. An effective strafing attack by 12 fighters could reduce the value of a ship's turret guns by a tenth for the rest of the day, or her secondary battery by a fifth, or her anti-aircraft battery by 30 percent. Nine fighters strafing a ship just before a bombing attack would halve her anti-aircraft effectiveness for that attack. Single fighters armed with light bombs and machine guns could also seriously damage destroyers.

Anti-aircraft rules were revised again in 1938, the new assumption being that airplanes would attack in divisions of six to nine aircraft, coming from one direction as rapidly as possible. "Although about 100 planes of different classes can be handled tactically in the same group when cruising, it is assumed that during the attack they will split up." Groups would be strafers, level bombers, dive bombers, and torpedo planes all attacking in separate groups. Only one group could attack one division of ships at the same time from one direction, though two could attack two separate divisions at the same time.

By 1939, the fleet was using drones as anti-aircraft targets, with startling and depressing effects. Existing fire-control systems seemed well suited to placing bursts within lethal range of aircraft, but the drones, which were considerably flimsier than operational aircraft, often survived just such explosions. The damage tables in the 1939 rules did not yet reflect this uncomfortable perception. The rules even allowed the game director to increase the effect of anti-aircraft fire by up to 30 percent in special or unusual circumstances. The 1939 tables were retained in the 1941 rules, the last peacetime edition.

Aircraft Navigation and Reliability
Aircraft navigation mattered in two complementary ways. First, it determined how far aircraft could fly and still find their way back to their carriers. This was always consid-

erably less than the radius of action based on fuel supply. For example, in a wash-up after a game in the early 1930s, a senior officer who had come from the fleet commented that airplanes flying more than 25 miles from the fleet were often lost. These airplanes could have been brought back from a much greater distance using radio, but no one was willing to risk the carriers that way. In the late 1930s, both the U.S. Navy and the Royal Navy developed carrier homing beacons which, in theory, solved this problem, but they required considerable expertise, and they were not always effective in actual operations. They did not feature in the evolving game rules.

Second, air navigation determined how effective scouting could be. The scout had to know where it was when it spotted an enemy. Otherwise, no follow-up attack would succeed. This problem was demonstrated, for example, at Midway, when the Japanese fleet was found by land-based patrol aircraft, but carrier attackers had difficulty finding it themselves, despite information available when they were launched.

An airplane generally had to estimate its position by dead reckoning based on its known performance and its compass heading. The main problem was prevailing wind, which would move the aircraft off course and might also affect its speed over the sea. That was only a minor problem over land, but the sea was featureless. To a limited extent, surface wind could be observed from wave action, but the wind at altitude might be quite different. A carrier could launch balloons to measure wind directly overhead, but the wind at various altitudes 50 or a 100 miles away often varied considerably.

Under the 1923 rules, aircraft dead-reckoning steering errors might be as great as 3 degrees if over 1,000 feet altitude, 5 degrees if over 3,000 feet, and 10 degrees if over 6,000 feet. The increase with height was because an airplane would estimate drift across its path by observing the movement of the surface of the sea or by using navigational float lights. Wind at altitude usually differed greatly from wind at the surface. Umpires would indicate whether pilots would have the benefit of wind indication using a balloon. Since War College problems generally involved scouting, such errors could have a considerable effect on the estimate and decisions of a commander receiving news from air scouts. In 1925, the 3-degree altitude was 2,000 feet. The 1926 rules eliminated these figures, but made the same point about dead reckoning.

None of the later rules quantified navigational errors, but in the game, aircraft were increasingly important for scouting. Their navigational errors could have considerable tactical and even strategic consequences, but they were not really taken into account. Games certainly did take into account poor performance by scouts and poorly framed messages, but they do not seem to have anticipated some of the gross failures marking actual combat—in the Solomons, for example (the Savo Island debacle began with failed air scouting).

The 1926 rules did include figures for the increased chance of breakdown due to continued flying without overhaul or during a lengthy flight. During a half-hour flight, the chance of breakdown was 1 in 200, but for a 12-hour flight (by a patrol plane, for instance) it would rise to 1 in 15. These rules also pointed to the loss of morale due to the load of continuous flying over time. The 1927 and later rules went into considerable detail both for the number of hours pilots could be expected to fly each day and for the effect of fatigue. For example, fighter pilots could not be expected to fly more than four hours a day except for additional time in one 24-hour period. Multipliers were used to capture various kinds of stress, such as the effect of overtime. For flight personnel, recovery from fatigue would require three hours for every effective extra hour (with the multipliers). A curve was included to show breakdown hazards due to increased flight time.

The 1931 rules added graphs of navigation, rendezvous risks and for chance of failure for long flights (up to 20 hours for patrol planes), and a multiplier for the chance of success depending on what percentage of a flight was made at cruising speed. The navigational graph gave the percentage chance of failure for both flight duration and distance from a ship; by this time, some shipboard aircraft had nominal ranges as great as 400 miles. For example, a 150-mile flight or four flight hours gave a 25 percent chance of failure—which was considered grossly optimistic in a game wash-up.

Operational losses were difficult to foresee and to calculate. In an effort to simplify the process, the 1936 rules gave loss penalties; once they added up to 100 percent for a squadron, it would permanently lose one of its aircraft. Carrier plane losses in a smooth sea were given as 0.5 percent (1 percent in a rough sea) per flight. Such figures covered take-off and landing crashes, collisions, crashes in thick weather or high winds, and forced landings due to running out of fuel. -

Notes

A Note on Sources

The Naval War College did not initially maintain its own historical archive. Instead, its papers were retired to regional branches of the U.S. National Archives. In the 1960s, they were brought back to the college to form its own collection. Most of the Naval War College publications—the pamphlets describing games and those describing lectures—seem to have survived. No correspondence describing the rationale for various scenarios seems to exist. Moreover, there is no reason to be certain that papers for all of the games have survived. For example, there is no trace in the college collection of informal games played to test various technical concepts in the 1930s. The evidence of such work is in OPNAV and other files citing college contributions to decisions about ship and weapon characteristics.

Descriptions of games vary. Generally surviving materials describe the scenario and also the staff solutions for the Blue and enemy "Estimates of the Situation." Students received the description of the problem and wrote their own estimates, but in order to play a game a single estimate had to be adopted for each side. It included the central decisions (in effect, commander's intent) that would govern further actions. Once the students received the relevant staff solutions, they were assigned subordinate commands. During the game, a record ("History of Maneuver") was kept on a move-by-move basis. Ultimately, there was a formal critique, which might be retained in written form. Many fewer histories and critiques than staff solutions have survived.

A second major source of material is the papers of the OPNAV War Plans Division in the National Archives division at College Park, Maryland (NARA II). This file includes lists of war-gaming material provided to War Plans by the Naval War College, as well as most (but not all) of this material, and what amounts to a library of lectures roughly parallel to those that survive at Newport.

A third important source is the collection of OPNAV and General Board files in the National Archives in Washington. It provides a few examples of direct influence by the Naval War College on specific technical and policy decisions. This evidence is largely lacking in Newport because the archive there seems not to include correspondence from the president of the Naval War College to CNO.

1. Naval Transformation

[1] The recent account of inter-war gaming at the War College, John M. Lillard, *Playing at War: Wargaming and U.S. Navy Preparations for World War II* (Dulles, VA: Potomac Books, 2016) describes gaming and some of its effects on participants, but does not connect gaming experience with wider issues of inter-war U.S. naval development, such as the rise of carriers. He seems most concerned to debunk a previous, apparently off-hand, dismissal of the games as ritualistic and unimaginative (or worse). The only other published book that is concerned with the war-gaming experience is Michael Vlahos, *The Blue Sword: The Naval War College and the American Mission 1919–1941* (Newport, RI: U.S. Naval War College, 1980). *The Blue Sword* is mainly concerned with how the War College helped weld U.S. naval officers into a coherent group sharing important ideas in common. It includes a list of the war games and points to the significance of Operations IV of 1933, the game that changed U.S. Pacific strategy. Dr. Vlahos is not concerned with other impacts of the games on the larger Navy. Neither author seems to have looked beyond the archives at Newport to examine the way in which the War College interacted with the wider U.S. Navy of the time, to the extent that Lillard seems unaware of the 1930 London Naval Conference, an important watershed for the War College as an advisory organization.

[2] For the War Plans Division, see, for example, Edward S. Miller, *War Plan Orange: The U.S. Strategy to Defeat Japan 1897–1945* (Annapolis, MD: Naval Institute Press, 1991), 17. A comparison between officers listed in the 1 April 1941 *Navy Directory* and the roster of War College graduates shows only one not listed in the War College roster: Rear Admiral W. R. Munroe, commanding Battleship Division 3 in the Pacific Fleet.

[3] The torpedo story has often been told. The magnetic exploder, which the U.S. Navy called the "Index Device," was intended to detonate the torpedo below the enemy's hull. It would blow in the bottom of a large ship, and it could snap a smaller ship in half; such detonators (often acoustic in modern torpedoes) are now quite common. The Royal Navy and the German navy developed their own magnetic exploders at about the same time as the U.S. Navy and all encountered serious problems. In the U.S. case, the problem was exacerbated by defects in the back-up contact exploder, which turned out to be too flimsy. The culprits were apparently a lack of pre-war testing (partly because the exploder was too secret to expose, and partly because testing was too expensive for a Depression-era fleet) and arrogance on the part of the responsible organization, the Bureau of Ordnance and its Torpedo Factory. This episode, which drastically reduced the efficacy of U.S. submarines before the problem was solved in mid-1943, explains why modern U.S. torpedoes are so often tested against full-scale targets.

[4] All forms of radio used before the late 1930s could be intercepted well beyond the horizon, sometimes thousands of miles away. During the 1930s, navies adopted high-frequency radio, which could often be detected an ocean away. The U.S. Navy seems to have led the shift toward it. The Japanese seem to have been the first to develop effective high-frequency radio direction-finders, which made it possible for them to associate messages with particular ships and then with particular movements. For its part, the U.S. Navy discovered this Japanese capability through code breaking, and it adopted security measures to frustrate Japanese exploitation of U.S. high-frequency radio. In turn, the U.S. Naval Research Laboratory was asked to develop a high-frequency direction finder—which it failed to do, claiming that no such device could work. OPNAV insistence on this project was a rare pre-war example of the operational Navy pressing the materiel bureaus (in this case, the Bureau of Engineering, which was responsible for the Naval Research Laboratory) to do what they claimed was impossible. For details, see this author's *Network-Centric Warfare: How Navies Learned to Fight Smarter Through Three World Wars* (Annapolis, MD: U.S. Naval Institute Press, 2009), 23 and 268–70.

[5] Covert U.S. observation of the 1930 Japanese maneuvers revealed just that. Worse, it appeared that the Japanese had found effective means of defeating the U.S. plan. Plans had to be completely revised. See the

official history of the pre-war development of the Naval Security Group (RG 457, Box 108, SRH-355, Naval Security Group History to World War II, prepared and compiled by Captain J. S. Holwick Jr., USN (Ret.), June 1971, 71–76). According to the Security Group history, the analysis of the Japanese maneuver was done by Captain R. E. Ingersoll, who was then in the OPNAV Division of Fleet Training (however, according to a 1943 Security Group history, he was attached to War Plans). The understanding of the Naval Security Group was that the framework of the U.S. war plan was changed as a result of Ingersoll's studies, but no trace of his analysis or report survived in its file copy of the 1930 report (page 75). Ingersoll had been head of the Intelligence Section of ONI (Op-16B), which was later the Special Intelligence Branch. According to this history, the report did not go to ONI because CNO Admiral William V. Pratt did not trust its acting director. For the 1930 maneuver, the Japanese managed to mobilize their fleet and its reserve so secretly that the U.S. naval attaché in Tokyo was unaware of it. Secret Japanese mobilization was a feature of many later war games. Reactions to the 1930 revelations included much more secure radio practices, including the "Fox schedule" broadcasts (all messages intended for individual ships were broadcast by a shore station, not on a ship-to-ship basis) and encrypted call signs. See Timothy S. Wolters, *Information at Sea: Shipboard Command and Control in the U.S. Navy, from Mobile Bay to Okinawa* (Baltimore: Johns Hopkins University Press, 2013), 142–55.

[6] In 1936, Sutherland Danlinger and Charles B. Gray published a book analyzing U.S. and Japanese naval strategies, *War in the Pacific: A Study of Navies, Peoples, and Battle Problems* (R. M. McBride & Co.), on just this basis.

[7] The Royal Navy, which had invented the aircraft carrier during World War I, lost control of its aircraft to the new Royal Air Force on 1 April 1918. British records show that the navy fought hard to regain control through the inter-war period, finally winning in April 1939 (although the RAF retained control of all shore-based aircraft). By the 1930s, British official writers considered the U.S. Navy far ahead of theirs. In the late 1930s, for example, the Admiralty tried to negotiate an exchange of aviation information with the U.S. Navy, and was much concerned that it might be giving away its technical secrets. The senior Admiralty technical advisor, Director of Naval Construction Sir Stanley Goodall, wrote that since the Fleet Air Arm was a constant source of anxiety and U.S. naval aviation was considered the best in the world, the Admiralty should gladly take whatever was offered. The inferiority of British naval aircraft is well known; British records show that those involved tried hard to avoid admitting it. See the author's *Fighters Over the Fleet* (Annapolis, MD: U.S. Naval Institute Press and Barnsley, UK: Seaforth, 2016) for examples in the both pre-war period and early in World War II.

[8] Each of the two carriers displaced about 36,000 tons. The Washington Treaty included a clause allowing for the addition of up to 3,000 tons in order to provide existing ships with modern protection against torpedoes and bombs. This was presumably an incentive to navies not to seek to build new ships that would be so protected. Curtis D. Wilbur, the Secretary of the Navy in office at the time of the Washington Conference, repeatedly said that there were 3,000 more tons "in the treaty." His successors were somewhat nervous about this interpretation, and General Board records reveal unsuccessful attempts to redesign out the 3,000 tons. Between wars, the carriers were always rated at 33,000 tons with a note that this was exclusive of special protection against bombs and torpedoes. An innocent reader might imagine that this was a matter of a few hundred tons or less. All calculations of remaining U.S. treaty carrier tonnage counted the two large ships as 33,000 tonners. By way of contrast, the later *Yorktown*, which was considered a very satisfactory carrier, displaced 20,700 tons, and the World War II *Essex* displaced 27,500. However, the designs of the big new carriers were relatively inefficient, so they operated only about as many aircraft as later ships—initially about 100, but fewer as aircraft grew larger. The Japanese *Akagi* and *Kaga* were roughly parallel conversions, of about the same size. The British converted the "light battlecruisers" *Courageous* and *Glorious* into 20,000-ton carriers with an official capacity of 54 aircraft (fewer in practice). The big U.S. ships were completed in 1927–28.

[9] The initial impetus for Japanese carrier aviation seems to have been a 1921 British mission led by Sir William Francis-Forbes (later Baron) Semphill; this was a logical choice given that the British had recently invented the aircraft carrier. See for example Mark R. Peattie, *Sunburst: The Rise of Japanese Naval Air Power 1909–1941* (Annapolis, MD: U.S. Naval Institute Press, 2001), 18–19. However, Japanese Fleet Admiral Osami Nagano stated in 1945 that "I think our principal teacher in respect to the necessity of emphasizing aircraft carriers was the American Navy. We had no teachers to speak of besides the United States in respect to the aircraft themselves and to the method of their employment…. We were doing our utmost all the time to catch up with the United States." This statement is from Scot MacDonald, "Evolution of Aircraft Carriers: The Japanese

Developments," in *Naval Aviation News* from October 1962. It was part of a series on aircraft carrier development. MacDonald did not give his source, and it was *not* in the transcript of Admiral Nagano's interrogation in volume II of *Campaigns of the Pacific War*, published in 1945 by the U.S. Strategic Bombing Survey.

[10] For an excellent account of the inter-war Navy, see Thomas C. Hone and Trent Hone, *Battle Line: The United States Navy 1919–1939* (Annapolis, MD: U.S. Naval Institute Press, 2013). The end date is 1939 rather than 1941 partly because with the outbreak of World War II in 1939 a powerful Atlantic Squadron, which later became the Atlantic Fleet, was hived off from the unitary U.S. Fleet.

[11] This may seem ludicrous given the World War II achievements of naval aircraft, but wartime experience actually backs it up. Between 1939 and 1944, only two modern battleships were sunk by air attack (another was sunk an anchor at Taranto, Italy): HMS *Prince of Wales* and the Italian *Roma*. The latter was sunk by a guided bomb that could not have been delivered by a carrier airplane (the battlecruiser HMS *Repulse* does not count as a modern battleship). During the same period, two modern battleships, *Bismarck* and *Scharnhorst*, were sunk by gunfire, as well as the older Japanese *Kirishima* (a modernized battlecruiser). A fourth, the Japanese *Hiei* (a sister ship of *Kirishima*), was disabled by cruiser gunfire and sunk by aircraft while immobile. It is, moreover, arguable that the only modern battleship sunk at sea by aircraft using weapons deliverable by carrier aircraft, *Prince of Wales*, fell victim to a lucky hit due to poor tactics on the ship's part (see the author's *British Battleships 1906–1945* (Annapolis, MD: U.S. Naval Institute Press, 2015) for details based on modern expeditions to the wreck). The first modern battleship sunk at sea by concentrated air attack that did not exploit some tactical error was the huge Japanese *Musashi*, in October 1944—and the number of aircraft involved would not have been imaginable before then.

[12] The law was first stated in 1916. The unstated caveat was that, as the Battle of Jutland showed, there was a limit to how many battleships could be concentrated usefully. That limit was, however, well above the strength of the post-1918 U.S. battle fleet.

[13] Even with infrastructure, forward basing was unattractive. Between the two world wars, the Royal Navy, like the U.S. Navy, envisaged a campaign against Japan culminating in a fleet engagement. Ideally, the British would have their fleet in theater at the outbreak of war, and the Singapore naval base was built up for that purpose. It had facilities far superior to those in the Philippines, but they were not enough. The British repeatedly avoided basing even their three battle cruisers at Singapore, usually citing the lack of local labor and poor facilities for the families of ships' crews.

[14] In the fall of 1939, the U.S. Fleet consisted of a Battle Force, a Scouting Force, and a Submarine Force. The Battle Force consisted of the fleet flagship plus four battleship divisions, four cruiser divisions, two destroyer flotillas, the carriers, and a force of light minelayers (converted destroyers) and minesweepers permanently based at Pearl Harbor because they had relatively short range. The Scouting Force consisted of four heavy cruiser divisions and the patrol planes with their tenders. Both the U.S. Fleet and the Asiatic Fleet had supporting fleet trains (tenders and oilers). Earlier, the ships based at Pearl Harbor had been designated as the Control Force. By late 1939, many of the carriers and cruisers were operating in the Atlantic. At this time, there were also a special service squadron operating in the Caribbean, and a force (40-T) left over from the Spanish Civil War. The ships that earlier had constituted the Scouting Fleet were now part of the Atlantic Squadron, built around the three oldest battleships.

[15] Figures from the Naval History and Heritage Command (NHHC) website, presumably as of the beginning of that fiscal year (1 April 1939).

[16] As understood during the inter-war period, the rules of trade warfare required that a raider visit and search merchant ships to determine whether they were carrying contraband and thus were subject to seizure or sinking. The crews of any ships sunk in this way had to be conveyed to a safe destination, explicitly not in lifeboats. That is why British merchant seamen from the ships sunk onboard by the German cruiser *Graf Spee* were onboard the warship when she fought her last battle off of Montevideo in December 1939. No such rule could apply to a submarine, which had no accommodation for captured crewmen. Ships owned by the enemy were automatically subject to seizure or destruction. So were ships in enemy convoys (convoy as a trade-protection technique was certainly considered in inter-war games and also in inter-war U.S. Navy thinking).

[17] It can, moreover, be argued that the successful U.S. submarine campaign was a variant of the earlier idea that a U.S. battle fleet would shield the smaller ships enforcing strangulation via blockade. It seems reasonable to argue that Japan was unable to adopt an effective convoy strategy to protect her merchant ships because the struggle with the main U.S. fleet absorbed the resources involved. The Imperial Japanese Navy was always painfully aware that should the main U.S. fleet be free of the threat of the Japanese fleet, the convoys and merchant ships would soon be swept away. That was what the British had argued after World War I: Namely, that the convoy strategy, which defeated the German submarines, succeeded because the British Grand Fleet kept the German High Seas Fleet at bay, preventing it from sweeping up convoys. On one occasion in November 1917 when British intelligence failed to counter a German sortie, two German cruisers wiped out a convoy. They were far more powerful than the two destroyers escorting it. U.S. officers had reason to remember because U.S. battleships seconded to the Grand Fleet were then assigned to cover convoys. Several games envisioned submarine operations—including some unrestricted ones—against Japanese trade as part of a diversionary raiding campaign (Operations I of October 1936 for the Class of 1937 and Operations III [Strategic] of November 1939 for the Class of 1940). Early post-1918 proposals by senior submarine officers that an unrestricted submarine campaign should be included in any Orange war plan were rejected on the basis of the way in which the German unrestricted submarine campaign had played out during the previous war. The manifest illegality of the German campaign had helped tip the United States into the war, with fatal consequences for Germany. Americans had reacted to the German campaign because the Germans had been unable to (and had made no effort to) avoid sinking neutral (American) ships.

[18] British records show that they were already preparing to ask for an arms-control conference. The four post-war British battle cruisers (cancelled due to the Washington Treaty) were justified to Parliament as bargaining chips in such a conference. A British official historian went so far as to suggest that the British triggered the Washington Conference. See the unpublished Cabinet history, Y. M. Streetfield, "Ten Year Rule 1919–1929," CAB 101/295 in the British Public Record Office, Kew. The British record of the Washington Conference includes remarks by the head of the Japanese delegation indicating intense Japanese public sentiment favoring some end to the expensive naval building program.

[19] Before World War I, the U.S. Navy regularly asked for scout cruisers, but Congress was willing to authorize only battleships and large destroyers. This record is slightly muddied by the General Board's decision a few years earlier not to ask for larger cruisers, because ten large armored cruisers had already been authorized. These and contemporary large cruisers were rapidly outmoded as fleet scouts because they offered little or no speed advantage over "dreadnought" battleships authorized from 1907 on. After that, the only satisfactory fleet scouts were battle cruisers and small fast cruisers called scouts. Because only three of them had been authorized, the pre–World War I U.S. Navy found itself using large destroyers as fleet scouts, an unsatisfactory practice. The pre–World War I U.S. Navy considered but rejected the construction of larger scouts, including battle cruisers. Designs for such ships survive in the National Archives files on U.S. Navy preliminary design, Entry 449 of NARA RG 19. The big 1916 shipbuilding program, which to some extent triggered the 1921 Washington Conference, included ten large cruisers and six battle cruisers that would have gone some way toward solving a tactical (but not strategic) scouting problem. However, cruisers had other roles. In the U.S. Navy, they were needed to beat off enemy destroyer (torpedo) attacks against the battle line. In the Royal Navy, cruisers were also considered an essential way of protecting global British trade. The British therefore argued throughout the inter-war period that they needed cruisers both for the fleet and for commerce protection, hence a much larger cruiser fleet than the U.S. Navy should want. Unfortunately, no U.S. Administration could lightly cede superiority in cruiser numbers to the British. To some extent this problem ruined a 1927 attempt by President Coolidge to extend the Washington Treaty to cruisers. In 1929–30, the British were willing to accept a form of parity because their economy could not support a program to replace their large but aging cruiser force. U.S. naval officers painfully aware of the lack of U.S. cruisers used the apparent imbalance in numbers to promote cruiser construction; Congress approved an eight-cruiser program in 1928 and then a 15-cruiser program. Stephen Roskill, *Naval Policy Between the Wars, I: The Period of Anglo-American Antagonism, 1919–1929* (London: Collins, 1968), 505–16, describes how U.S. naval officers broke up the conference. It seems to this author that the U.S. officers at Geneva feared that any limit on cruisers would encourage Congress not to authorize any of them. If that is true, the much smoother progress of the subsequent London Conference (1930) can be explained by the Navy's reaction to a Congressional agreement to buy badly needed cruisers for the U.S. Navy. The need for strategic scouting in any war with Japan was probably obvious

to many U.S. officers, but war gaming would have reinforced that understanding. Roskill emphasizes the fundamental difference between the U.S. Navy, which needed large cruisers for strategic scouting in the western Pacific, and the Royal Navy, which needed large numbers (hence smaller) for trade protection.

[20] The Washington Treaty limited the total tonnage of battleships and aircraft carriers. Arms controllers wanted to extend those limits to all other types of warships and indeed to other arms. The 1930 treaty extended the limits on total displacement to cruisers, destroyers, and submarines. The Japanese also managed to extract an increase in their ratio. That Japan received the short end of the 5:5:3 tonnage ratio (factors of 5 for the United States and the United Kingdom) and this became a rallying cry for Japanese nationalists. The ratio was variously justified on the ground that the Japanese would be defending their home waters, whereas any British or U.S. attackers would lose much of their fighting power en route there; that the Royal Navy and the U.S. Navy had to defend broader areas, including the Atlantic; and that the British Empire required additional defenses. Later, it was claimed that nothing short of a 2:1 ratio could have insured U.S. success against Japan, so that the 5:3 ratio sufficed for Japanese defensive needs. U.S. records of the evolution of the U.S. position, which embodied the 5:5:3 ratio, do not explain its origin. It was certainly not the result of gaming or of advice by either the Naval War College or the General Board. Nor did the War College provide advice during the run-up to the 1930 conference (or, for that matter, the 1935 conference that framed the 1936 treaty). By the early 1930s, the U.S. view was that Japan was building every ship allowable under the treaties, whereas the United States was not. The pre-1939 U.S. naval rearmament acts were justified publicly as a means of creating a modern U.S. *treaty* navy, even though the 1938 Vinson-Trammell Act was signed after treaty limits on total tonnage had been abandoned and Japan had withdrawn from the treaty system in 1937. That this language had to be used as late as 1938—after Japan had invaded China and had sunk the U.S. gunboat *Panay* there—shows the enormous strength of public sentiment that considered naval arms control a guarantee of peace. The treaty language implied that ships would be replaced when they were over-age in treaty terms. In 1936, when the first new U.S. battleships were approved, U.S. battleships were mostly over-age, but existing carriers were not.

[21] This is evident from the Pratt papers at the War College and also from General Board papers on arms control at the National Archives. Admiral Pratt was the General Board's advisor on naval arms control during the conference.

[22] The Naval War College archive (RG 8) contains a 1935 paper replying to a CNO request for advice as to what building program to pursue once the existing treaties expired at the end of 1936 ("Building Program on Expiration of Washington and London Treaties," UNC 1935–26, RG 8 Series 1, Box 37, Folder 8. CNO (Op-12) sent a request for advice to CinC U.S. Fleet and to the president of the War College dated 1 April 1935, and on 6 July repeated his request. CNO's assumption was that all restrictions would be lifted. In fact, the London Naval Treaty of 1936 maintained (in some cases reduced) limits on individual ship size and armament, but it eliminated all restrictions on total numbers. The study was sensitive because there was enormous political pressure to maintain and even to tighten overall limits (the U.S. delegation came to the 1935 London Conference carrying just such a policy, even though by this time it was obvious that the slide toward World War II had begun). The War College reply, signed by its president, Admiral E. C. Kalbfus, was dated 1 October 1935. It contrasts strongly with the game-based analysis of future cruiser construction completed in 1930, and described below. The 1935 War College letter emphasizes the political requirement that any program had to be acceptable to the public, and therefore framed it in terms of the 5:5:3 ratio adopted at Washington in 1921 and embodied in the later London treaty. It had to be accepted that the public would oppose any building program that might be construed as the beginning of an arms race, although the public might be persuaded to accept something more than 5:5:3 if that could be shown to be necessary. Since ratios had become acceptable, the War College suggested that in future they should apply to numbers of ships rather than to tonnage, as that would make it possible to increase the efficiency and strength of the Navy. The president of the War College did point out that although a 5:5 ratio with Red was satisfactory, 5:3 (against Orange) was inadequate for a trans-Pacific offensive; a ratio of at least 2:1 was needed. Since the public would reject any such ratio, the best the War College could offer was to seek an effective ratio of that sort through superiority of individual ships. Whether or not CNO agreed, that approach became impossible once the 1936 London Naval Treaty maintained limits on individual ship size and, indeed, reduced the allowable size of cruisers. The existing numbers of foreign battleships precluded any great increase in numbers, but by 1937 the United States would have seven over-age battleships, and it might be possible to replace five of them. The War College did propose that

three of the five should be battlecruisers (a type the United States lacked), matching the three Japanese battle cruisers, "in view of the long period of cruiser and commerce warfare that will probably precede a decisive engagement.... If for any reason our air forces were unable to operate in an area into which enemy battle cruisers had projected themselves, the latter would be a formidable menace to such light forces, convoys, and shipping of ours as might be there...in effect [this exchanges] a tactical advantage during a possible general engagement for strategical and tactical advantages on perhaps numerous occasions during the pre-battle period." To some extent this reflected war-gaming experience, in which most of the action during a trans-Pacific advance occurred before any main fleet battle could occur. It may explain the decision to adopt high speed for the first U.S. post-treaty battleships of the *North Carolina* class. The president of the War College certainly did not envision high speed for any other post-treaty battleships: He would have been satisfied with 21 knots, as in existing U.S. battleships. As for carriers, current Orange numbers (four built and two building) would justify ten U.S. carriers on the basis of the 5:3 ratio. At the time, the United States had four built, two building, and one authorized, the existing four including the obsolete *Langley*, which would be reduced to seaplane tender status when the authorized carrier *Wasp* was built. All the new carriers should be of the heavy offensive type (*Yorktown* class). During the rest of the pre-war period, the United States would lay down one more carrier (*Hornet*), a repeat *Yorktown*. Note that carriers were listed sixth in priority, after battle cruisers, battleships, submarines, destroyers, and light cruisers, and before minelayers and auxiliaries—hardly the sort of strong endorsement implicit in earlier War College comments on building programs. The War College paper included a recommendation for 12 squadrons of patrol planes (16 aircraft each) at the outset of an Orange war, many more being needed later on. This figure was based implicitly on gaming experience rather than on a numerical comparison with the Japanese (Orange). However, the discussion of cruiser numbers appended to the letter was decidedly not based on gaming. Instead, it was based on a breakdown of the U.S. fleet, plus some ratios between numbers of major units and desired numbers of cruisers. For example, it cited Admiral Jellicoe's World War I precept that there should be five cruisers for every three battleship (giving 25 cruisers), one cruiser as flagship of each destroyer flotilla (two), and two for each carrier (12 for six carriers). Past studies were cited as the basis for a requirement for another 21: nine to protect focal areas (trade protection), six to protect lines of communication, and six to attack enemy trade. Other figures could be derived from numbers the Japanese already had, this total being 53 compared to the 60 obtained from the breakdown of tasks.

[23] Discussions of U.S. ships and the policies that created them are based mainly on the author's series of illustrated design histories of U.S. warships: carriers, battleships, cruisers, destroyers, submarines, small combatants, and amphibious ships and craft, all based on formerly classified U.S. official records, including records of policy makers. All were published by the U.S. Naval Institute Press. I have also published a parallel series of British design/policy histories covering carriers (and their aircraft), battleships, cruisers, destroyers, and lesser escorts, both before, during, and after World War II, again based on official formerly classified papers.

[24] The fatal hit, which was the first the Japanese made, was near the point at which the ship's outboard propeller shaft emerged from her hull. In retrospect it seems that this hit in turn was made because the ship's commander waited too long to begin the standard defensive measure of combing the torpedo tracks. It would not have been logical for the pre-war Navy (and its competitors) to abandon the battleship as a capital ship—certainly not before 1943, when the Germans demonstrated that large *land-based* bombers armed with guided bombs could sink the Italian *Roma* and disable the British *Warspite*, *Roma* being a fully modern battleship. Sinkings at Pearl Harbor do not count because the ships sunk were not maneuvering and also because they were not at a high degree of readiness, with many watertight doors open. The fleet at anchor was subject to a much higher percentage of hits than would have been the case in the open sea, and the lack of watertight integrity magnified the effect of the hits. The author has elaborated the *Prince of Wales* argument, using modern examinations of the wreck, in his book *British Battleships 1906–1945* (Annapolis, MD: U.S. Naval Institute Press and Barnsley, UK: Seaforth, 2015). The sinking of the accompanying battlecruiser *Repulse* was much less of surprise; she was relatively weakly protected, and had a weak anti-aircraft battery. British records show that the sinking of the *Prince of Wales* was a great shock, as she was a modern ship whose torpedo protection should have been proof against the Japanese torpedoes.

[25] Games certainly taught U.S. officers that carriers could be sunk by fast battleships or even by cruisers, particularly if they did not use their aircraft effectively. This helps explain why Admiral Halsey did not detach his battleships when he took his carriers north at Leyte Gulf. It also raises the question of why the

commander of the carrier HMS *Glorious* allowed his ship to be attacked by the two German battleships *Scharnhorst* and *Gniesenau*, which sank her off Norway in 1940. The implication of the loss of the British carrier would seem to be that the British were unable to teach their officers, particularly carrier commanders, about the facts of fleet life.

[26] The connection between the air-defense role in the passage through the Mandates and the anti-air capability of the new destroyers is explicit in General Board hearings that established their basic design features. The dual-purpose guns were also justified by their ability to deliver plunging fire in support of troops landing on fortified islands, another requirement underpinned by gaming. The new high-angle 5-inch/38-caliber guns appeared for the first time in the *Farragut* class, the first post–World War I destroyers, which were designed toward the end of the period of War College influence on warship designs. There is no direct evidence that the heavy torpedo batteries of later destroyer classes resulted from Naval War College arguments, but their adoption did coincide approximately with the War College's accepted recommendation that torpedo tubes be removed from cruisers. That left destroyers as the only remaining source of torpedoes in night surface actions.

2. The Naval War College and Gaming

[1] See, for example, Miller, *War Plan Orange*, 15. An account of the inter-war General Board, emphasizing its role in formulating official U.S. Navy policy and also in determining ship characteristics, is John T. Kuehn, *Agents of Innovation: The General Board and the Design of the Fleet That Defeated the Japanese Navy* (Annapolis, MD: U.S. Naval Institute Press, 2008). Dr. Kuehn emphasizes the board's continued espousal of a mobile fleet repair facility, particularly one or more towable floating dry docks. The contention of the present work is that the board became interested in such facilities because war games showed how vital they would likely be. For details of General Board discussions of ship characteristics, see the author's series of illustrated design histories of various types of U.S. warships, all published by the U.S. Naval Institute Press.

[2] This account of the 1908 controversy is based on John B. Hattendorf, B. Mitchell Simpson III, and John R. Wadleigh, *Sailors and Scholars: The Centennial History of the U.S. Naval War College* (Newport, RI: Naval War College Press, 1984), 61–64. As a junior officer, Sims became a disciple of British gunnery innovator Captain Percy Scott, RN; his own work on gunnery in the U.S. Navy led President Theodore Roosevelt to make Sims inspector of target practice. Sims was an early supporter of the all-big-gun battleship espoused by British First Sea Lord Admiral Sir John Fisher. His reports in the naval attaché series at the U.S. National Archives (RG 38) suggest that Sims enjoyed special access to the Royal Navy—to the revolutionary battleship HMS *Dreadnought*, for example. He was widely considered a strong anglophile. Before World War I, he made a speech before a British audience promising that if there was a war, the United States, as an offshoot of Britain (part of "greater Britain") would come to the aid of the British.

[3] That is, the first U.S. battleship with all big guns and steam-turbine propulsion (the earlier *South Carolina* had the guns, but not the turbines).

[4] As indicated in the 1913 Admiralty report of the committee assigned to consider the organization and training of the Royal Naval War College, held by the Admiralty Library in Portsmouth (Royal Naval Historical Branch). It includes an account of the U.S. Naval War College as it then existed, written by Captain C. F. Sowerby, RN (page 42 of the British report). He began with the three-month summer conference, "which they are very anxious to keep [as such]. To this conference are sent all sorts of questions by the Navy Department. War plans are here discussed, elaborated, and in all probability practically settled. Schemes for maneuvers, strategical and tactical questions are brought up, in fact everything in connection with the handling and movement of the fleet. Admirals commanding are in constant communication with this conference. Lectures are given on naval history, strategy, tactics, international law, gunnery, torpedo, engineering etc. A great stress is laid on the intimate connection of gunnery and tactics, thus the design of new ships forms a topic of discussion. War games are of course played. This course is not compulsory, but officers are at times told off to attend. This college is commanded by an Admiral, who has a staff of generally one captain, five or six commanders or captains." The U.S. Navy had, as far as was known, no war staff, and hence no chief of war staff so-called "and it would seem the admiral commanding the War College was the nearest

approach to it they have; he was frequently at Washington." At this time the Secretary of the Navy had four aides, including an aide for operations whose office would soon be transformed into the Office of the Chief of Naval Operations. Captain Sowerby wrote that the aide for operations was practically in the position of the Royal Navy's First Sea Lord, with the War College and the director of naval intelligence subordinate to him. Sowerby added that a few officers, generally commanders and lieutenants, stayed on at Newport through the winter working on various problems, attending lectures, and writing essays; as of a year earlier (1912), it was hoped that those who showed special aptitude would be allowed to stay on another year. "Whether this was with the idea of making them a war staff is not known." This report is not otherwise identified. A typescript version is designated T.21605.

[5] This was more radical than it may appear. The old Bureau of Navigation was responsible for naval personnel, but its chief was also the Secretary's primary advisor for operations. (The bureau was also the conduit for inputs from the fleet. For example, before 1909 its files show requests (which were often accepted) for modifications to U.S. submarines. The result of the reform was that the Bureau of Navigation was reduced to personnel functions, although for many years the Bureau of Naval Personnel was styled BuNav rather than, as later, BuPers.

[6] Spector, *Professors of War*, 142–43, and Hattendorf et al., *Sailors and Scholars*, 82–84. Daniels also apparently rejected Fiske's arguments for increased investment in naval aviation. Fiske had proposed the torpedo bomber in 1912.

[7] These were hardly junior officers. One was Captain Sims, who had helped ignite the 1908 battleship controversy. The other three were on the War College staff: Captain J. S. McKean, Commander Yates Stirling Jr, and Captain E. H. Ellis, USMC. Ellis was later a pioneer exponent of amphibious warfare in the Pacific. McKean would later head the OPNAV material division and ended up as a Vice Admiral. Stirling was later rear admiral and chief of staff of the U.S. Fleet. At this time, the War College president was a captain. Of six officers in the second long course (1912–13), two were in that summer conference (1913), which seems to have been a very limited affair (a total of only five officers). The full two-committee summer conference was revived in 1914. The 1914 academic year was the first not to feature a summer conference, and also the first to feature an enlarged long course (13 officers).

[8] According to Hattendorf et al., *Sailors and Scholars*, 79–80, the first two long courses culminated in the four-month summer conference. This was not the change ordered by Secretary of the Navy Josephus Daniels, who succeeded von Lengerke Meyer after Woodrow Wilson became President. After a June 1913 visit, he ordered the college to provide three different courses. An elementary course (three weeks) was intended to teach junior officers tactics. A four-month "preparatory course" would add the first part of the long course (Hattendorf et al., *Sailors and Scholars*, 80, see it as the short course or renamed summer conference). The long course became the 16-month War College course, consisting of a four-month preparatory course followed by 12 months of advanced work. There was also to be a correspondence course. Each of the two longer courses was to have 20 students, and faculty was to be drawn from graduates. Ronald Spector, *Professors of War: The Naval War College and the Development of the Naval Profession* (Newport, RI: Naval War College Press, 1977), 125–26. In July 1913, Daniels personally ordered 50 Atlantic Fleet offices to attend a two-week elementary course. The War College considered this too short, and President Rodgers replaced it with a correspondence course. It was also difficult to arrange two four-month preparatory courses, as in each case officers had to be detached temporarily from ships that were thinly manned. Rodgers asked Daniels to issue a general order eliminating the two short courses in favor of the single long course. Aid(e) for Operations Rear Admiral Fiske refused to endorse it, as Rodgers would soon be relieved by Rear Admiral Austin M. Knight. Knight then endorsed Rodgers's draft, and it was issued in January 1914. The new standard course would last 12 months, and two such courses (15 students each) would be enrolled each year: one in January and the other in July. Hattendorf et. al., *Sailors and Scholars*, 80–81. The correspondence course was formally established by a 1 April 1914 general order.

[9] During this period, American destroyers were seen primarily as ocean-going torpedo craft, which would seek out enemy capital ships and attack en masse. At this time, the U.S. Navy had the most powerful destroyer torpedo batteries in the world. Sims remarked later that although the theory was well understood, no one in his command had any idea of just how the destroyers were to find their targets or of how they were

to attack. This was not a unique problem; a few years later, the Royal Navy found that it also had no idea how destroyers operating independently were to find and attack major enemy units, particularly at night. Sims's remarks were reprinted as "Naval War College Principles and Methods" in U.S. Naval Institute *Proceedings*, March–April 1915, 390–91. For his earlier skepticism, see Hattendorf et al., 76–77. Moreover, Hattendorf et al. point out (page 88) that Sims was allowed to choose a staff (previously the destroyer flotilla had not had one).

[10] Sims was appointed president on 16 February 1917 and promoted to rear admiral on 19 March.

[11] The War College program for the January and July classes of 1917 survives as file UNT in NWC RG 8 Series 1. It is mainly a diagram of which problems were to be solved. In January, the new class would have a week of reading, after which it would have two problems (I and II), then a scouting and screening exercise, followed by tactics classes at the start of February, followed by Problem III and then scouting and screening again, then more tactical classes in March, followed by Problem IV and then Maneuver I (presumably a larger-scale game). After that, tactics, maneuvers, and solutions would be interleaved. The curriculum explicitly includes a logistics game (VIII) and at least three chart maneuvers (IX, X, and XI). Between January and July, time would be divided between strategy (101 periods), tactics (109), scouting (68), and logistics (14), with similar but slightly different apportionment between July and January.

[12] After the war, the U.S. Navy published the papers of the Planning Section: Officer of Naval Intelligence Historical Section, *American Naval Planning Section, London* (Washington, DC: GPO, 1923). According to the preface of this book, contributed by Captain D. W. Knox, all planning for the U.S. naval force in Europe initially came from Washington. Sims protested that he needed a planning section in London because it had much better access to timely information, not only about Allied naval activities and requirement, but also about the enemy. According to Knox, the London planning group was authorized by CNO Admiral Benson after he visited London in November 1917. At that time, the Admiralty staff was being reorganized and a planning section had just been created. In 1923, Knox was in charge of the Naval Library and Records and of the U.S. Navy's Historical Section; he was the direct predecessor of later chiefs of the Naval Historical Center and of the current Naval History and Heritage Command. He was one of the first two members of the London planning section.

[13] The planning group was initially a section within the Operations Division of the Royal Navy's War Staff. This staff had been created to meet a political demand. In 1911, Prime Minister Herbert H. Asquith asked his army and navy chiefs what they expected to do in the event of a war with Germany, which seemed entirely possible. He rejected the "secret" naval war plan proposed by his First Sea Lord, and he accepted the view of his Secretary of State for War that the Admiralty needed reform in the shape of a war staff like the army's. No one pointed out that the navy already had a staff in the form of the naval intelligence department, and the new war staff was formed almost entirely from members of the earlier department. This demand was the immediate basis for appointing Winston Churchill to the Admiralty; unfortunately for the historian, a case can also be made out that the entire drama was window-dressing intended to move Churchill out of the really sensitive position of Home Secretary (in charge of internal security), in which he had been less than successful. The move led to reorganization of the Royal Naval War College to support the new war staff. In army terms, the war staff was intended to help implement command decisions, not to ask wider-ranging questions, and a war staff education on the army model was very much about how to turn high-level decisions into detailed orders. Just as the army had its own staff college to produce staff officers, the navy was to use its Royal Naval War College (created in 1900) to produce officers for the new war staff. That was quite different from the earlier concept that, like Newport, the Royal Navy War College was intended to produce better general-purpose officers. The origin and role of the pre-1911 Royal Navy War College were described in a 1909 paper written after a visit by Rear Admiral R. P. Rodgers, USN, at that time Naval War College president, in the NWC Archive (RG 8, Box 6, coded DNT). Rodgers reported that the war course having proved its value, it was broadened and the War College made an adjunct of the naval intelligence department, "which in England largely fills, under the First [Sea] Lord, the duties of a General Naval Staff." In 1906, the war course was renamed the Royal Naval War College and detached from the Greenwich Naval College to Portsmouth. Under its first president, Captain H. J. May, RN, the War College used its gaming capability to answer questions such as the relative value of speed and armor. Its answers were not entirely accepted, and it seems that such issues were not raised after May died in 1904. The answers and the reactions are in ADM

1/7597, "Exercises Carried Out At the Royal Naval College, Greenwich," dated 7 May 1902 (the war college conclusions were also printed as naval intelligence department papers). Later evidence of pre-1914 British naval tactical development seems to reflect the experience of full-scale exercises and post-exercise analysis; war gaming or table-top exercises are never cited in the surviving papers. There is, however, evidence that the Royal Navy continued to war game, as some pre-1914 rules have survived.

[14] See William N. Still Jr., *Crisis at Sea: The United States Navy in European Waters in World War I* (Gainesville, FL: University Press of Florida, 2007), 43. Still first quotes Admiral Lord David Beatty, CinC Grand Fleet, complaining in April 1917 of the need for a more effective executive body including a planning body, but he then emphasizes the role of Lloyd George in forcing reorganization. He states further that Lloyd George identified planning as the weakest feature of the Royal Navy. It seems likely that the planning group was created in response to outside pressure rather than to a demand from within the Royal Navy. The evidence is that U.S. officers attached to the London planning section of Sims's staff considered the British planners weak and unwanted.

[15] CinC U.S. Atlantic Fleet to Secretary of the Navy (Operations), 11 October 1917, page 6 of a general report by fleet commander Admiral H. T. Mayo of a visit to England and France in August–September 1917 in OPNAV Confidential file C-47-9, NARA RG 80. "Until very recently there has been no planning section, or was there any definite body of men charged with the function of looking ahead, or even of looking back to see wherein lay the causes of success or failure, nor any means of furnishing the heads of the Admiralty with analyses and summaries of past operations in order that decisions as to continuing old operations or undertaking new ones might be reached with a due sense of 'perspective' both as to past operations, and as to the co-ordination of new operations in a general plan."

[16] This was not an idle question. In November 1917, two German light cruisers wiped out a British convoy headed to Norway. The last German fleet sortie, in April 1918, was intended as a convoy raid. However, German intelligence was poor: There was no convoy to raid. The planning section idea was imaginative, but the German ships did not have the required range, nor were the Germans willing to make what amounted to a suicidal gesture. However, ocean raiding of exactly this type was of considerable interest to the Germans before World War II, and it was practiced during the war.

[17] Still, *Crisis at Sea*, 45. Sims first requested a formal planning staff in October 1917. Secretary of the Navy Josephus Daniels opposed the idea in line with his general rejection of the creation of any kind of general staff. However, he referred the question to Admiral Benson, who included it in his list of items to discuss in England in November 1917. Benson endorsed the idea, arguing (presumably based on Sims's ideas) that a U.S. planning group could sell U.S. ideas to the Admiralty. It was assigned office space adjacent to that of the British planning staff. The first two members of the planning group were Captain Frank H. Scholfield and Captain Dudley W. Knox. Comments by U.S. officers in the classified OPNAV records (RG 38 in NARA) suggest that Still greatly overrated the power of the British planners.

[18] See for example Hattendorf et al., *Sailors and Scholars*, 112–13. The War College resumed its pre-war practice of running two overlapping one-year courses, with classes entering in June and December. Beginning in 1920 with the Class of 1921 (also described as the Course of December 1921), there was a single senior course each year, typically consisting of about 30 officers, many of them captains (later some rear admirals). The principal departments were Command, Strategy, and Tactics, Command being headed by the college's chief of staff. The college also ran a correspondence course. Both the Strategy and Tactics departments ran their own games. Beginning in 1924, the college added a junior class (lieutenant commanders and lieutenants) with its own staff and, in some cases, its own games. In theory, the junior class was intended to prepare officers to command major ships; the senior class was intended to prepare them for fleet and major fleet organization command. This book concentrates on the senior games. The NWC archive includes some junior class games, the avowed purpose of which is often to familiarize officers with the character of different ships in the fleet.

[19] According to a 1930 lecture on OPNAV organization, the original legislation provided for 15 officers to assist the CNO. In December 1918, a total of 267 officers were assigned to OPNAV. Given the scope of OPNAV tasks, an act of 12 August 1918 provided for an Assistant CNO (ACNO). OPNAV contracted

post-war, so that on 1 November 1930 it had 84 officers exclusive of those under instruction or on temporary duty. Lecture, Rear Admiral W. H. Standley (ACNO), "The Functions of the Office, Chief of Naval Operations," delivered at the Army War College on 5 September 1930, in OPNAV War Plans Division records, NARA II.

[20] Sims, "The Practical Officer," 2 December 1919, RG 12, Box 1, Folder 20, NHC, 25-6.

[21] Address to the Class of June 1919 (the first post–World War I class), NWC Archives, RG 8, file U-N-T 1919/136 in folder UNT 1919–1920.

[22] Sims's triumphant return to Newport is described in Hattendorf et al., *Sailors and Scholars*, 116–17. Sims's recommendations were reviewed and cut back by CNO Admiral W. S. Benson, who had worked with and against Sims during World War I.

[23] Hattendorf et al., *Sailors and Scholars*, 122–25, describe the controversy. Presumably Sims's underlying theme was that Daniels had unduly politicized the Navy, and that before the outbreak of war he had far too enthusiastically followed President Wilson's pacifist policy. It is difficult to separate Sims's attack on Daniels from the larger fight between the insurgents who had led the Navy in European waters and those who had shaped the fleet before the war.

[24] In theory, because it directly advised the Secretary of the Navy, the General Board was lead naval agency during the various arms control negotiations of the inter-war period. It might be imagined that because the War College alone had the technique to compare different fleets, it should have been deeply involved, but that does not seem to have been the case. The General Board framed a U.S. Navy position prior to the Washington Conference, but it was overridden by President Harding and the State Department. Important clauses in the 1930 London Naval Treaty, particularly provision for the flight-deck cruiser, were the personal responsibility of Rear Admiral William V. Pratt, who was U.S. naval advisor. He may have been inspired in part by U.S. needs demonstrated by war gaming. U.S. CNO Admiral William H. Standley seems to have been instrumental in negotiating the 1936 London Naval Treaty (the original U.S. negotiating position, developed by the State Department, was completely overcome by events). War College analysis may have helped the General Board head off a British attempt in 1927–33 to cut the size of battleships to cut their cost. The British attempt is evident in British records, in the treaty records of the General Board, and in General Board file 420-6 (battleships).

[25] The existence of the struggle can be inferred from intense controversies over what may now seem minor issues of ship design. The General Board won the fight over cruisers in 1920; its terms are clear from the General Board cruiser file (420-8). General Board files (420-2) include Sims's recommendation for the U.S. 1918 program. It is essentially a list of the types of ships then being built for the Royal Navy, adapted to U.S. weapons and therefore somewhat enlarged (they included a U.S. equivalent to the then-revolutionary fast battleship/battlecruiser *Hood*). The cruisers would have been equivalent to British fleet cruisers of the time, the C and D classes. Sims and his London planners brought the cruiser idea home with them. Meanwhile the General Board espoused big long-range cruisers armed with 8-inch guns for operations in the Pacific. The issue had an unstated international dimension. In 1919–20, the British were trying to avoid a costly increase in cruiser size. Sims and his planners were, in effect, supporting them—and, it might be argued, working against the new kind of war plan the U.S. Navy required for the Pacific. The General Board won this fight. It lost the next major fight, over battleship design in the mid-1930s. The fact of the creation of the OPNAV War Plans Division clearly shows the loss of the General Board's war planning role. The power of the inter-war General Board is probably somewhat exaggerated for historians by the high quality of the records it left (now in the National Archives in Washington), compared to the much sparser and less-organized records left by various parts of OPNAV.

[26] Inter-war presidents were Rear Admirals William S. Sims (16 February–28 April 1917 and then 11 April 1919–15 October 1922); Clarence S. Williams (1 November 1922–5 September 1925); William V. Pratt (5 September 1925–17 September 1927); Joel R. P. Pringle (19 September 1927–31 May 1930); Harris Laning (16 June 1930–14 May 1933); Luke McNamee (3 June 1933–15 June 1934); E. C. Kalbus (15 June 1934–15 December 1936 and 30 June 1939–16 June 1942); and Charles P. Snyder (2 January 1937–27 May

1939). Pratt was CNO between September 1930 and June 1933, having served as CinC U.S. Fleet between appointments as president of the War College and CNO.

[27] The extent of its information-gathering is evident in the surviving Research and Intelligence record group (RG 8) in the War College Archive. Most of the files are in the same form as the ONI (largely attaché) series held by NARA in Washington in RG 38. There was a rub. Apart from code-breakers, the inter-war U.S. Navy had no source of secret intelligence about other navies. That is evident from the ONI and War College intelligence files, for example in the very limited information gathered about the main potential enemy, Japan. This problem is indicated indirectly by the pride with which the code-breakers pointed to the one piece of technical intelligence they obtained, the trial speed of the Japanese battleship *Mutsu*.

[28] This change is evident in the War College's roster of students and instructors.

[29] These comments surfaced when the U.S. Naval Institute published the 1942 edition of *Sound Military Decision*, which it called the "Green Book," in its "Naval Classics" series (October 1992). According to Hattendorf et al., *Sailors and Scholars*, 155–61, when he became President of the War College in 1934, Rear Admiral E. C. Kalbfus devoted considerable energy to total revision of the pamphlet on the "Estimate of the Situation" on the theory that its logic was the key to successful naval decision-making. At this time the Navy was beginning to expand, diluting the War College's influence except through its correspondence course. It seems arguable that in Kalbfus's view the college's publications thus became much more important, since most of the fleet could not benefit from actual gaming. Hattendorf et al. see *Sound Military Decision* as the only handbook for decisions available to World War II naval officers. This is not a point made by Hattendorf et al. in their description of Kalbfus's effort.

3. War Gaming and War Planning

[1] Unsigned booklet, "The Naval War College," 10 June 1932, NWC RG 4, Box 55, folder 1. The booklet was intended as an introduction to new officers. It was probably written by the War College President, at that time Rear Admiral Harris Laning. Laning made this point particularly clearly in his memoir (Harris Landing with introduction by Mark Russell Shulman, *An Admiral's Yarn* [Naval War College Historical Monograph No. 14, 1999]), 262: "student officers train for high command in war by actually applying their knowledge of fighting to war situations; in war, no one condition is quite like another, and what might win in one case could bring complete disaster in another."

[2] For example, the Naval War College archive contains a lecture delivered before the 1913 Summer Conference by Captain William McCarty Little, "The Philosophy of the Order Form," in RG 8, Box 20, coded XCDO. At the time, McCarty Little was an important figure in the War College and he was instrumental in having the first four students assigned to the long course. His point is that the order form describes a complete course of action, hence can guide subordinates developing their own orders. This order form is equivalent to the later "Estimate of the Situation."

[3] Spector, *Professors of War*, is particularly critical of what he sees as the narrowing of the War College in his final chapter, "Epilogue: The Long Twilight," which covers essentially the same period as this book. He argues that, by the early 1930s, strategic problems were merely tests of isolated parts of larger plans (the account here of the impact of the "big game" of 1933 suggests the opposite). An alternative view would be that this shift reflected the maturation of the Orange War Plan, the basic ideas having been worked out.

[4] Spector, *Professors of War*, 117–18, credits introduction to Captain McCarty Little and his younger colleagues. He considers the origins of the system obscure, but does note a system of the same name used by the Army War College, extensive notes having been taken by CDR William L. Rodgers during his tour there in 1907–1909. Spector credits Captain McCarty Little with taking up an English translation of German General Otto von Griepenkerl's *Letters on Applied Tactics* as the basis for a new type of instruction. He formed a committee with Lieutenant Commander C. T. Vogelgesang and Marine Major John H. Russell (a future Marine Corps Commandant). Little made the important leap to a two-sided game.

⁵ The opposing argument is that too much initiative can be deadly. In framing instructions before Jutland, Admiral Sir John Jellicoe was clearly worried that a battle might become a melee, in which case his own ships might find themselves shooting at each other. That had nearly happened at Tsushima in 1905. Japanese commander Admiral Togo had divided his fleet into two divisions. In the smoke and confusion, one nearly engaged the other. We might now say that this clearly depends on how well everyone involved has situational awareness. In an army, an over-energetic junior officer might advance so far that his flanks would be exposed, and an alert enemy might well exploit that. A transition to objective-oriented orders remains a current goal in army training.

⁶ The British ships were the large "Town"-class cruisers, which they felt compelled to build instead of the far more affordable *Leander* class. The U.S. and Japanese ships were the *Brooklyn*s and the *Mogami*s. As recounted later, the U.S. Navy decided to build larger rather than smaller cruisers on the basis of War College analysis of what the other two navies (particularly the Japanese) were likely to do. The War College analysis is described in greater detail below. British reasoning (and shock) is evident in various Admiralty documents I have used for my history of British cruisers. British decisions on all types of major surface ships during the inter-war period seem marked by wishful thinking and mirror-imaging. This is particularly obvious in the discussion of what battleships the Japanese were likely to build once they withdrew from the treaty system (announced in 1934). The British badly wanted to limit the size of future battleships so that theirs would be affordable. They converted this wish into a totally false assumption that the Japanese could not go much beyond the earlier 35,000-ton limit. The reality was the 62,000-ton *Yamato*. The War College technique was an antidote; at the least, the War College would have asked what alternatives the Japanese might choose. Admittedly the U.S. Navy also did not guess the Japanese would go as far as they did, but in 1934 there was a serious attempt to guess what an unlimited Japanese battleship would look like.

⁷ Spector, *Professors of War*, 119, claims that the "applicatory method" was revolutionary in that it made the creation and use of doctrine possible.

⁸ Jellicoe's "Grand Fleet Battle Orders" survive in several files in the Archives at Kew, in versions reproduced post-war for instructional purposes (ADM 116/1341–1343). These orders were supplemented by "Grand Fleet Gunnery Orders" and "Grand Fleet Torpedo Orders," as well as by "Grand Fleet Secret Orders." The apparent bulk of the "Grand Fleet Battle Orders" was increased by inclusion of brief comments on current experience, for example, on the Battle of Dogger Bank. Jellicoe's successor Admiral Beatty renamed the orders "Grand Fleet Battle Instructions," but they were still the sort of detailed orders Jellicoe had issued. It is not clear whether the Royal Navy ever issued an overall tactical handbook between wars, although it did publish a document called "The Tactics of the Gun" (no copy of which has surfaced) and some other detailed handbooks. They contrast with the standardized formations and general tactical instructions issued by the Office of Fleet Training and by the U.S. Fleet itself between wars. After Jutland, the Royal Naval College tried and failed to find a tactical solution to the German turn-away. Gaming would probably have shown that the notion of a submarine trap, which so occupied Jellicoe and which resulted in the construction of the fast British K-class submarines, was impossible due to the limits of communications. Remarkably, the idea that the British would lead them into a submarine or mine trap appears in the 1914 German tactical orders, which the British captured and distributed in translation in October 1914. The 1915 version of these orders has survived in a collection of German tactical documents in the Admiralty Library in Portsmouth, and the January 1914 version (in translated printed form) is in the Public Record Office.

⁹ Jellicoe's view that decentralization required training is evident in a memorandum on plotting as a basis for divisional tactics, prepared by Captain F. C. Dreyer on the eve of war, a copy of which survives in the Backhouse Papers in the Royal Naval Museum, Portsmouth. Backhouse, who was later First Sea Lord, was Jellicoe's flag secretary. Probably the strongest attack on the social system that placed obedience and deference to authority uppermost is Andrew Gordon's *Rules of the Game: Jutland and British Naval Command* (London: John Murray, 1996).

¹⁰ It is striking that only one of the surviving U.S. games (Tactical Problem VI for the Class of 1922) simulated Jutland, although many classes heard or gave lectures about the battle. The British inter-war senior course apparently devoted two weeks each year to Jutland, showing officers what had happened and what would change using modern methods and weapons. The great difference between Jutland and most of the

big battles simulated by the War College was aircraft. Jutland involved a single British scout airplane, which had no real impact because its radio report never reached the flagship. Inserting mass air forces would have changed the battle out of all recognition. Records of the British presentation have not surfaced, but it seems unlikely that aircraft were a major feature of the modernized version of the battle. In the British case, the sole game played by the senior war class was a one-week campaign game simulating a war in the Far East. Again, no detailed accounts seem to have survived, so it is not possible to say how much aircraft figured in this game. It was probably a simulation of the movement of the fleet to the Far East as a prerequisite for the expected fleet action there. Since the British fleet would have steamed through the Indian Ocean, it would have been subject to submarine attack and to submarine-laid mines, but not to the sort of large-scale attacks that figured in U.S. war games simulating the progress of the U.S. fleet to its own expected fleet battle.

[11] Both pre-war and wartime tactical handbooks survive in the National Archives at College Park.

[12] Hattendorf et al., *Sailors and Scholars*, 72.

[13] The General Board records at NARA include a file (General Board 418) on the development of Battle Plans Nos. 1 and 2. It begins with a 1903 file produced by the Naval War College, giving extracts from the report of a special board on tactics. The paper is marked "Archives of U.S. Naval War College," but it is not clear that a copy survives in Newport. On 4 June 1904, the General Board formally requested a study of battle plans. The existing signal book allowed a senior officer to maneuver the growing U.S. fleet, but he needed guidance as to what he should do in combat. Secretary of the Navy Morton ordered the War College to form the summer course officers (rear admirals and captains) into a board to answer the three tactical problems the General Board posed. The General Board includes the result: a 1 July 1906 tentative battle plan (Battle Plan No. 1) prepared for testing by the fleet. The file also includes a 3 August 1906 letter to Commander Niblack (who had written about the War College and officer education) from British Admiral Charles Beresford. An attached memo recounted that when Beresford met Admiral Dewey on 4 May 1905, he "stated that Admirals in the British Navy get no training in handling fleets before they attain flag rank, and, after attaining that rank, some have little time and inclination to handle them. He found no fault with the method of dealing with strategical questions, but believes that this lack of experience in handling tactical questions practically is a source of great weakness to the Navy." Beresford was proud of forcing his subordinates to practice tactics, but pointed out that the Royal Navy had no regulation requiring such exercises: "[S]ome Admirals do not maneuver their squadrons beyond the mere passage from port to port." He wanted Dewey to support his drive for such a regulation. For his part, the Secretary of the Navy ordered all commanders in chief afloat to test and report on the new battle tactics which had just been prepared by the General Board in conjunction with the War College. The General Board file includes these reports; Battle Plan No. 1 was accepted. It and Battle Plan No. 2 were embodied in the Atlantic Fleet war orders when the United States went to war in 1917.

[14] The texts of the lectures have survived, and many of them offer insight into the typical views of the inter-war Navy. Because the game material is massive, and hence relatively inaccessible, many writers have taken the lectures as the core of the War College course. For example, the lectures present what seems to us a very conservative view of the coming technology of aircraft. If, however, the view is Sims's or Harris's—that the students would learn by doing—the details of the lectures are almost irrelevant.

[15] Post-game discussions by War College President Sims were concerned entirely with the art of tactical decision-making. As head of the Tactics Department, Harris Laning introduced comments on the tactical lessons of the games.

[16] Gaming in tandem with the 1927 Fleet Problem is described by Lillard, *Playing War*, 83–84. Presumably to avoid political complication, the enemy was called Black, although the composition of the enemy fleet indicated that it was British. Black was usually reserved for Germany, the most likely pre-1914 enemy.

[17] The simpler pre-1917 games were often played in the fleet; in fact, one of the surviving sets of rules was promulgated not by the War College, but by the Atlantic Fleet. They were seen as means for the fleet to educate itself and also to undertake tactical development.

18 The War College Archives contain several early sets of game rules. The 1901 rules contain both tactical (including one-on-one) and strategic rules, a total of 21 pages. The 1905 rules (coded XMAT, in NWC RG 35) are more elaborate, 25 pages long (after an introduction), with special curves to make it possible to take maneuvers into account. An attempt is made to indicate fire effect at increasing range. NWC RG 4, Box 90, Folder 7 (XMAC) is a more elaborate set of tactical war game rules, with appended tables of ships of the U.S. and German navies and merchant fleets, issued in 1912. It amounts to about 61 pages including the extensive tables. The inter-war rules included no such tables; all ship data and fire effect data were separate. NWC RG 8, Box 90 contains Lieutenant Commander H. E. Yarnell's account of "Naval Chart Maneuvers 1916," meaning a dissertation on war gaming as it was then practiced (it is coded XMAC Accession 1916/174 in the War College archive). Yarnell discusses the mechanics of gaming as it was then practiced, with rules for items such as radio messages. No copies of the much-improved rules issued in 1919–22 have surfaced at the War College, but some of the record groups involved are disorganized and may contain them. The record for 1923–41 in RG 8 is fairly complete.

19 War College files (RG 4, publications) include a massive treatise issued by the Intelligence Department in June 1931, *Construction of Fire Effect Tables 1922; Notes on Revision of Fire Effect Tables Up to March 1930; The 1931 General Revision of Fire Effect Tables*. Formulas and figures were adjusted to match new data. Material on the construction of the 1922 rules is coded XOGF (RG 8, Box 96, Folder 7).

20 Rear Admiral Bruce McCandless, USN (Ret.), "The *San Francisco* Story," in *Proceedings* of U.S. Naval Institute, November 1958.

21 For example, in a 1939 game (Operations VI, Class of 1939) *Hood* (CC-4 in the Royal Navy list) was credited with a lifetime of 17.7, compared to 16.1 for a *Royal Sovereign*, 16.6 for a *Queen Elizabeth*, and 19 for a *Nelson*. U.S. figures were 18.5 for a *Maryland*, 17.7 for a *California* (which had the same armor), 17.3 for the modernized *New Mexico*s, and 17.2 for *Pennsylvania*. Only the *Nevada*s (16.2) and older ships were as weak, in these figures, as older the British battleships. All figures were in the common denomination of 14-inch hits.

22 Laning, *An Admiral's Yarn*, 327.

23 "Superiority of the British Over the United States in the Fighting Strengths of the Capital Ships (Excluding Aircraft Carriers) Resulting From the Treaty Limiting Naval Armaments and Study and Computation of the Relative Fighting Strengths of the British and United States Navies," a letter from the President of the War College. NWC RG 8 Series 2, Box 97, folder 5, coded XOGF/Safe. The letter of transmission to the Secretary of the Navy is dated 13 October 1922; it refers to an earlier submission dated 12 October, which is supplemented by the current one. A much more elaborate comparison (with the same conclusion) was produced in 1924. The War College Class of 1923 gamed an engagement between the U.S. and British fleets as its Tactical Problem II. During the Washington Conference, the General Board evaluated national naval strengths on the basis of age and displacement, concluding that the ratio of British superiority was only 1.107. Sims's analysis showed a crushing British superiority: an effective ratio of 1.44:1 at 15,000 yards, but 2.56:1 at 24,000 yards.

24 Lieutenant Commander H. R. Thurber, *Suitability of Naval War College Instruction in Gunnery With Reference to Tactics*, in folder NC (for War College), but in a War College jacket coded XOGF 1937-104, in the CNO/SecNav classified files (NARA RG 80). Thurber was particularly worried by what he called the range-band theory. Certainly it could be argued that armor and shell performance would define ranges at which some ships could fight without being penetrated, but Thurber showed that the data from which the tables were built was much less than credible. Data on foreign deck armor was nearly nonexistent. What little there was, was grossly inconsistent. The width of the range bands implicit in War College tables was 2,000 to 6,000 yards, but the endless uncertainties Thurber cited would swamp that. For example, Thurber estimated a difference of 9,000 yards in penetration between old and new armors. Grossly uncertain information on Orange ships such as heavy cruisers did not help. Were the belts on those cruisers three or four inches thick? The difference was equivalent to 6,000 yards for a U.S. 8-inch gun. Was it two inches thick? The actual figure was about four inches, but no one knew that before 1945. Uncertainty as to deck thickness in the rebuilt Japanese *Kongo* class battle cruisers equated to 10,000 yards (four versus seven inches over magazines, the

true figure being 4.7 inches). It is not clear how seriously Thurber was taken, but the survival of his report in the CNO/SecNav classified file suggests that he was. Thurber's analysis seems to have been the result of his experience as a member of the War College class of 1936. He retired as a vice admiral.

[25] War College archives include several studies taking supposed British means of increasing gun elevation without modifying turrets into account, and the issue is also prominent in General Board records (mainly 420-6). Bulges were anti-torpedo structures built outside a ship's hull, typically including air spaces. The Washington Treaty certainly allowed their installation. Looking back, it is difficult to see why a fleet would deliberately flood bulges, thus reducing its immunity to torpedo fire. The British did not change their own interpretation of the treaty and therefore did not increase the elevation of guns onboard ships they modernized in the late 1920s and early 1930s, but it is entirely possible that the chronically underfunded Admiralty was using the treaty as an excuse to avoid expensive gun-mounting modification. For the U.S. side of the controversy, and the U.S. battleship modernization program, see the author's *U.S. Battleships: An Illustrated Design History*. For the British side, and for arms control negotiations, see his *British Battleships 1906–1945*.

[26] In 1936, the U.S. communications intelligence organization broke a message reporting the trials of the battleship *Mutsu*, which was now credited with a speed of over 26 knots (the previous accepted figure had been 23.5 knots). We now know that this was also the original design speed of the ship, unreported for some reason through the 1920s. The real surprise, not known until late in World War II, was that reconstruction of the earlier Japanese battleships had given them much the same speed as the *Mutsu*, which meant that the Japanese fleet could have crossed the "T" of the U.S. fleet unless it was stopped in some other way. Through the war, U.S. official handbooks credited the Japanese battleships with much lower speeds (so did the British). Presumably the high speed of the *Mutsu* did not figure in these handbooks because of the way it was discovered. None of this turned out to matter during World War II, because (as the War College forecast) the Japanese withheld their battle fleet pending attrition of the U.S. fleet (which did not, in the event, succeed). The most visible effect of the discovery of the speed of the *Mutsu* was redesign of the *South Dakota* class. There is no evidence that the British did even this well.

[27] Laning, *Admiral's Yarn*, 263–64. Sims told Laning to incorporate what was being learned in his thesis on battle tactics, which became a student handbook.

[28] Van Auken had been a gunnery officer during World War I; he had run afoul of his fleet commander by writing that failure to install a new fire-control system would doom the fleet in battle. He spent the post-war decade in the naval education system, attending the Naval War College as a member of the Class of 1927. His outspokenness shows in some of his written comments explaining war game lessons. Van Auken's World War I story is told by C. C. Wright, "Questions on the Effectiveness of U.S. Navy Battleship Gunnery: Notes on the Origins of U.S. Navy Gun Fire Control System Range Keepers," *Warship International* 41/3 (September 2004). After leaving the War College, van Auken commanded the battleship *Oklahoma* for a year.

[29] U.S. war planners were certainly aware that they needed large numbers of auxiliaries to transport an expeditionary force to the western Pacific, and, by 1929, printed War Plans documents included a list of required conversions (this document is available as a microfilm at NHHC). They included aircraft carriers converted from liners (XCV) and seaplane tenders converted from merchant ships (XAV) as the only ways of increasing sea-based air power in a treaty-limited era. It also included armed merchant cruisers (XCL) to harass enemy trade and to protect U.S. sea communications from enemy attack. The XAVs were dropped in 1935 both because there were too few fast U.S. liners suitable for conversion (particular ships were earmarked and preliminary plans drawn) and because conversion would take too long—at least 180 days. The conversions envisaged at the time were quite elaborate, resulting in ships resembling existing fleet carriers, but slower. The need for fast naval auxiliaries was explicit in public explanations of the need for the Maritime Commission, created by the Merchant Marine Act of 1936. It was headed by former U.S. Navy Chief Constructor Rear Admiral Emory S. Land.

[30] War College records (RG 8) include a brief description of the 1933 design of a battle cruiser as a candidate for construction once the battleship-building "holiday" initiated by the Washington Treaty (and extended by the 1930 London Naval Treaty) expired at the end of 1936. Up to that time, proposals for such battleships had envisaged conventional low-speed designs (the same was true of the British). War games highlighted

the operation of the Japanese battle cruisers, not only in fleet action, but also in cooperation with Japanese carriers and in raiding operations against vital U.S. sea communications to the western Pacific. The General Board absorbed the idea. Its proposed battleship sacrificed firepower for speed. CNO preferred a different balance, a somewhat slower ship with more firepower. He won the fight because at the crucial time he was also acting Secretary of the Navy. Even so, the resulting ships, the *North Carolinas*, had a rated speed six knots faster than their predecessors. They seem to have been conceived as a fast wing for the future U.S. battle line, because the next class of battleships, the *South Dakotas*, were to have traded speed (reduced to 24 knots) for better protection. That is not evident from what happened; the *South Dakotas*' speed was increased (via a radical redesign) because U.S. Navy signals intelligence picked up evidence that the rebuilt Japanese battleship *Mutsu* was a good deal faster (over 26 knots). During World War II, a U.S. Navy summary of the achievements of signals intelligence singled out this information as its most important pre-war coup, although that is open to question. Unfortunately, correspondence files that might have demonstrated a link between the 1933 sketch design and General Board policy do not exist; the General Board files (in NARA RG 80) show the results of its thinking, but generally not its sources.

[31] This change is evident in War College advice proffered in 1930. Unfortunately, the War College archive apparently does not contain communications from the college president to OPNAV and other naval offices (or, for that matter, correspondence relating to the scenarios chosen for war games) because it was reconstructed in the 1960s from records that had gone to various Federal records centers. In this book, War College influence has been traced using mainly the records of OPNAV and of the General Board.

[32] Series II of the War Plans Division (OPNAV Strategic Plans Division) files in NARA II is "Naval War College Instructional Materials 1914–1941," beginning with Joint Army-Navy Problem (Overseas Expedition) of 1923 in Box 14; the next in the series is Stratedgical Problem I (Joint Problem) of 1924. Boxes 28–34 are studies conducted at the War College. Another War Plans Division file is the receipts for war gaming material received from 1937 onward. These studies range from fleet tactics to the strategy of economic warfare.

[33] The U.S. Navy's views as to which games mattered are reflected in the requests for gaming material from the War Plans Division. Its surviving archive for 1936–41 includes numerous Orange-Blue games, but only one Red-Blue game, from 1938. That game was presumably intended to support a study conducted that year of which British bases the United States should seek to obtain for hemispheric defense. The only other game in these files involving Red was a 1941 Atlantic exercise in which the United States and Britain were allied against Germany.

[34] Van Auken went further in his formal commentary on the big 1931 Blue-Red (i.e., against the British) game (Operations III for the Class of 1932). He considered it essential to game and plan a cooperative war by both countries against Japan and its allies. He repeated his comment the following year in commentary on a Blue-Orange game (Operations II for the Class of 1933).

[35] It also helped that the United States began naval rearmament in 1937 with the expiration of total tonnage limits imposed by the Washington Treaty and the London Naval Treaty of 1930. However, the U.S. rearmament program was dwarfed by that of the British.

[36] Operations Problem II-1927, with introductory remarks dated June 1927 by Captain H. E. Yarnell, who wrote that "it is a situation which represents very closely what would happen under actual conditions." NWC RG 4, Box 34, Item 1285-F.

[37] By 1929, War Plans documents included the proviso that if the Japanese did not see sense once the blockade began, it might be necessary to bomb Japanese cities. It was widely believed that the flimsy Japanese cities were unusually vulnerable to bombing, particularly using incendiaries. This possibility did not affect planning or gaming because the same islands that had to be seized to support an effective blockade could also be air bases. The main effect of the possible change was to add another type of required auxiliary, an aircraft transport for army bombers. No invasion of Japan was contemplated because it was considered most unlikely that a sufficient army could be raised and transported across the Pacific. See Miller, *War Plan Orange*, 150–52 and 162–64 for the use of air power; elsewhere in the same chapter, he estimates the numbers of troops the Army thought would be needed.

[38] Many accounts of the disappearance of Amelia Earhart, the famous aviator, connect her flight with a Navy-sponsored reconnaissance project. Whether or not that is true, it reflects the well-known and badly frustrated Navy interest in the islands. Post-war investigation suggests that the islands were not fortified, but that facilities such as seaplane ramps were built.

[39] The U.S. demand for high performance is evident in BuAer's draft fighter history (through 1945) in the files of the Ships Histories Branch at NHHC (formerly the Aviation History Branch). The demand for performance is discussed as the key factor in U.S. naval fighter development at least from the early 1930s, resulting in pre-war designs such as the Buffalo, Wildcat, and Corsair. It was in striking contrast to the consistent British view that low performance was acceptable because British carrier aircraft would face only other carrier aircraft. The Admiralty position is described in, for example, a proposed Fleet Air Arm handbook produced in 1936 (ADM 116/4030). When the handbook was circulated to senior officers for comment, Admiral Sir Dudley Pound, then Home Fleet commander, asked pointedly how a concept to defeat Japanese carriers applied to the more likely war in European waters against land-based aircraft. The answer from the director of Naval Air Department was simply that this possibility had not been considered. The main British inter-war scenario involved a decisive battle between a British fleet sent east to Singapore and a Japanese fleet coming from the north. The British would not meet Japanese aircraft en route to Singapore, and they assumed that the only aircraft they would have to fight before and during the decisive battle would be Japanese carrier aircraft. It is not clear what they planned to do after the decisive battle, when they would have to operate near Japan in order to impose an effective blockade.

[40] Miller, *War Plan Orange*, generally distinguishes between "thrusters" and "cautionaries" who moved the war plan back and forth.

[41] This shift is described by Miller, *War Plan Orange*, 139–42.

[42] In Box 19 of the War Plans Division series, NARA II. This game does not figure, at least explicitly, in Miller, *War Plan Orange*. He sees the shift based on a series of developments, including a new assumption that the Japanese would fortify the Mandates and improvements that made aircraft much more effective. However, the War College had been ascribing improved performance to aircraft for some time. Miller also attributes the shift to a change in personnel from "thrusters" to "cautionaries." It is impossible to say just how the 1933 Operations IV game influenced thinking in the War Plan Division, but it seems that, at the least, anyone rejecting "thrusting" could use it as powerful ammunition. Note that this is the *only* war game described in detail by either Vlahos, *The Blue Sword*, or Hattendorf et al, *Sailors and Scholars*.

[43] MacArthur was involved because U.S. war plans were joint documents approved by a Joint Board established in 1903. The Army view, reflected in the first volume of its official history of the Pacific War, was that U.S. Pacific strategy was about defending or else regaining the Philippines. That was very different from the consistent Navy view that Japan had to be defeated, disgorgement of a conquered Philippines being one of several consequences. See Louis Morton, *Strategy and Command: The First Two Years* (Washington, DC: Government Printing Office, 1961); the first chapter describes pre-war U.S. strategy. It is possible that Morton's concentration on the Philippines is connected with General MacArthur's insistence that the Philippines were the appropriate objective in October 1944 (the Navy would have preferred to seize a base in Taiwan or on the Chinese coast). Morton describes the Army position: Although the Japanese could occupy most of the Philippines, it would be worthwhile to hold Manila Bay as a fleet base, or even to hold just Corregidor to deny that base to the Japanese. This view does not correspond to the Navy view well before 1934 that Manila would surely fall and that the fleet would have to go to an alternatate port in the southern Philippines. The situation was further complicated by congressional action in March 1934 promising the Philippines independence in 1946. Morton does not mention MacArthur's acceptance of the change in the accepted Orange war plan, which dropped any attempt to reinforce the Philippines in wartime (it is dated to 1936). After the war plan had changed, Army students at Newport commented that the change had been inevitable, as they had doubted that the islands could be held. They seem not to have realized that the key point (at least to the Navy) was that even if part of the Philippines could be held, the fleet almost certainly would not have made it there intact enough to defeat the Imperial Japanese Navy—which was necessary in order to go to the next step, the blockade. In effect, the students' comment demonstrated how wide a gulf existed between the services'concepts of a future Pacific war.

44 It might be argued that the war planners could simply add up opposing forces to compare them, or that they could follow the earlier path of the "applicatory method" and lay out likely enemy courses of action for comparison with U.S. courses of action. Simply adding up forces would not have sufficed to predict success or failure; a campaign was just too vast for that. A force ratio might be used to predict the outcome of a single battle, but not in detail. It could not predict how many ships would emerge how badly damaged, for example. All of the war planners had been indoctrinated at the War College to see a two-sided game as far better than the predictions a single planner or committee might make. A single planner or even a committee was just too vulnerable to wishful thinking. Even after the War College had lost much of its influence, the war planners continued to use its gaming data; they still wanted to avoid pitfalls. Moreover, gaming made it possible to tease out unexpected possibilities. For example, during a 1932 game (Tactical VI, Class of 1932) a student suggested a carrier air raid on Tokyo as a way of concentrating Japanese attention. The idea was incorporated into a report on interesting developments in gaming, but it was not played out at the time—only because the game had no mechanism for testing it against the team playing the Japanese.

45 The key advantage of seaplanes was that their tenders were not limited by treaty. It probably helped that Congress did not consider them combatants. The seaplane carriers were modified tenders (AV). Their designs incorporated carrier-type catapults. The aircraft they would launch were intended to provide Marines with bomber support from lagoons near a contested island (they retained the AV designation). This concept explains experiments with floatplane versions of dive bombers and probably also fighters. It was dropped soon after war broke out, and the Marines could rely on numerous escort carriers. The new idea of a fully mobile seaplane force was symbolized by the transfer of seaplanes from the Base Force at Pearl Harbor to a new Patrol Wings Scouting Force. In 1939, there were only two modern tenders, the *Wright* and the ex-carrier *Langley*, but new ships were being built. Fourteen old destroyers had just been converted as prototype AVPs, and nine minesweepers were also reclassified as AVPs. In 1940, the first purpose-built seaplane tender, *Curtiss* (AV-4), was completed.

4. War Gaming and Carrier Aviation

1 The closest any World War I naval airplane came to participation was scouting during the Battle of Jutland. It had no impact because the radioed report from the airplane was not transmitted onward by its carrier. At the end of the war, the British had assembled a torpedo-bomber strike force, but the war ended before it could see combat. Moreover, the strike force was intended to attack the German High Seas Fleet in its harbor, not at sea.

2 During World War I, U.S. naval aviators had considerable access to British thinking, and they wrote extensive reports, which among other things helped convince the U.S. Navy that it wanted carriers. Direct contact with the British continued for a time after the war, as the British hoped that the United States would continue to help maintain the peace. Thus, U.S. files include contemporary plans of British carriers and, in 1922, the British provided the U.S. Navy with details of its arresting gear. British files show that most exchanges stopped in 1920, when it was clear that the United States would not be joining the League of Nations. One great irony of the inter-war period is that the relatively large tonnage allowed for carriers under the Washington Treaty—135,000 tons for the United States and the United Kingdom—was the result of a British initiative based on perceived fleet requirements, taking into account the very limited number of aircraft the British believed that any one carrier could operate. In 1921, no other navy saw the need for a substantial carrier force, and the U.S. Navy would probably have been satisfied with the conversion of the two battle cruisers.

3 The first approach may have been a set of data provided by Director of Naval Aviation Captain T. T. Craven dated 17 June 1919, replying to a 4 June letter from Lieutenant Commander H. M. Lammers. Land (wheeled) aircraft would fly only from flying-off platforms on decks and turrets; they could not land back onboard. Craven did not allow for fighters, but only for large (patrol) flying boats, spotters, torpedo bombers, and reconnaissance aircraft. Of these, the spotters were the fastest, at 90 miles per hour; the slowest were the torpedo bombers, at 70 miles per hour. Servicing time between flights was two hours for large aircraft or one hour for smaller ones. Endurance was six hours for a reconnaissance plane, four for a spotter, and three for a torpedo bomber. Aircraft could spot single ships 30 miles away and close formations 40 miles away. Max-

imum radio range for a land plane was 75 miles (they could receive messages at 250). There was no estimate of hitting probability under various conditions. Craven suggested that the War College should pay attention to lighter-than-air craft (blimps and rigids) that it had neglected in the past. NWC RG 8, Box 77, folder 6, Accession 1919/561, coded XAEG.

[4] The June 1923 maneuver rules are in NWC RG 4, Box 18, folder 20; they are marked 803-B.

[5] NWC RG 8, file XMAR (1935-12) dated 26 February 1935, "Manuever Rules Pertaining to Aircraft and Airplane Characteristics—Revision of." Games often involved smoke screens laid by aircraft; King drily remarked that students at the War College were much better equipped than pilots to lay them correctly.

[6] The most striking exception was the British torpedo bomber–spotter–reconnaissance airplane (TSR), the Swordfish. In games, it was accorded rather better performance than it had.

[7] These were not new categories, although the designations were new. Converted carriers figured in 1920 war games under the designation JC (J for carrier, C for conversion). This was a natural category; nearly all the existing British carriers were conversions of one kind or another. The first British carrier with a full-length flight deck, HMS *Argus*, had been converted from an incomplete liner laid down for an Italian company. It seems unlikely that the XOCV designation implied the later idea of converting liners upon mobilization to expand the U.S. carrier force.

[8] Initial estimates of bomb accuracy were optimistic. For airplanes at up to 1,500 feet, it should take two bombs to make one hit on a large or intermediate ship, such as a battleship or cruiser; up to 4,000 feet, four; up to 8,000 feet, six; and up to 12,000 feet, ten (12 for the intermediate ship). Airplanes dropping bombs from below 500 feet would be destroyed by their blast. The 1924 rules recognized that these figures, taken from target practice, would likely be cut sharply in reality. For example, when bombers were opposed by considerably superior forces, hitting would decrease by 40 percent. The umpire would assign further decreases due to target maneuvers or speed or anti-aircraft fire. Figures became more realistic as experience accumulated; they were based on results in the annual bombing exercises. The 1925 rules assigned a 40 percent chance of hitting (i.e., less than before) to a bomb dropped from 1,500 feet; 15 percent from 8,000 feet, and 10 percent from 10,000 feet, in each case against a large ship. Anti-aircraft fire was expected to force aircraft to attack from higher altitude and thus to lose accuracy (in 1931, the minimum bombing altitude was 2,000 feet, and at that altitude a bomber would achieve only 26 percent hits against a non-maneuvering ship). The penalty exacted by opposing fighters increased (a superior number of fighters could reduce bomber effectiveness by 60 percent). In 1936, bombing figures were reduced by a quarter to take into account the "fog of war." Horizontal bombing was accorded greater accuracy in 1937, probably because the U.S. Navy now had a precision bomb sight (the famous Norden bombsight). Now, a bomber was accorded 8 percent hits from 9,000 feet (but much less against a fast ship). There was also an uncertainty in the effect of hits in the 14-inch shell currency used by the War College. In the 1933 rules, a 1,000-pound bomb hitting a battleship was equivalent to 0.58 shell hits (0.75 against a carrier or cruiser), but in the 1937 rules a 500-pound bomb hitting above water was considered equivalent to a 14-inch shell, and a 1,000-pound bomb was considered equivalent to two 14-inch hits.

[9] Thus, in the 1925 rules the chance of being wrecked while landing in daylight was given as 1/30 rather than 1/20 (1/20 rather than 1/10 in darkness). In 1927, rules were made more elaborate, the chance of being wrecked landing on a carrier depending on the type of airplane and the weather. In smooth weather, a fighter had a 1/25 (0.04) chance of being wrecked on landing, but in rough weather this rose to 1/10. Figures were again reduced in the 1929 rules, to 0.03 onboard a carrier in day under all sea conditions (0.06 at night). Comparable figures for a landing field ashore were 0.001 and 0.03. In 1933, the chance of being wrecked by landing on a carrier was reduced to 0.01.

[10] NWC RG 4, Box 29, Folder 9.

[11] The game did not include individual decisions by fighter pilots; it is not clear who realized that the observation planes were only a feint, and that not all of Blue's fighter pilots should (or would) have been drawn in.

[12] The post-game analysis pointed out that the barrier was effective only against aircraft coming from an expected direction. Gaming was the way to guess unexpected tactics.

[13] This presumably meant that more aircraft were being brought up from hangar decks and prepared on deck (e.g., by having their wings unfolded). In 1924, the U.S. Navy had no idea of how that would work.

[14] The game lasted this long because the Blue commander failed to take advantage of the situation and turned away from Surigao.

[15] Laning noted two important tactical innovations that year. One was the sustained Blue air operation; he described the careful refueling arrangements as "building for the future." The other was the circular formation, which the inter-war Navy later adopted, invented by Roscoe MacFall. Reeves was chosen as Laning's successor as head of the Tactics Department. Given Laning's generally disparaging remarks about the tactical choices made by other players, it seems very unlikely that he considered anyone except Reeves worth choosing.

[16] NWC RG 4, Box 26, Folder 2. In this game, Red's fleet headed for the Caribbean to attack Blue trade routes. Blue's objective was to prevent Red from establishing itself in strength there. The objects of the Tactics Department were to explore (1) how a fleet (in this case Red) could best conduct its train; (2) how the Blue fleet could deal with Red's superior speed and firepower; (3) how Blue could overcome Red's superiority in cruisers (in this case, 24 Red versus 10 Blue); (4) whether, given the same total number of aircraft, the two big U.S. carriers were superior to six smaller Red carriers; and (5) whether cutting the Red fleet by four capital ships would overcome its superior fighting strength. The game did not answer (1), because Blue never tried to destroy Red's train, despite its advantageous position. It did show that neither side could easily deny the other air spotting. Blue was able to keep spotting even after Red destroyed the flight decks of its two carriers and thus (in theory) gained "control of the air." The only consequence of losing the flight decks was that Blue spotters could not land back on the carriers to refuel, but by then the gun battle was over. Blue had many more destroyers than Red. An important cruiser role was to break up destroyer torpedo attacks on the battleships. In this case, Blue had so many more destroyers that some of them could be devoted to sinking Red cruisers (at the end of game Blue had more cruisers than Red). The post-game critique pointed out that the Red commander evidently did not give sufficient consideration to the question of which side had air control and thus could use airplanes for spotting. It made no provision for the possibility that Blue—and not Red—would be able to spot from the air, which in turn would determine which side could fire at greater range. "As this was a question yet to be decided by the outcome of the air engagement, it should have received a most prominent position in the Red commander's battle plan." Red certainly intended to seize air control at the outset and directed that every effort be made in that direction, but he should have reckoned with the possibility that this plan would fail. Blue made no provision for using his air forces to observe enemy movements, yet this was vital from the moment of contact. "All movements of enemy forces, their disposition, the launching of air attacks, should be promptly known." The critique, dated 14 January 1925, was unsigned, but was probably written by Reeves as head of the Tactics Department.

[17] Treaty rules included a proviso that new ships in categories whose total tonnage was limited could be replaced only to replace over-age ships in those categories (the treaty included rules as to lifetime). *Langley* was not counted in future projections because, as an experimental ship, she could be discarded at any time to be replaced by a new ship (or a fraction of one).

[18] "Use of 10,000 ton Cruisers for Carrying Aircraft," in General Board 420-7 file 1925–1931 (NARA RG 80), from War College President Rear Admiral C. S. Williams (writing as a member of the General Board) to Senior Member, General Board, 23 April 1925. This letter rejects the idea of a combination cruiser-carrier, a concept that nonetheless survived into the 1930s.

[19] General Board file 420-2 Serial 1313 dated 30 March 1926 in NARA RG 80 file 420-2 1926–1927).

[20] NWC RG 8 Series 1, Box 37, folder 6: UNC 1925–1927. The cover letter from the President of the War College is dated 11 April 1927, coded UNC (1927-83).

[21] Summary of officers' comments by Lieutenant Forrest Sherman attached to the letter from Admiral Pratt, dated 1 December 1926. The officers were Captain J. W. Greenslade, Captain S. R. Moses, Captain B. B. Wygant, Commander R. R. Stewart, and Lieutenant Forrest Sherman. Moses pointed to the advantages of converting a suitable merchant ship into a medium carrier, noting that authority had already been given to convert a second experimental (training) carrier comparable to *Langley*. Inputs were letters from the bureaus of Aeronautics and Construction and Repair. Aeronautics strongly favored smaller carriers. Sherman pointed to experiments he had conducted on scale models of the flight and hangar decks of the *Lexington* and *Saratoga* using paper airplanes to show problems that could arise if aircraft spotted on deck did not correspond to changing requirements (one conclusion was that the elevator arrangement of these ships was defective). He pointed out that fighters spotted aft could take off over heavier bombers spotted forward. His main conclusion was that the ships could not operate as many aircraft as were planned.

[22] Light bombing meant the new tactic of dive bombing, which in 1926 involved fighters carrying 25-pound bombs. These weapons were sufficient to wreck a flight deck and send a carrier back to a shipyard. In his summary comments, Sherman proposed building a non-watertight false deck 2.5 feet above the real deck. This idea does not appear in the approved characteristics for *Ranger*, but she had exactly the sort of easily repaired flight deck Sherman envisioned, the only difference being that there was no heavy structure immediately below it. Subsequent U.S. carriers through the *Essex* class had this type of flight deck. Sherman also called for the largest possible hangar. Having visited the *Lexington*, he was impressed by the potential fire hazard in a hangar, and proposed that the hangar of any new ship be open and well ventilated. He mistakenly believed that the British had drawn the same conclusion from their earlier carriers, including HMS *Furious*.

[23] It is not clear why Sherman did not realize that the British had no choice; these were the only two existing hulls they could convert under the limits in the Washington Treaty.

[24] As areas, they were proportional to the square rather than the cube of dimensions. Displacement (volume) was proportional to the cube.

[25] General Board 420-7 (Serial No. 1362) file 1926-31, Characteristics dated 1 November 1927 (NARA RG 80). These characteristics were formally approved by the Secretary of the Navy on 7 November.

[26] Memorandum dated 17 October 1927 in General Board 420-7 file 1926–1931 (NARA RG 38). Horne called for a program of one 10,000-ton carrier specifically to support the battleships' spotting planes (36 spotters and 36 fighters onboard); one 13,000-tonner (32 knots) for light cruiser planes (54 observation and 54 fighters); and four 14,000-tonners (29 knots) specifically to support bombers (32 bombers, 48 fighters). At this time, the only design on offer was *Ranger*, which was sometimes credited with capacity for 104 aircraft.

[27] *Ranger* was authorized under the FY29 program and ordered on 1 November 1930. Her design was revised after the two big carriers entered service. She had been conceived as a flush-decker, but their operation demonstrated the importance of an island, and one was belatedly added. How late in the process that came is evident from the fact that she retained the tilting funnels required by her original flush-deck design; trunking them into the island would have been too expensive.

[28] Until April 1939, the Fleet Air Arm was part of the RAF, and the RAF leadership did not want to divert scarce resources to it. However, the Royal Navy paid for the airplanes and the pilots, and it does not seem to have understood that it needed so many more of both. This was partly, though not completely, reflected in a failure to equip carriers with aircraft on anything like a U.S. scale.

[29] The London Naval Treaty of 1930 tightened the definition of a carrier to include any ship intended to launch aircraft. In addition, it included limits on the total tonnages of cruisers, a category not limited by the Washington Treaty (except that maximum tonnage was 10,000 tons). As a consequence, installing a flight deck on an existing battleship or cruiser would move it into the limited carrier category. Pratt extracted a limited exemption for air-capable cruisers in connection with the primary British objective at the conference, which was to end construction of heavy (8-inch gun) cruisers. The British thought that any navy building a light (6-inch gun) cruiser would necessarily limit its size and that therefore the cost of maintaining their own cruiser fleet would be reduced. This did not turn out to be the case.

[30] "A Tactical Study based on The Fundamental Principles of War of the Employment of the Present Blue Fleet in Battle showing Vital Modifications Demanded by Tactics," February 1925, in NWC RG 8, Box 92, folder 2.

[31] Reeves never wrote down these conclusions; they are deduced from the conclusions included in the relevant game and from what Reeves then did as Commander, Air Squadrons, Battle Fleet. Had the results of the 1925 game not been available, it might have seemed that Reeves would have been motivated by the need to concentrate large numbers of torpedo bombers against initial targets in the enemy's battle line. In game after game from 1923 on, the question was whether the enemy's flight decks should be the initial target.

[32] Thomas C. Hone, Norman Friedman, and Mark D. Mandeles, *American and British Aircraft Carrier Development 1919–1941* (Annapolis, MD: U.S. Naval Institute Press, 1999), 40.

[33] Reeves's new technique is described in the annual report of Aircraft Squadrons, Battle Fleet for 4 September 1926 through 30 June 1927. The Naval War College archives contain a two-page excerpt from the *Weekly News Letter of the U.S. Battle Fleet* for 29 August–4 September 1926 reporting that nine aircraft landed on in an average of one minute and 16 seconds each; "it is now possible with a squadron trained and flying frequently to fly a squadron on and off at intervals not exceeding one minute thirty seconds." This item is coded XMAR and UANOP/1926-136 in NWC RG 8.

[34] These figures were attained when *Langley* was fitted with both transverse and longitudinal arresting wires. The latter were intended to keep an airplane from slipping to one side (the first British airplane to land on a carrier was lost on a second attempt when it was blown over the ship's side). The longitudinal wires were eliminated in 1929, considerably reducing the landing interval. When he discussed aviation at Newport on 2 August 1929, Captain John Halligan Jr., commanding officer of the carrier *Saratoga*, said that once these wires were gone he expected the landing interval to be reduced to 40 seconds. He had recently carried out a dramatic air attack on the Panama Canal as part of Fleet Problem IX. NWC RG 8 Guest Lectures file, Halligan folder, Lecture File (53). For this operation, *Saratoga* had 110 aircraft onboard, more than 90 of them assembled (she was carrying *Langley* aircraft as well as her own). Halligan said that her flight deck could accommodate more than a hundred aircraft ready for takeoff, including a squadron of heavy bombers. No foreign carrier had anything like this capacity.

[35] The author remembers the late senior British naval constructor D. K. Brown asking him how American carriers with smaller hangars than their British counterparts could operate so many aircraft (he sent a letter tabulating relative hangar areas). The British and the Japanese were both forced into using multiple hangars (stacked one atop the other) on carriers, with all the attendant operational problems of complicated elevators. The British became aware of U.S. practices during the 1930s, but by that time the Royal Navy found attempts to increase the rated number of aircraft per carrier blocked by the Royal Air Force. When he designed the two-hangar *Ark Royal* in 1934, the director of naval construction pointed out that, using U.S. practices, she could have stowed all her aircraft on deck, considerably simplifying the design. *Ark Royal* was fitted with a barrier, but she did not employ U.S. deck-parking practices. They were introduced in other British carriers during World War II. The Japanese never employed deck parks, for all that they claimed they were following U.S. examples in carrier operation.

[36] Reeves was still commander of Air Squadrons, Battle Fleet. In his August 1929 address to the Naval War College, Captain Halligan (commanding officer of *Saratoga*) quoted Reeves "whose work can truly be described as brilliant:" *Saratoga* had been unopposed even though she had had to approach the target much more closely than she would have in wartime. In war, she would have launched 220 miles from her objective, and thus probably would have been unopposed and undetected due to her high speed. In the exercise, *Saratoga* was escorted only by the cruiser *Omaha*. That suggested to Reeves that attack groups should be organized around a carrier, cruisers (armed with 8-inch guns), and destroyers. They could avoid detection thanks to their speed, and they could punch through an enemy's scouting forces to "arrive at a point from which [they] could launch an air attack which could be stopped only with the greatest difficulty." Reeves pointed out that such employment demanded more U.S. cruisers, a category in which the U.S. Navy was deficient.

[37] Flight-deck cruisers vanished from war games after Admiral Laning left the War College in 1933 (which coincided with the departure of their great sponsors, Admiral Pratt as CNO and Admiral Moffett at BuAer

[due to his death]). Probably, Pratt's departure in June 1933 was the key event. The concept survived as late as 1940.

[38] Memo from Admiral Mark L. Bristol, who headed the General Board, dated 20 July 1931, in General Board file 420-7 on characteristics of the new carrier.

[39] The War College letter dated 30 July 1932 is in RG 80 CNO/SecNav file (SC) CV; it replies to a CNO letter dated 12 May 1932 enclosing General Board letter 420-11(SC) CV of 7 May 1932. The Naval War College letter refers further to an August 1931 War College letter NC-3(1930-168) discussing aircraft characteristics. A General Board (SC) CV 420-11 letter dated 20 July 1932, which is in the SecNav/CNO CV file (classified correspondence), connects the War College letter with a proposed alteration which would give *Lexington* and *Saratoga* greater bomb and torpedo stowage. The War College had failed to answer specific questions about how much such stowage was needed, and the question was urgent because the modification was to be included in the coming year's budget. The questions were also circulated within the fleet. The surviving 420-11 file includes correspondence on bomb and torpedo stowage dated July 1932, beginning with the statement that stowage in *Lexington* and *Saratoga* was clearly inadequate—a conclusion that could certainly be drawn from the War College letter, but which the ships' commanding officers had reached the previous year after the Fleet Problem.

[40] The War College letter went on to describe lessons of Blue-Red games after observing that "the chances of war appear most remote between Great Britain and the United States." Flight-deck cruisers and cruiser aircraft had proved very valuable in scouting along distant British trade routes. In games against a Red force trying to establish a base near Nova Scotia or in the Caribbean, Blue carriers with both shore and carrier-based aircraft were valuable in reducing or destroying the Red air force, "upon which so much depends for Red's service of information. Just as in the Blue-Orange problems, a great need has been shown for aircraft forces in strategical operations in addition to the demands of the major engagement."

[41] The games showed that one side or the other generally gained control of the air over the engaged battle lines and thus denied air spotting (i.e., long-range fire control) to the other; but the games also showed that ships had to close to less than 20,000 yards to secure decisive results before they exhausted their ammunition. At such ranges, air spotting was not needed. In some cases battleship spotting aircraft were launched prematurely, either due to a misunderstanding of the situation or a deliberate enemy feint intended to weaken Blue spotting and long-range fire. Without flight decks nearby, the spotters could not refuel and be serviced; the battleships certainly could not stop to recover them. They had to be sacrificed. "On the other hand, it has been shown that, with our spotting planes on carriers with flight decks damaged *before* an engagement, the battle line opens fire under a severe gunnery handicap at extreme ranges." War game rules offered considerable incentive to open at favorable ranges defined by the relative vulnerabilities of the two battle lines, a factor always included in tabulations of relative capability. A 1937 analysis cited elsewhere in this book emphasized the unreality of this "range band theory," because it was based on very incomplete information and ignored many significant combat factors.

[42] Although in principle a technical bureau like the Bureau of Ordnance, at this time BuAer was also responsible for pilot training and for air tactics (the relevant arm OPNAV arm, Op-05, was not created until 1943). Hence its more strategic view of carrier operations, which seems to have taken into account typical war plans affected by War College gaming.

[43] These comments were made in connection with a newly proposed 25,000 ton cruiser-carrier armed with 8-inch guns. BuAer pointed out that within the available tonnage only two could be built, costing one flight deck and aircraft capacity. Two *Lexington*s and five *Ranger*s would have supported 635 aircraft; the new distribution offered 553. With two 25,000-ton carriers in place of the three now planned, the number would fall again, to 429; and instead of seven carriers there would be only five. The fleet air force would be reduced by a third compared to the original plan. The current London Naval Treaty of 1930 allowed for construction of as many as eight flight-deck cruisers.

[44] *Wasp* had an armor deck but no torpedo protection; she was designed for later addition of side armor, if and when the treaty limit was abandoned. That was not done. She was slower than the *Yorktowns*, but faster than *Ranger*. Given the fixed total carrier tonnage available, weight growth in *Ranger* while she was being built

squeezed down the tonnage available for *Wasp*. *Wasp* was not included in the FY34 program, but rather in FY35 (she was launched only in 1939). Another year's delay would have carried her clear of the treaty limit on total U.S. carrier tonnage.

[45] 27 December 1927 memo by Admiral Reeves in the CNO/SecNav classified files (NARA RG 80), Folder CV 1926–31.

[46] *Essex*-class machinery was designed to sustain 20 knots astern to support landings over the bow, into the forward arresting gear. It is not clear whether the *Yorktowns* had a similar capability. The hangar-deck catapult was eliminated when additional anti-aircraft guns were installed blocking its opening, but before that, it was onboard several *Essex*-class carriers. Pre-war game rules eventually counted the usual two flight deck catapults, but not any hangar deck installation. The *Yorktown*-class hangar was partly open so that aircraft could be fueled in it. Partial closure was considered desirable to deal with high wind and foul weather.

[47] The C&R letter referred to Comairbatfor letter S1-1/FF2-3(1775) of 5 May 1932. Unfortunately, this letter has not survived in the CNO/SecNavy correspondence file in RG 80, folder CV 1932. The detailed letters from Air Squadrons Battle Force that have survived from this period do not mention survivability.

[48] BuAer seems not to have appreciated the implications of quick repair, because in a 30 July 1934 note to the Secretary of the Navy (Aer-E-34-SCP CV5&6/S12 in CNO/SecNavy classified files in NARA RG 80 File CV July–Dec 1934) on a proposed further lengthening of flight decks, it referred to the need to be able to operate with part of the flight deck shot away (paragraph 6), a possibility that explained the need for two-way arresting gear and the ability to steam astern at high speed (19.6 knots). "Under this damage condition the length of the remaining intact deck is a measure of the carrier's residual fighting power which further emphasizes the importance attached to obtaining the maximum possible flight deck length...." The General Board approved an additional 65 feet, but no more because additional length added weight, which had to be subtracted from the third, smaller, carrier (*Wasp*).

[49] The design of the British "armored deck" carriers reflected the utterly different British operating practice. These ships were conceived when the British faced the prospect of attack from land bases in the Mediterranean and in the North Sea. Up to that point, they had had the same idea as the U.S. Navy: They could gain air superiority by hitting the enemy's carriers first. Unfortunately, as the U.S. Navy already well understood, land bases could not easily be neutralized by air attack. The British idea was that the hangar could be armored so that a ship's aircraft could survive an air attack inside it, then emerge to operate (the sides of the hangar were armored against cruiser fire). The ends of the flight deck were not armored. The cost of such protection in terms of numbers of aircraft was high: On about the same tonnage, *Illustrious* had only about half as many aircraft as the unarmored-deck *Ark Royal* although she displaced slightly more. Armor also made for a shorter flight deck, which during World War II turned out to be more dangerous when the ship operated high-performance U.S. aircraft. Moreover, despite the considerable ingenuity involved in the British design, it could withstand only the 500-pound bombs contemporary British dive bombers could deliver. The armor was defeated by German Stuka dive bombers dropping much heavier bombs. It would also have been defeated by contemporary U.S. dive bombers, which normally carried 1,000-pound bombs.

[50] Under the last of the inter-war treaties (the London Naval Treaty of 1936), the allowable tonnage of individual carriers was reduced from 27,000 to 23,000 tons. The design displacement of the *Essex* was 27,500 tons. This provided enough weight to add an additional armored deck on the hangar deck. It protected the ship's hull from the effect of a bomb passing through the flight deck (the *Yorktowns* had only a single armor deck, over their machinery). This arrangement was chosen in preference to a proposal to armor the flight deck—not so much to protect the flight deck as to protect the ship underneath it. A June 1939 study of a modified *Yorktown* showed that flight-deck armor would have weighed 1,460 tons, reducing aircraft complement by at least two thirds to about 25. Even then, numerous openings in the deck, for example for elevators, would compromise any protection it offered. An increase in the thickness of the protective deck deep in the ship to resist a 1,000-pound bomb dropped from 10,000 feet was considered more practicable; it would cost 600 tons. The next step, once war broke out and tonnage limits were abandoned, was the two-deck solution.

[51] General Board 420-7 of that date, in NARA RG 80 General Board files.

[52] *Wasp* was delayed to FY35. Construction of the two *Yorktowns* was partly funded under the National Industrial Recovery Act rather than the usual Navy appropriation.

[53] General Board documents list *Hornet* as a single carrier in the second 1939 Deficiency Bill, and in the spring of 1938 this ship was described as delayed. *Hornet* was actually ordered on 30 March 1939, the *Yorktown* design being adopted to avoid delay.

[54] GB 420-2 Serial 1803. In an accompanying paper (GB 420-2 Serial 1790), written at about the same time, the General Board pointed out that the two *Yorktowns* and *Wasp* were about to enter service, offering the possibility of comparison between different carrier sizes and also "opportunity for better determination of the requisite fleet carrier strength. The Board feels that this experience and information should be at hand before determination of future carrier building, and believes it of sufficient value to warrant omitting carriers from the 1940 program. The Board at present contemplates including two 20,000 ton carriers in the 1941 program."

[55] General Board study of the FY40 program dated 25 April 1938, responding to the Secretary's request for guidance, 5 January 1938. There was certainly a prospect that Navy fighters were about to become fully competitive with land-based aircraft. The Grumman F4F Wildcat prototype had already flown and the Corsair, for a time the fastest fighter in the world, was about to win the 1938 fighter design competition. Basic policy was to maintain the force ratios embodied in the earlier treaties, presumably because the 5:3 ratio over Japan had strategic significance. War gaming showed, however, that the ratio that mattered was the ratio between total Japanese and American air arms, not the numbers of carriers. At this time, Japan was believed to have at least five, but possibly more, carriers. That was correct: Japan had the huge *Akagi* and *Kaga*, the small prototype *Hosho*, the small *Ryujo*, and the recently completed *Soryu* (a near-sister was under construction). It was apparently not known that the Japanese had built auxiliaries and subsidized merchant ships specifically for wartime conversion into carriers.

[56] Enclosure to GB 420-2 dated 5 April 1938 based on the Vinson-Trammell Act and the 1938 supplementary Vinson-Trammell Act; it included 4 rather than 2 carriers because some would be over-age (hence replaceable) during the ten years. The plan was based almost entirely on shipyard capacity. The Bureau of Aeronautics was not represented; the plan was submitted by Construction and Repair, Engineering, and Ordnance. A 6 April 1938 cover letter by the Secretary of the Navy approved the program as a basis for budgetary planning and particularly the procurement of shipbuilding facilities. In this sense, battleships and carriers were almost interchangeable. An attached memo from the General Board states that "in view of the arising questioned value of aircraft carriers relative to increasing improvement in characteristics of land planes, it is believed that if more carriers are desired they should be built as soon as possible," hence in FY41. The four other carriers replaced *Lexington* and *Saratoga* in tonnage terms. No tactical rationale was given. A General Board memo on the program was dated 25 July 1938 (GB 420-2 Serial 1802). It listed the 1939 carrier (*Hornet*) only because construction had been authorized; "the Board is not in favor of constructing this carrier at this time, and has stated in previous correspondence that it contemplates including a carrier in the 1941 program. If a carrier is to be built at this time, the Board believes that it should be …a repeat design of the *Enterprise-Yorktown* as it appears extremely doubtful that an efficient all-purpose carrier can be constructed on less than approximately 20,000 tons." This would seem to undercut the roughly contemporary statement that carrier construction was being suspended to await experience with two sizes of carrier.

[57] GB 420-2 Serial 1860 dated 30 June 1939. *Yorktown* displaced about 20,000 tons. This paper gave crude initial characteristics, including armament (not aircraft capacity, however) and speed (35 knots).

[58] GB 420-2 file 1939. The supporting analysis is in a paper signed by R.S. Crenshaw, dated 25 September 1939, which refers to earlier "Two-Ocean Navy" programs. Numbers were based on the number of battleships in the two oceans needed to provide at the minimum a parity of *battle-line* strength *at any time* in the area of operation [emphasis in the original]. Numbers of other types were set by the need to support the battle line and then by the need to carry out other necessary operations. It was assumed that Japan would ultimately have 12 modern battleships. Germany and Italy together might have 16, although they would maintain only 8 active ones. That gave a requirement for 16 battleships in the Atlantic and 20 in the Pacific, in each case to maintain a 5:3 ratio—a total of 36 U.S. battleships. The corresponding figure for carriers

was 18. Note that the initial "Two-Ocean Navy" paper in this package was dated 2 May 1939, before the outbreak of war, but after it must have been obvious that something was about to happen. It set out various ratios of superiority which would provide security in various parts of the Atlantic and Pacific.

5. The War College and Cruisers

[1] The General Board argued that cruisers were, if anything, more important now that battleship numbers had been cut drastically by the Washington Treaty. Battleships were now too valuable to risk in many secondary roles, which in future would be filled by cruisers. This was not a bad prediction; during World War II, cruisers often acted, in effect, as junior battleships. Cruiser roles were varied enough that it was impossible to set appropriate numbers. Rather than any particular operational or tactical rationale, the General Board proposed as a policy that the 5:5:3 ratio in capital ships should apply to all other classes. Quite aside from its simplicity, this type of calculation was easiest to justify politically, as it did not invoke any particular naval scenario. During the Washington Conference, the General Board recommended that the lifetime of cruisers be set at 15 years, which made all existing U.S. cruisers over-age and hence due for replacement (the ten 1916 "scouts" were then under construction). The British adopted the same figure in 1925, and it was embodied in the 1930 London Naval Treaty. In a 7 April 1923 summary of its preferred future building program (General Board 420-2, no serial number given), the General Board gave the cruiser strength of the "strongest power" (Great Britain) and "the second strongest power" (Japan) and stated that 20 10,000-ton cruisers were a bare minimum for the United States. If more such ships were laid down abroad, the weakness of the U.S. Navy would worsen. In a follow-on submission dated 18 April 1924 (Serial 1205), the board called for the construction of 16 10,000-ton cruisers, given British and Japanese programs. Congress authorized eight ships in 1924, construction of which was to be spread over four years. On 3 April 1926 (420-2, Serial 1271), the General Board reported again on cruiser policy, pointing out that the United States currently had 11 large and 11 small obsolete cruisers. It did not point to any particular required number, but asked simply that they be replaced. In a 1927 submission, the General Board called for parity with the British; as of 1928, they would have a total of 333,380 tons of under-age cruisers, including those under construction. The U.S. cruiser figure was 95,000 tons (including the first two U.S. 10,000-ton cruisers), leaving a deficit of 238,380 tons—equivalent to 24 of the new cruisers. This seems to have been the origin of the call for 24 cruisers, including 6 left over from the 1924 authorization. Similar reasoning applied to Japan led to a call for 22 new cruisers. These figures were produced as part of an attempt to frame a 20-year building program. The General Board file includes a 27 September 1926 letter from the president of the War College (Admiral Pratt) to Rear Admiral A. T. Long of the General Board, but it is concerned mainly with the advantages of describing the program as replacements rather than as increases in the Navy. That led to the creation of a list of the old cruisers, which would be replaced by new ones. The issue, as stated by the college, was whether the obsolete ships should be retained for their value in wartime or retained until replaced, or scrapped to clear the list of obsolete tonnage. The old cruisers might retain value as convoy escorts and as flagships for base forces. In an 11 October 1926 memo for the president of the Naval War College, Captain H. E. Yarnell pointed out that the old cruisers were too vulnerable to underwater damage to be effective escorts in wartime, and that to retain them as possible flagships of base or subsidiary forces was hardly justified. They might have some value as arguments for their replacement, if legislators could be informed as to their condition and their limited remaining value (otherwise "to many people, including Congressmen, a ship is a ship regardless of age)." Pratt doubted that they would count in a future armament conference, since he thought that tonnage ratios would be determined more by the needs of the countries involved than by anything else. Pratt wrote across the bottom that the memo had been "discussed with the Staff and with the President [himself] and is agreed to." A 27 September 1927 General Board submission to the Secretary of the Navy (General Board 420-2 Serial 1358) recommended a five-year program whose size was based on that of the British cruiser force, to provide 13 8-inch gun cruisers by 1932 with another 15 building, and a total of 33 such cruisers by 1936. This paper responded to a 20 September 1927 request by Secretary Curtis D. Wilbur for a five-year program based on the 20-year program he had already approved. The only reference in this file to war gaming is a statement, in a December 1927 version of the program, that "every war game, whether played at the War College or carried out in practice on the high seas [i.e., in Fleet Problems], emphasizes the need for an increased number of vessels of the cruiser type. The number called for is the result of the experiences of war

and annual post-war combined fleet exercises and represents the conclusions of commanders-in-chief afloat and the General Board." However, it seems clear from the other General Board papers that the figures used were based on the political argument of parity with the British (and the 5:3 ratio against the Japanese). They were what the General Board thought Congress might be convinced to provide, not the larger number the fleet needed. Military considerations entered in the choice that only the largest (10,000-ton) cruisers were acceptable. In January 1929, when the 15-cruiser bill (HR 11526) was going through Congress, the General Board attacked a proposed amendment that would have cut it to five cruisers ("inadequate even for training") and would have reduced the ships to 7,000 tons (which would have precluded mounting 8-inch guns). It characterized the latter as "a step backward of ten years" leaving a ship "inferior in some respects to the converted merchant vessel." This was the 4th Endorsement of GB 420-2 Serial 1400. The proposed amendment also called for the President to convene a further naval arms conference to be held in Washington on or before 1 January 1930. The 15-cruiser act approved on 13 February 1929 called for construction of five cruisers in each of FY29–31.

[2] They seem to have kept changing the terms of any agreement, so that no British offer was ever acceptable. Prior to the Geneva Conference, the U.S. Navy's General Board reviewed U.S. naval policy on the basis of parity with the Royal Navy; it called for construction of a total of 24 8-inch cruisers. As a first step, the General Board wanted Congress to provide money to build the eight cruisers authorized in 1924. For the role of the U.S. officers at Geneva, see Stephen Roskill, *Naval Policy Between the Wars* (London: Collins, 1968), I, 505–13. The British were never able to extract an explanation of why the U.S. Navy wanted large cruisers, because it was politically impossible for the U.S. Navy to point out that it was designed specifically to fight Japan. Ironically, the Royal Navy was also being designed to fight Japan, but its representatives apparently could not tell the U.S. officers that. Because this point could not be discussed openly, the British were unable to offer a compromise that would allow the U.S. Navy its large cruisers. British internal naval papers emphasize the desperate need the British felt to cut the cost of naval construction. A British compilation of American press material gathered during the run-up to the 1930 conference (ADM 116/2686) reveals claims that the Geneva conference failed because of effective lobbying by William S. Shearer, who was hired by the major naval shipbuilders. President Hoover was portrayed as outraged and determined to overcome the "big Navy" lobby in the interest of peace (the articles include his proclamation of the Kellogg-Briand Pact, which "outlawed" war). In all of these articles, the British and not the Japanese figure as possible future maritime enemies, even though all contemporary internal U.S. naval papers emphasize the Japanese as the only likely future enemy.

[3] The one naval officer who was allowed to attend as a delegate, Admiral of the Fleet Sir John Jellicoe, did so as governor-general of New Zealand. He was the one who pointed out that under the terms of the treaty, the British would be accepting a reduction below the 70 cruisers he considered essential to safeguard the sea traffic of the empire (it is not clear how he reached this figure). Ultimately, the treaty called for a reduction to 50 by the time it expired in 1936, although some cruisers scheduled for scrapping were actually retained. In the case of cruisers, parity was difficult to define because individual cruisers differed so much in armament and in other qualities; the U.S. tried to create an appropriate "yardstick."

[4] Pratt was supported by a naval technical staff of twelve, including Rear Admiral Moffett, Rear Admiral J. R. P. Pringle, Rear Admiral H. E. Yarnell, and Rear Admiral A. J. Hepburn, as well as four captains. They included Captain A. H. Van Keuren of the Bureau of Construction and Repair, who produced sketch designs to test the practicality of various treaty proposals (copies of these designs survive). Pringle was president of the Naval War College. Moffett was the long-serving chief of the Bureau of Aeronautics. Hepburn was chief of staff of the U.S. Fleet. Yarnell was chief of the Bureau of Engineering (responsible for engines), but in 1928 he had commanded the carrier *Saratoga* during her surprise attack on the Panama Canal during the Fleet Problem, and in 1932 he would execute a similar attack by both fast carriers against Hawaii during Fleet Problem XIII (at that time he would be serving as CinC Asiatic Fleet).

[5] See, for example, Stephen Roskill, *Naval Policy Between the Wars* II (London: Collins, 1976), 67.

[6] General Board treaty papers (GB 438 series) include a discussion of "parity" in naval air power, GB 438-1 (Serial No. 1464). It was based on a proposal by Rear Admiral Moffett, chief of the Bureau of Aeronautics, that comparisons should be based on the flight-deck areas of the various fleets rather than on carrier

displacements. In this connection, he considered the huge *Lexington* and *Saratoga* poor bargains, as two carriers of half their displacement offered real advantages. Moffett proposed moving these two ships to the "experimental" category so that they could quickly be replaced by smaller units. The General Board pointed out that the United States had consistently maintained that tonnage was the best and most equitable basis for limitation. Moffet's comment presumably reflected the new carrier operating practices that Reeves had developed. The board added that allotting different nations specific carrier tonnages offered equality of opportunity, not outcome, in the way the tonnage was used. Looking forward to the London Conference (the General Board letter was dated 6 January 1930), the General Board added that if the conference sought to reduce total carrier tonnage below 135,000, the United States should insist on moving the two big carriers into the experimental (i.e., instantly replaceable) category—but not otherwise. The two carriers were acceptable only if enough tonnage was left over to buy a sufficient number of carriers. Moffett did not know (or preferred not to acknowledge) that the British equated carrier capacity to hangar size, so he produced a table crediting British carriers with far more aircraft than they ever carried—for example, the *Furious* and her two half-sisters were given 72 aircraft, whereas their rated capacities were 36 for *Furious* and 54 for *Courageous* and *Glorious*—and even that was considered extreme. His British figure totaled 348, compared to 216 (including 36 for the tiny *Langley*) for the United States. Because most of the British carriers fell under the "experimental," hence instantly replaceable, category, the British could in theory build seven 13,800-ton carriers compared to five for the United States, giving them a total of 669 versus 565 aircraft, a 20 percent advantage. Moffet's ideas had no impact on the initial U.S. treaty proposal, which was drafted by the State Department without reference to the General Board, but he certainly had an impact at the conference, where he was a member of the staff working with Admiral Pratt.

[7] The 6-inch gun was generally considered the largest whose shell, weighing about 100 pounds, could be handled by one man. Although specially built warships would have power ammunition handling, a converted merchant ship would not. In theory, a 6-inch gun would weigh about 42 percent as much as an 8-incher, and a mounting would be correspondingly lighter. Mountings would not be quite as much lighter, but the British designed a 7,000 ton 6-inch cruiser (*Leander* class) they considered satisfactory. It was generally assumed that the cost of a ship was proportional to her displacement. The treaty referred to 6.1- rather than 6-inch guns, because the French used that caliber. However, in the end the French rejected the treaty because it would have prevented them from building what they considered sufficient numbers of large destroyers (*contre-torpilleurs*) armed with 5.5-inch guns. These ships would have come under the cruiser category.

[8] Cruisers were assigned a lifetime; under the London Naval Treaty of 1930 they could not be replaced until they were over-age.

[9] As a dramatic gesture supporting the conference, both Hoover and McDonald suspended or cancelled new cruisers. Hoover was attacked for violating the law passed by Congress. Although McDonald's cancellations were not entirely reversed (three of the five British cruisers were built), Hoover found himself ordering all five of the projected FY29 ships (Cruisers 32 through 36). All 15 of the 1929 cruisers were eventually built.

[10] The 6-inch gun tonnage was adjusted during the conference, which ultimately allowed the United States to build three more 8-inch gun cruisers, but required it to delay laying them down. The initial U.S. position is described in a British memo dated 5 February 1930, in the collection of confidential conference memoranda in British files (ADM 116/2746).

[11] The General Board 420-2 file (naval policy, including building policy) includes a 28 May 1930 letter (coded Aer-GB CV) from Rear Admiral Moffet, chief of the Bureau of Aeronautics, pressing for more carriers. Moffett claimed that a U.S. proposal that landing decks be allowed on all combatant ships was defeated, leading to the proviso that landing decks could be placed only on a quarter of total cruiser tonnage (amounting to 80,850 tons of cruisers), provided that they were not designed primarily as carriers. Moffett also claimed that it was only with the greatest difficulty that an attempt to cut the carrier allowance (of which the United States had used only 66,000 of the 135,000 tons) was headed off. He expected further attempts at any future conference, hence wanted the Navy to build as many air-capable ships as possible before any further naval arms control conference could meet. That included five 13,800-ton carriers, to be laid down in 1931–34.

[12] General Board file 420-8, folder on flight-deck cruisers dated 1934 includes notes from Admiral Moffett's diary and papers dated 16 June 1933 and initialed G.W.D. On January 25, 1930, Moffet agreed to the tentative U.S. negotiating position except that he pressed for authority to use battleship and cruiser tonnage for aircraft. He furnished Secretary of the Navy Adams a copy of secret papers regarding aviation, "especially urging an option be obtained to use cruiser tonnage for aircraft." At a 30 January 1930 conference of delegates, he read a paper arguing that the United States was at a disadvantage because the huge *Lexington* and *Saratoga* had consumed so much of the allowed 135,000 tons of carriers; the British would have more flight decks. He opposed any reduction in overall carrier tonnage. He was willing to stop the construction of 10,000-ton carriers, but wanted five for the United States; the total 135,000-ton allowance was not enough. Moffett also pointed out that the British, but not the United States, had merchant ships suitable for conversion to carriers. A practical way out would be to allow transfer of tonnage from other categories to carriers. When a delegate suggested simply avoiding the question of carriers altogether, Moffett said that the conference was intended to extend parity (with the British) across the board, and that under current conditions the United States would not have parity. Moffett suggested that each country have the option to use cruiser tonnage, or at least some of it, for 10,000-ton carriers, with guns of 6.1-inch or smaller caliber. Another possibility would be to allow each nation to build carriers instead of battleships; Moffett "said that battleships are obsolete vessels, that neither the Congress nor the people would authorize building any more of them, but that they would authorize building of carriers and therefore authority to transfer battleship tonnage to carrier tonnage would result in more carriers." On 1 February, Moffett proposed that the right be obtained to transfer up to a sixth of allowed tonnage between the categories of carriers, surface craft, destroyers, and submarines, but this idea was rejected. On 3 February, he opposed the proposal to charge all carriers of 10,000 tons or less against cruiser tonnage: "neither the Delegation nor the advisors appreciated the value of aircraft except Admiral Yarnell, and that all hands were thinking in terms of surface ships. He said that the surface ships we had would last until the next war and then that everything would be aircraft." On 6 February, he remarked that the delegates seemed to have come to the conference determined to secure an agreement regardless of what sacrifices that might entail. The next day, Secretary Adams said that he was right: It was necessary to get an agreement—"we could not go back without one." Adams did say that he would prevent any cuts in the 135,000-ton carrier tonnage allowed in the Washington Treaty (British Prime Minister MacDonald had proposed to cut it to 100,000 tons, and there was also a British proposal to cut it to 80,000 tons). Moffett said that he had done all he could to get more carriers, but would have to be satisfied with holding the line at 135,000 tons. Adams also said during the conference that in future the United States might want to build carriers instead of cruisers. By 18 February, Moffett was frustrated. Not only the delegates, but also Admiral Pratt rejected all attempts to allow battleship or cruiser tonnage to be used instead for carriers, and the draft treaty also prohibited 10,000-ton carriers. "My only consolation is that I have done my best." Yet another possibility was to give the Japanese their desired larger (than 5:3) ratio in cruisers in exchange for more U.S. carriers. On 11 March, Pratt announced that he had recommended that aircraft carriers under 10,000 tons be charged to the cruiser category (he stated that all of the naval advisors had agreed to this before leaving Washington). The British and French agreed (the Japanese were doubtful). On 24 March, Moffett wrote in his diary that "aircraft will settle next war. Don't care how many surface vessels we build. The more the other nations build the less money they will have for aircraft." On 21 March, he had asked Captain van Keuren, the ship designer, to produce a sketch of a cruiser with 6-inch guns and a landing deck; the tentative design was finished the next day by Mr. Bates, the chief civilian ship designer. Moffett remarked that such a ship would violate the principle of purity of type and the principle that aircraft carriers should remain out of range, but "I think, however, we may meet the opposition to aircraft carriers and especially the argument that aircraft carriers must be accompanied by a cruiser. It might be useful in the transition period between aircraft carriers and cruisers." He told Admiral Pratt that such a ship might be produced outside the carrier tonnage—that it would help solve the problem of insufficient carrier tonnage. Secretary Adams was unenthusiastic, and Yarnell said that guns would interfere with carrier functions. Pratt was enthusiastic. On 1 April, Moffett was designated to return home, leaving Pratt to handle the possibility of transfer of tonnage. According to his diary, the U.S. delegates wanted to limit aviation at sea as much as possible. Pratt failed to get the 10,000-ton carrier but promised to push for the right to put flying decks on all types of ships should the United States wish to do so. Moffett said that it was not what he had wanted, but might be satisfactory. In a later note, Moffett wrote that the proposal had met with great opposition, leading to the compromise allowing 25 percent of cruisers to be fitted with flight decks.

[13] General Board (NARA RG 80) file 420-7, file 1925–1931, letter coded SC138-7.

[14] The Bureau of Aeronautics disagreed vehemently, in a 22 March 1926 letter in the same General Board file. It saw the Construction and Repair studies differently; according to the letter, it appeared that a fairly satisfactory ship could certainly be built, to carry 24 torpedo bomber/scouts or 64 fighters. This file includes the Bureau of Construction and Repair report on alternative carrier designs. *Langley* (12,700 tons) did not count in the U.S. total at this time because she was experimental.

[15] Letter from RADM C. S. Williams, writing as a member of the General Board, to the Senior Member of that board, 23 April 1925, in General Board 420-7 file 1926–1931. Williams was President of the War College at the time.

[16] Letter dated 19 September 1930, NWC RG 8 file UNC-1929–1935 (Box 37, folder 8).

[17] No such calculation appears anywhere in U.S. records; the point is that the stated demand for about 24 large cruisers was well below what might have been calculated.

[18] Letters quoted in the report on the analysis of torpedoes in cruisers and destroyers, file XTYB (1930-170). The letters themselves have not surfaced either at Newport or in the National Archives in Washington.

[19] GB 420-8 of 11 December 1930 in War College file 'Determination of Characteristics of Cruisers,' file XTYB 1930-157 and 1931-38 NWC RG 8, Box 113. The duties were (a) screening; (b) supporting and opposing attacks by light forces including aircraft; (c) scouting; (d) operating offensively on lines of communication; and (e) operating defensively on lines of communication. The ships should be evaluated in night action, day action, and low-visibility day action. This file includes a 25 November 1930 letter from Rear Admiral Rock (chief of C&R) to Admiral Laning (War College) enclosing a statement to the General Board, and commenting that the cruiser the Navy would ultimately want would be one that could "stand up and fight and not be in imminent danger of being put out of action in the first attack, due to absence of necessary protection."

[20] At this time, the new-construction program was CL 32–38, with the nature of CL-39 not yet established. CL-32, -34, -36, and -37 were immediately reordered to the new CL-38 design. The new design was so attractive that it was adopted for three of the five ships of the FY29 program (cruisers 32–36): the cruisers *New Orleans* (CA-32), *Astoria* (CA-34), and *Minneapolis* (CA-36) as well as the heavy cruisers of the FY30 program (CA 37-39). The other two FY29 ships, *Portland* (CA-33) and *Indianapolis* (CA-35) had been ordered from private yards, hence any radical design change would have been very expensive. Their designs were modified to considerably improve protection. In 1930–31, these changes had not yet been made, so the new design was referred to as CL- (or CA-) 38 (CL-37 was to be ordered from a private yard). Plans were drawn up to build CL-39 as a flight-deck cruiser, but that did not happen. In effect, the War College argued that the United States should abandon building 8-inch cruisers *assuming that the flight-deck cruiser was adopted*. The decision *not* to adopt the flight-deck cruiser thus guaranteed that the United States would build the three permitted 8-inch gun cruisers (CA-39, -44, and -45).

[21] The initial War College reply to the question, posed by the Secretary of the Navy, was dated 6 January 1931; it was filed with XTYB (1930-167). Six alternatives were suggested: 10,000 tons (12 to 15 guns, protected against 8-inch fire between 12,000 and 24,000 yards); 9,000 tons (12 guns, similar protection), 8,000 tons (ten guns, protection against 6-inch guns at the stated ranges), 7,000 tons (nine guns, 6-inch protection), 6,500 tons (nine guns, slightly better than 5-inch protection), and 6,000 tons (nine guns, 5-inch protection). All alternative ships were credited with the same speed (32.5 knots) and with the usual torpedo armament (six tubes), but with different numbers of guns and with different levels of protection. Radius at 15 knots was given as 10,000 nautical miles for the 8,000- to 10,000-ton ships, 9,500 nautical miles for the 7,000-ton ship, and 9,000 nautical miles for the two smallest ships.

[22] "Blue Cruiser Requirements," War College file XTYB/1931-53. The destroyer section included British destroyers, but only after the Japanese entries.

[23] It seemed unlikely that the Japanese would transfer tonnage from the destroyer to cruiser categories:"[O]n account of the peculiar geographical situation of Orange ... it would seem that destroyers and submarines—all smaller types—are more valuable to Orange in a defensive role than larger types."

[24] Cruisers were considered key to trade protection, and the only way to get even second-rate ones quickly would be to convert merchant ships. The standard for suitability was gross tonnage above 4,000 and speed above 15 knots. Japan had 26 such ships, compared to 83 in the U.S. merchant fleet. None of these ships was converted for cruiser duties during World War II, but converted cruisers figured in U.S. war plans. Note that submarines, which would have access to the "safe" trade routes, did not figure in this analysis.

[25] Alternatives were (1) 10,000 tons with 8-inch protection and 15 6-inch guns; (2) 9,600 tons with 12 6-inch guns and 8-inch protection; (3) 8,350 tons with 12 6-inch guns and 6-inch protection; (4) 7,175 tons with nine 6-inch guns and 6-inch protection; (5) 6,000 tons with nine 6-inch guns and 6-inch protection; (6) 6,000 tons with six 6-inch guns and 6-inch protection; numbers denote scheme numbers. These alternatives were defined by the General Board and based on design studies by the Bureau of Construction and Repair. Naval War College game rules made it possible to compare these alternatives and to conclude that (2) offered greater fighting strength as well as advantages, important in a long overseas campaign, of seaworthiness, sustained speed, and habitability.

[26] We now know that the Japanese considered and rejected a flight-deck cruiser as an alternative to building the carrier *Soryu*. The War College logic would not have predicted Japanese construction of any additional carriers.

[27] Two types were compared, one protected against 6-inch fire (24 aircraft) and one protected against 8-inch fire (12 aircraft). The latter was considered at least a match for an 8-inch gun cruiser (actually superior at some ranges). The ability of the flight-deck cruiser to cover large areas while scouting and the effect of dive bombing by its aircraft made it far superior to an 8-inch gun cruiser for scouting and also for offensive screening, commerce destruction, and convoy work. Its 6-inch guns would be most effective against destroyers in a formation.

[28] If the United States built 15 heavy cruisers, they would mount 137 8-inch guns against 104 Japanese. Of the U.S. cruisers, five would be adequately protected against 8-inch fire, the rest insufficiently protected; but none of the Japanese ships were so protected (eight were protected against 6-inch fire, four not at all). In a fight between these two groups, the United States would have an even chance of victory and probably a slight edge. However, in a fleet action "it is impossible to say positively that any number were adequate."

[29] The last of seven French 8-inch cruisers was *Algérie*, approved under the 1930 program; the Italians, their rivals, built their last 8-inch gun cruiser *Pola* under their 1930–31 program.

[30] It did not help that current policy was o use four 8-inch cruisers as flagships. Battleships were considered unsuitable because they were tied down in the battle line; a flagship should be smaller and faster, with adequate communications and space for an admiral and his staff. By 1939, perhaps because cruisers were needed so badly for other roles, the battleships were again flagships (of the fleet and the Battle Force). Heavy cruisers served as flagships of the Scouting Force and of the cruiser components of the Scouting Forceand a heavy cruiser was flagship of the Asiatic Fleet.

[31] There would be a zone of ranges at which the 6-inch gun cruiser could not return 8-inch fire. However, at long ranges (22,000 to 27,000 yards), 8-inch accuracy would be poor, particularly since the ships were heavy rollers, and inside 22,000 yards many existing heavy cruisers could be penetrated by 6-inch fire. The new heavily protected 8-inch gun vessels were a different proposition: They were immune to 6-inch fire outside 8,000 yards (outside 10,000 at a 90-degree target angle). Such a ship could keep a 6-inch gun cruiser under fire from 29,000 down to 8,000 yards before suffering penetrating hits.

[32] The requirements set by war against Red were mentioned only in passing. "We rest on firm ground in demanding parity with Red for it may be assumed that Red will maintain a safe margin of superiority over all others."

[33] The U.S. analyst assumed that the British would continue to build 6,500- ton cruisers (*Leanders*), which was certainly their intent as they emerged from the 1930 London Naval Conference. However, when the U.S. Navy and the Japanese decided to build large 6-inch cruisers, the British felt compelled to follow suit (theirs were the "Town" class). They also invoked an escalator clause to retain some cruisers due to scrapping, citing the worsening political situation. At the 1935 London Naval Conference, the British tried to avoid continued construction of unaffordable large cruisers by capping individual cruiser tonnage at 8,000, the result being their "Colony" class—which turned out to be ruinously limited in growth margin. The United States designed a class of 8,000-ton cruisers, but war broke out (and the treaty limits lapsed) before they could be laid down.

[34] After the London Conference, Admiral Pratt claimed credit for a clause that created an unlimited category of "sloops" displacing up to 2,000 tons and armed with guns of up to 6.1-inch caliber. In theory, they could take over the anti-raider role. The U.S. Navy built two of them, *Erie* and *Charleston*, armed with 6-inch guns. The French, who did not sign the London Treaty (but largely abided by it) laid down ten *Bougainville* class "colonial sloops" (of which they completed eight) armed with 5.5-inch guns. The British built less-powerful sloops, conceived as anti-submarine and later as anti-aircraft escorts, in quantity.

[35] Conclusions drawn in the War College paper were somewhat ambiguous. The optimum for a Blue-Orange war was given as four flight-deck cruisers, six large (9,600 tons) and six small cruisers (6,500 tons) plus the ten *Omaha*-class cruisers. For Blue-Red, it was four flight-deck cruisers, two large (9,600 tons) and 18 somewhat smaller (8,250 tons) cruisers, the *Omaha*s being discarded. In both cases construction of 8-inch gun cruisers ended at 15. A compromise scheme would be four flight-deck cruisers, nine large (9,600 tons) and 12 smaller (6,500 tons) cruisers. It seemed that although 6,500 tons might be too little, a good flagship could be built on 7,125 tons. Building four such flagships would reduce the 6,500-tonners to 12 and the 9,600- tonners to six.

[36] That is, the U.S. Navy, or at least the War College, expected the Japanese to emphasize the torpedo, both to wear down the U.S. fleet as it proceeded west through the Mandates, and in any fleet action. This was entirely without knowledge of the primacy the Japanese were assigning torpedoes, or of their work on the Type 93 ("Long Lance") oxygen-powered torpedo. At this time, the U.S. Navy was working on longer-range torpedoes for battle-line work. CinC U.S. Fleet recommended a torpedo with a range of 15,000 yards and called for high speed (the General Board wanted at least 35 knots) in destroyers and destroyer leaders. In August 1931, new settings gave torpedoes the desired long range at 27 knots (they could reach 7,000 yards at 42 knots). It had to be assumed that other navies were making similar improvements. The U.S. Navy had been interested in long-range destroyer torpedo attacks on battleships since before World War I, and its destroyers had the heaviest torpedo batteries in the world (12 tubes in four triple mountings) and a technique designed to allow them to fire all of their tubes in a single salvo (curved-ahead fire). The War College pointed out that the role of large long-range torpedo attacks had changed radically since World War I. At that time, fire-control systems required ships to maintain steady courses as they fired. Mass torpedo attacks could force battleships to maneuver and thus ruin their fire control solutions. However, modern U.S. and foreign fire-control systems could cancel out the effects of such maneuvers. To be effective, attacking destroyers had to ensure that their torpedoes actually reached the enemy battle line, disrupting it either by hitting or by forcing ships to maneuver so wildly that their fire-control systems could not cope.

[37] This was based on 5.75-inch belt armor, which the General Board later reduced to five inches.

[38] A slightly later War College account pointed out that World War I experience had shown the value of protection (Admiral Jellicoe cited the loss of his battlecruisers at Jutland as proof that firepower without protection might well be pointless). However, that had been obvious since 1916, and only in 1931 was the U.S. Navy suddenly interested in heavily protected cruisers. It could be argued that foreign navies were reaching the same conclusions, reflected in the redesign of the French and Japanese cruisers (the British cruisers that would have been built had the 1930 conference failed were also heavily protected). The new Japanese *Mogami*s were thought to be well-protected. Just why this was believed is not clear, as it was also claimed that the existing (and much larger) Japanese heavy cruisers were poorly protected.

[39] The tone of the War College memoranda suggests that the idea was very much C&R's own initiative, but that its chief thought he had support elsewhere in the Navy. At this time, the British were promoting the idea that battleships should be pruned back to 25,000 tons and 12-inch guns. The U.S. Navy rejected the idea, partly on the ground that small ships could not resist unlimited weapons such as bombs and torpedoes. Captain Thomas Withers made the interesting point that were cruisers to become the new capital ships, the United States would find itself with a battle line both much slower than the enemy's and badly outranged (the U.S. Navy would have 6-inch guns on the new cruisers C&R advocated, the enemy 8-inch guns). The situation would be hopeless. At present, the U.S. battle line was slightly slower than that of the probable enemy (presumably Japan), but it was not outranged. Moreover, powerful and fast light U.S. forces could overcome the enemy's slight speed advantage. If the battle lines consisted of cruisers, the enemy would be as fast as the U.S. light forces, and these would be compelled to remain within the support of the U.S. battle line, possibly even behind it. The 6-inch main battery was apparently a necessary trade-off for the sheer weight of armor C&R envisaged. A few of the Naval War College staff did accept the idea that cruiser speed could be reduced to some extent. However, any U.S. cruiser might have to escape battlecruisers. War College games had featured cruisers running down enemy carriers. In 1932, Orange was credited with one 28.5-knot carrier and Red had three 31-knot carriers. Fast liners suitable for conversion to carriers were being built. All of these factors made 29 knots the effective minimum cruiser speed. At this time, the fast-liner argument applied only to Japan, but it is not clear whether this was known.

[40] C&R offered a range of possible cruisers, of which Alternative A was the 10,000-ton 8-inch gun cruiser approved in June 1931 (CA-37–38). The bureau's preferred design was C, with a maximum speed of 27.5 knots; F had a speed of 29.7 knots.

[41] This might mean analysis requested by either the CNO or the General Board, or both. The result was "Analysis of Use of Torpedoes in Light Cruisers and Destroyers in Major Actions on the Game Board," NWC Archive, RG 8, Box 113, file XTYB (1930-170). The writer was Commander A. G. Kirk, USN, later a major World War II amphibious commander. This document begins with a compilation of staff views on cruiser characteristics. Its main content is the War College letter to CNO dated 2 April 1931, which refers to several earlier letters sent in January 1931.

[42] Commander Raymond Spruance, then on staff, examined the speed issue in a letter to the War College chief of staff. He thought that the game board, on which battles began with all forces in position, distorted the perception of the value of speed that seagoing officers understood. A cruiser should be fast enough to avoid heavier and more powerful ships and sufficiently protected to engage other cruisers and, with impunity, lesser ships. Spruance rejected the 6-inch gun 10,000-ton cruiser because he doubted that all that tonnage would offer much additional ability to stand up to capital ship gunfire, or to dive bombing, or to torpedoes. Numbers mattered, and many cruisers would be needed to screen a large convoy and its escort of heavy ships as it steamed across the Pacific. A cruiser had to have an ample speed margin over the capital ships with which she was working, or against which she was fighting. A 27.5knot cruiser would be good for about 25 knots at sea, and it was well known that the last two knots of a ship's speed were typically obtained only under perfect conditions and after some time had been spent working up to full power. A 25-knot cruiser would have no real margin over British 23-knot battleships and she would be slower than the 26-knot Japanese battlecruisers. Nor would she have a sufficient margin over 21-knot U.S. battleships. The projected U.S. destroyers (*Farragut* class) were designed for 35 knots, and should be able to keep station at 30; how could a 27.5-knot cruiser back them up?

[43] Some of the points raised by the War College were based on fleet experience, as gathered by the War College to support gaming, rather than on gaming itself. One was that because the 1.1-inch shell would explode only on contact, pilots would not know they were under fire (most shells would miss), hence would not be deterred. The existing 5-inch gun might be less effective, but the effect *on a pilot* of its bursts would be considerable. It might suffice to break up an attack. World War II experience confirmed this perception; the 1.1-inch gun had to be given tracer ammunition to make pilots aware they were under fire (the British had similar problems with their pom-pom). Captain van Auken, who was responsible for analysis of game lessons, pointed out that C&R was blithely sacrificing a known anti-aircraft weapon that might require 1.1-inch or .50-caliber guns to supplement it, when and if these latter guns were successfully developed. He was the only one of the staff who pointed out that the 5-inch guns were needed to deal with heavy bombers, submarines, and close-in destroyers under reduced visibility or at night.

Notes

[44] "Torpedoes on Light Cruisers," item XTYB/1931–32 in NWC RG 8, Box 113, folder 8, a memo from the President of the War College to the General Board, 16 January 1931.

[45] After the Battle of Midway, the cruiser *Mogami* suffered serious damage from explosion of her torpedoes. According to the War College paper, "only recently has the College learned that the war heads of unfired torpedoes are likely to be detonated."

[46] Post-game discussions of Operations III (1932) showed that the question was generally how effectively torpedoes could be delivered. After this game, for example, Rear Admiral Halligan, a student who had come from the carriers, said that there was no question that torpedoes were more effective than bombs as a way of slowing down enemy battleships, but the issue was how well they could be delivered. A destroyer captain admitted after the same game that after all his experience, he had not the slightest idea of how to make a successful destroyer attack, night or day, and he thought (incorrectly) that the British were thinking of abandoning destroyers and torpedo attacks as a battle tactic. Simulated night action showed how sudden contacts could lead to instant torpedo firing; cruisers in the Blue screen were sunk by Red destroyer torpedoes. This particular game suggested that night destroyer torpedo attacks would be ineffective, but it was clear that this conclusion might be nothing more than a consequence of the rules. For example, destroyers trying to penetrate a screen were unable to get through by evasion. The Research Department, which was responsible for updating the rules, concluded that they needed review, including comparison with the main known case of night destroyer attack, the night after Jutland, as well as experience in fleet maneuvers and in night anti-destroyer gun practices.

[47] Typical inter-war Japanese torpedo batteries onboard destroyers were three triple or two quadruple tubes. The first U.S. destroyers built during the inter-war period (*Farragut* class) had two quadruple tubes, but that increased to three such tubes in the *Mahan* class, and some ships had four quadruple tubes—the heaviest torpedo battery in the world. The British typically used two quadruple tubes. As for cruisers, like the Americans, the British removed torpedo tubes to gain weight for aircraft and anti-aircraft guns when they modernized some heavy cruisers, but they retained torpedo tubes in new cruisers.

[48] The games began with Operations III (November 1931 for the Class of 1932): NWC RG 4, Box 58, Folder 17. The Red Situation is in Folder 18.

[49] The game began at the highest level, in which a situation developed that forced the Joint Army and Navy Board to update the Red War Plan. That gave students a chance to consider national strategy (in the set-up there was no U.S. government guidance). It was assumed that the war would be ignited by a clash between Red and Blue economic interests; at the time, the favored explanation for World War I was that it had arisen out of the clash of British and French versus German economic interests. In much the same way, it was assumed that war might break out against Japan due to that country's need for an exclusive market in China. Neither assumption would probably be accepted now; the Japanese policy of ejecting Westerners from Asia seems in retrospect to have been a form of a much earlier kind of nationalism. The description of the European situation was typical of the description of European politics in war game set-ups until the outbreak of World War II. As the 1930s progressed, the set-ups often interpreted appeasement as preparation for a deal in Latin America at the expense of the United States. The set-up emphasized the mutual hatreds and suspicions left by the World War, complicated by economic problems (the Great Depression). "For Red, with a trying situation at home, and the influence of Purple [Russia] close at hand, the internal political condition is far from satisfactory. Gold [France] is the historical enemy of Red, though recently an ally, but the economic advance of Gold since the war, combined with the economic decline of Red, has led to estrangement. Gold fears that Red resents her recently acquired importance in the European picture. Black's [Germany] sympathy leans towards Red. Blaming Blue for keeping victory from her grasp in the World War, she would favor the defeat of Blue as leading to an adjustment of war debts and perhaps of territory. Maroon [Spain] and Portugal would line up with Red. The first is a rising industrial power, strongly nationalistic with a natural antagonism to Blue because of Blue immigration and Latin American policies. The defeat of Blue would give Maroon an opportunity for trade expansion and perhaps for imperialism. Portugal has always benefitted from Red's friendship and economic assistance and would unhesitatingly return past favors by benevolent neutrality. Spain (sic) also will favor Red and probably extend benevolent neutrality." The attitude of sovereign members of the Red Commonwealth was uncertain. Emerald (Ireland) would be sympathetic to Blue, but might be restrained by proximity to Red. Crimson (Can-

ada) would be deterred by the likelihood that she would be invaded and occupied, and would probably lose her independence. If she stayed out of the war, Blue would have to watch the border, and by her neutrality Crimson could benefit Red by freely supplying food and essential raw materials. Newfoundland was not then part of Canada, so it could provide Red with a base. The issue of a neutral Canada presaged an important Cold War question: what to do in a global war if Cuba remained neutral. War might allow Orange to seek a dominant position in the Far East, leaving Scarlet (Australia) in trouble, particularly if Red were so weakened as to be unable to protect her. In Ruby (India), Red already faced an independence movement; in the event of war Purple (Russia) might try to ignite an open rebellion. Fear of this situation might be a considerable factor in Red relations with Orange. Orange would find her opportunities far too good to ignore. She would particularly want Blue to be crippled, because if Blue emerged victorious she would gain so dominant a position that Orange expansionism would be stopped, at the least for a very long time. Orange expansionism would be opposed by Yellow (China) and Purple and Scarlet, but Yellow (China) and Purple were both too weak and disorganized to be effective. Blue faced hostility in Central and South America, albeit somewhat veiled; assistance would not be available. Venezuela in particular was sympathetic to Red and would probably assist her in establishing the Red fleet in the Caribbean. All of these ideas translated to statements of relative Red and Blue strengths and weaknesses, the standard prelude to decision making in a war game. Elements of Red strength included her "astute diplomacy" and her control (through her empire) of some of the principal sources of raw materials. The linkage between a European scenario and the Far East made this Red-Blue war much more complex than a Blue-Orange war; there was a real question of how much of the Blue fleet would have to be retained in the Pacific. It is not clear how seriously any of this should be taken; any observer would know how difficult it was to write a realistic scenario for a war with Great Britain. U.S. naval strategists knew as much.

[50] NWC RG 4, Box 57, Folder 33.

[51] To simulate poor visibility, only umpires and movers were permitted in the game room. Every effort was made to expedite action, moves being made rapidly without scoring gunfire. All involved thought that the game was a good simulation of a night destroyer attack.

[52] In all later games, bomb hits had less effect.

[53] NWC RG 4, Box 61. The corresponding Tactical III game is in NWC RG 4, Box 62, Folder 27.

[54] This was the last strategic game to use the *Akron* for scouting, as she was lost on 4 April 1933 with great loss of life, including Rear Admiral Moffett. Her sister *Macon*, which entered service in June 1933, figured in a 1935 strategic game (she was lost that year). U.S. Navy interest in such dirigibles persisted as late as 1940. In the 1934 game, Blue had two rigid airships (*Akron* and *Macon*—ZRs), each with five fighters onboard. They were assigned to the Bissagos Base Group, to warn of any approaching enemy forces. That effectively kept them out of the game. One sent to observe the Cape Verde Islands found nothing.

[55] The Research Department considered Red's situation analogous to that of British Admiral Torrington in 1690. He was in the Channel facing a superior French fleet, and an important British convoy was homeward bound from Gibraltar. The enemy had landed in Ireland, where a British army was fighting. To safeguard communications with the army in Ireland, a detachment of the British fleet was placed in the Irish Sea. Torrington chose to hover near the enemy on the defensive, fighting only when he had a fair chance of winning. He considered that if he lost he would lose sea control permanently; he was powerful enough that the French could not chance an attack on him. At Beachy Head, Torrington refused to give the French the chance of a real decision by disengaging when the wind dropped. He retired slowly to the east, drawing the French to Dover and making it possible for the convoy to reach port. He was also drawing the French fleet away from the vital line to Ireland. The French gained almost nothing before they had to retire to Brest. Torrington was court-martialed but acquitted, and eventually his conduct was approved. In 1779, another inferior British fleet found itself in a similar position, and again the decision was to hover near the superior French force without accepting a decisive battle. On the basis of this experience, Red's choice should be to take the active defensive, ready to attack whenever presented with a local advantage, using his air and submarines offensively against the Blue battle line at once, but preserving his own battle line strength. He should also withdraw to pull Blue away from the critical Red trade routes (and toward reinforcements heading for Red). "The Red plan should be characterized by finesse, the Blue by determined attack and follow up."

[56] NWC RG 4, Box 66, Folder 3. The discussion at the critique is in NWC RG 4, Box 66, Folder 18.

[57] This balance reflected real carrier operating practices, as British carrier capacity had been published. Red had 54 observation aircraft on carriers (all U.S. observation aircraft were on battleship and cruiser catapults). Blue had 162 fighters versus 120 for Red, 144 versus 84 heavy dive bombers, 96 versus 8 scouts, 90 versus 84 torpedo bombers. Heavy dive bombers could deliver 1,000-pound bombs, the type used at Midway. Blue had two very large carriers (*Lexington* and *Saratoga*), four smaller more modern ones (like *Ranger*), and the small, slow *Langley*. *Langley* was assigned to the Train, the others to the Covering Force at a distance from the Train. Two carriers were each assigned to task groups: (i) Cape Verde Raiding Group, (ii) Carrier Scouting Group, and (iii) Striking Group. Those in the Cape Verde Raiding Group were to join the Striking Group (initially the two very large carriers) after completing operations in the Cape Verde Islands. That made four carriers assigned offensive roles. The Research Department critique noted that of these the carriers in the Cape Verde Raiding Group "did nothing which could not have been done by the cruisers' planes. They scouted where it was intended to make landings." During the early part of Blue's advance toward the Azores, the Scouting Group carriers maintained a patrol around the Blue Covering Group. Later, they were ordered to join the Covering Group, which thus included all six fast Blue carriers. This order aborted a planned attack by these two carriers on the Red battlecruisers. Red had eight carriers: four of about 22,500 tons (54 aircraft each), one of about 14,500 tons (30 aircraft), two of about 11,000 tons (36 aircraft each), and one of 11,000 tons (30 aircraft). The four large carriers corresponded to the three converted ships and, presumably, *Eagle* (which carried considerably fewer aircraft in reality). The 14,500-tonner was presumably *Argus*. One of the 11,000-tonners was presumably *Hermes*. The number of Red carriers made it possible to split them up: three in the Red Main Body, three in the Raiding Force, and two in the Control (shipping protection) force. In effect, the Main Body carriers were reserves. The Control Force carriers patrolled the trade routes between Red and Gibraltar. Red patrol planes out of Madeira found the Blue forces, and they were tracked by observation aircraft from the Red carriers. Red launched strikes against the Blue Covering Force and the Blue Main Body. After these strikes, the Red carriers were assigned to resist a large Blue air attack.

[58] A separate critique by the Research Department, dated May 1934, is in NWC RG 4, Box 66, Folder 20.

[59] A discussion at the critique brought out the War College's optimism regarding aircraft operations. The War College rules based scouting range on fuel capacity, but in practice it depended on how well pilots could navigate. Captain J. O. Richardson (later CinC of the Pacific Fleet) said that no cruiser plane had ever flown farther than 75 miles and returned by its own navigation, but someone pointed out that in Fleet Problem XIII, the carriers *Lexington* and *Saratoga* maintained a search pattern in all directions out to 125 miles in daylight, and recovered all their aircraft. These figures were considerably less than the search ranges used at the War College.

[60] There is a separate tabulation (January 1934) of 'Casualties from First Air Attacks.' (NWC RG 4, Box 66, Folder 17). The summary of "outstanding facts and conclusions" is in RG 4, Box 66 Folder 19.

[61] The General Board 420-8 file on Flight Deck Cruisers dated 20 February 1934 included as its first item a précis of General Board papers on the subject dated 26 January 1934. According to this paper, on 18 December 1930 CNO Pratt directed the War College to compare 8-inch and 6-inch flight-deck cruisers. The War College reported favorably on 2 January 1931, and on 16 January the Bureau of Construction and Repair provided the General Board with "spring styles" (preliminary sketches) of flight-deck cruisers as a guide to writing characteristics, the broad requirements reflecting a particular spring style, from which a design would be developed. A conference on the subject was convened on 19 January, attended by the Assistant Secretary of the Navy for Aviation, the chiefs of the relevant bureaus, and the director of War Plans. On 20 January, the General Board disseminated a paper comparing various types of cruisers and concluding that, following the lead of the War College, an experimental flight-deck cruiser should be built. Two days later, CNO commented on General Board characteristics for conventional cruisers. He wrote that the War College studies showed that a flight-deck cruiser would be a better all-round cruiser than any conventional one, armed with either 6- or 8-inch guns. CNO offered his idea of characteristics in a 22 February paper. On 28 January 1931, Secretary of the Navy Adams wrote that the present need was to develop plans for a flight-deck cruiser "since Congress has indicated its intention to authorize this type of ship, and it is doubtful if authorization bill will extend to pure 6-inch cruisers." Two days later, Adams formally approved the ten-

tative characteristics laid down by the General Board. At the end of June 1931, the General Board formally approved including one flight-deck cruiser in the FY33 program then being laid out.

[62] Contract plans were detailed enough for shipyards to use them as a basis for preparing bids that would be used to select a builder to receive a contract, hence the name (the only part not laid out in detail was the machinery spaces). This set may have been the only one produced between the world wars for a ship ultimately *not* contracted for.

[63] The 1934 General Board folder on flight-deck cruisers includes a paper written by Admiral Clark (General Board) dated 4 February 1931 (headed Office of Chief of Naval Operations, not General Board) for CNO but never signed or submitted it. It pointed to flaws in the War College studies supporting the flight-deck cruiser. The War College based its evaluation on the result of actual dive bombing with 500-pound bombs from 1,000 feet, assuming that one 500-pound bomb was equivalent to a 14-inch shell (an 8-inch cruiser shell was equated to 0.4–0.5 the value of a 14-inch shell). The War College also assumed that 12 aircraft could dive-bomb simultaneously. However, it was now generally accepted that aircraft would attack from 2,000 feet, halving the number of hits, and that attacks would be made in waves of three or two waves of six, giving a ship's anti-aircraft guns longer to fire. That would roughly halve the number of hits, so the overall effect would be reduced by three quarters. Moreover, the fighter-bombers envisaged as the equipment of the flight-deck cruisers had not yet dropped any 500-pound bombs in tests. The War College also assumed that with any bombers still onboard a flight-deck cruiser was superior to a conventional gun cruiser, because even if the flight-deck cruiser was sunk, the bombers would still be airborne and could still sink the gun cruiser. Clark also pointed out that Pratt was due to retire on 28 February 1933, hence could never be held responsible for the construction of any flight-deck cruisers in the FY33 or later programs. Only with the last cruiser design had U.S. 8-inch gun cruisers been properly armed. If the U.S. Navy did not build the last of the 18 heavy cruisers allowed under the London Treaty, the number with satisfactory armor would be reduced from five to two (at this time it was assumed that CA-32–36 would be only lightly armored, CA-37 being the first with satisfactory protection). Moreover, aside from the *Lexington* and *Saratoga*, future carriers would be unarmored (*Ranger* was still assumed to be the standard), hence would need heavy-cruiser escorts.

[64] In General Board 420-8 folder on flight deck cruisers dated 1934.

[65] Letter dated 26 February 1936 in CNO/SecNav classified files, NARA RG 80 Folder CC to CF (CF was the new designation for the CLV).

6. Downfall

[1] Hattendorf et al, *Sailors and Scholars*, 149. They quote Rear Admiral F. B. Upham, chief of the Bureau of Naval Personnel (which took over the War College), who argued that the War College was "primarily a technical school for the training and education of line officers," hence no different in principle than the Naval Academy and the Naval Postgraduate School. Upham had apparently made this case in a 6 August 1932 letter to CNO Admiral Pratt. The October 1934 action date is confirmed by the Navy directive issued at that time (information supplied by Curtis Utz, NHHC).

[2] Based on the reference in Hattendorf et al, *Sailors and Scholars*; they would almost certainly have found anything written somewhat later. It also seems clear that the remaining War College files give little sense of the laboratory role. Hattendorf and his colleagues were very thorough in their search of the War College archives, but they did not take it into account, particularly in the story of the College in the late 1930s. Most of the material used in this book to demonstrate the laboratory role is to be found outside the War College.

[3] This memo is coded NC3/XPOD(1934-34) E-ELM/(0). It is not in the classified SecNav/CNO file in NARA, but it is in the General Board 438 file. It is dated 27 February 1934. It seems to be the only War College paper in the General Board file on the London Naval Treaty of 1936, which is otherwise quite voluminous.

Notes

⁴ If the British insisted on a change, McNamee offered a counter-proposal: The huge British merchant fleet offered many candidates for conversion into wartime cruisers. He suggested that every three tons of merchant ships of over 10,000 tons and 20 knots' speed be equated to one ton of 6-inch gun cruiser tonnage.

⁵ In his *Naval Policy Between the Wars*, II, Stephen Roskill makes it clear that the British moved toward rearmament when the Japanese in North China treated British businessmen with contempt, making it clear their ultimate goal was to clear the Western powers from the Far East. The General Board was certainly aware that this was the Japanese goal; it was included in the report requested by the Secretary of the Navy. It was not in McNamee's paper, but a Japanese policy to end the Open Door policy in China was often used as a motivator for Japanese aggression in war games. By 1934, the British had abandoned the "Ten-Year Rule" (the concept that defense policy could be based on the assumption that there would be no war for ten years). The most obvious symbol of rearmament was that a committee formed to prepare a position to be used at the (abortive) League of Nations Disarmament Conference at Geneva (1932–33) was converted into the Defence Requirements Committee to determine how to fill holes in British defense.

⁶ The General Board treaty file (438) includes an extensive memo on the conversations with the British (GB 438-1 dated 30 July 1934).

⁷ The main remnant of the negotiations was acceptance of a Japanese proposal that each signatory (which in the end did not include Japan) notify the others of new construction long enough in advance to avoid surprises that might trigger an arms race. After 1936, there were attempts to convince the Japanese to sign the new treaty, the main provisions of which were limitations on the different classes of warships. In particular, the new treaty set the maximum battleship gun caliber at 14 inches unless a non-signatory (i.e., Japan) failed to agree. In that case, the maximum would be 16 inches. A further clause allowed for escalation in the size of battleships in the event that a non-signatory failed to agree to the 35,000-ton limit. Japan refused to agree to both provisions.

⁸ The General Board's file on the London Conference includes a list of acceptable U.S. positions. The need to defend the current 35,000-ton limit on battleships was emphasized. A British proposal to limit carriers to 22,000 rather than the previous 27,000 tons was considered acceptable. As for cruisers, "as a maximum concession and if necessary to obtain agreement, the Delegation may agree to a limitation on the number of 10,000 ton 6-inch cruisers for the United States and the British Empire and to an approximate 7,500 ton limit on additional 6-inch gun cruiser construction *unless an abnormal increase be proposed in the latter sub-category* [emphasis in original]." This was written before it became obvious that there would be no limits on total tonnage in any category. This paper was dated 4 October 1935; it was approved by the Secretary of the Navy on 14 October.

⁹ Kalbfus's comments are in "Building Program on Expiration of Washington and London Treaties," File UNC/1935-26 in NWC RG 8 Series 1, Box 37, folder 8. He was responding to a CNO circular letter sent on 1 April 1935 and to a 6 July 1935 second request, "as the Navy Department wishes to continue its study of this question." The original letter pointed out that there might no longer be restrictions as to types of ships, number and size of guns, and total tonnage either by categories or global. The two addressees were CinC U.S. Fleet and the president of the Naval War College. CNO Admiral Standley emphasized that the fact that such a study was being made "should not go beyond those officers who are required to assist in the preparation of the replies." The study was very secret because at this point those in London were still talking about ratios of tonnage between the major powers. Standley guessed correctly that the talks would collapse. He enclosed a detailed questionnaire. Questions included how many additional carriers should be built and also the minimum number of patrol planes that should be available at the outbreak of an Orange war, and their characteristics. All other combatants were included, and Standley also asked whether new types of ships were needed.

¹⁰ General Board 420-2 dated 18 March 1938. At this point, the program included four more carriers as replacements for *Saratoga* and *Lexington*, to be built under the FY45 and FY46 programs. The General Board was far more interested in battleships. General Board hostility to intensive carrier construction continued into World War II.

[11] The *Iowas* benefitted from an escalator clause in the 1936 treaty, activated when the Japanese definitely refused to be bound by its limits. No one was aware that the Japanese were about to build the huge *Yamato* class (about 62,000 tons) armed with 18.1-inch guns.

[12] When the bill was pending, it was simply a 20 percent increase, which would have meant an additional 27,000 tons. Including some tonnage still available, the United States could have built either one 30,000-ton carrier or two 15,000-tonners (but the 30,000-tonner would have been illegal). Presumably the figure actually adopted was intended to provide two more fleet carriers.

[13] The British, who could hardly be considered air-minded, had six fleet carriers on order in 1939. They were intended to supplement six existing fleet carriers. The Japanese ordered two large carriers (*Shokaku* and *Zuikaku*) under their 1937 program, at which time they had six carriers built or building. Another carrier (*Taiho*) was ordered under the 1939 program; however, the Japanese also had auxiliaries and subsidized liners intended specifically for conversion to carriers in wartime, a total of seven ships. Against that, in 1940 the U.S. Navy had six fleet carriers with another under construction and another (*Essex*) planned.

[14] The Germans used civilian aircraft as prototypes of bombers (to evade treaty restrictions), but there were entirely legitimate examples of conversions. In the late 1930s, the standard U.S. Army medium bomber was the Douglas B-18, a military version of the high-performance DC-2/3 airliner.

[15] General Board file 449 (aviation) includes a 30 October 1930 memo from Commander, Carrier Division One (Rear Admiral F. J. Horne), to Commander, Scouting Fleet, describing naval air missions, beginning with three long-range missions: strategic scouting, patrol, and long-range heavy bombing, all of which would be performed by large flying boats. Commander, Scouting Fleet agreed that large long-range flying boats were needed. Horne apparently had in mind something substantially larger than existing flying boats. A March 1931 endorsement by BuAer supported him, but cautioned that really large flying boats were too expensive to buy in any numbers.

[16] The 1933 competition is described in a history of pre-war seaplane development by Lee M. Pearson, the BuAer historian, in the November 1952 BuAer *Confidental Bulletin* (NHHC Aviation History Branch). Requirements included a 3,000-mile range at 100 miles per hour, a maximum speed of at least 150 miles per hour, a normal stall speed of 60 miles per hour, and a maximum weight of 25,000 pounds. BuAer chose Consolidated's design, which promised a maximum speed of 170.8 miles per hour and a range of 3,920 miles. The contemporary Army Air Corps Martin B-10A was credited with a maximum speed of 178 miles per hour (at sea level, more at altitude), and its range was far shorter. A summary of U.S. naval aircraft characteristics produced in 1933 (in the General Board 449 file, folder for 1931–34) shows that the main advantage enjoyed by the PBY (at that time designated P3Y-1) was its high speed: 162 miles per hour, compared to 134.7 for the P2Y-2, which had only slightly less range (2,900 versus 3,010 miles). The same technology of powerful air-cooled engines and lightweight structure made it possible for the Imperial Japanese Navy to develop land-based bombers capable of reaching Pearl Harbor from Japan, in the form of the G3M ("Nell").

[17] The new bomber designation applied on 21 May 1936 seems to have reflected a change in operating concept. It is sometimes suggested that the key difference was the advent of a precision bomb sight (the Norden bomb sight, developed by the Bureau of Ordnance). Bombing certainly mattered at the time. The General Board 449 folder for 1934 includes a 1 August 1934 BuAer report on the comparative values of small single-engine carrier-based aircraft and multi-engine (two- and four-engine) patrol planes. Its charts offer data on both scouting and bombing missions. Without bombs, a carrier scout bomber could scout out to 859 miles; a two-engine flying boat, which would cost three times as much, to 3,010 miles, and a four-engine flying boat (four times the cost) to 3,800 miles. Cost per scouting mile was somewhat less for the flying boats. For the maximum range (834 miles) of the carrier plane, the flying boats could carry more bombs: 5,500 pounds for the two-engine flying boat, 8,500 for the four-engine, versus 1,000 pounds for the carrier bomber. The flying boats offered considerably greater striking power, based on maximum bombing range and bomb load at three-quarters speed: 4,000 pounds of bombs for the four-engine flying boat at 2,250 miles' range; 2,000 pounds for the two-engine flying boat at 2,150 miles; and 1,000 pounds for the carrier bomber at 608 miles' range. The charts gave no hint of relative survivability in the face of enemy fighters. This comparison covered

only the scouting and long-range bombing missions. The 1935 General Board aviation folder (449) recounts the first test of torpedoes from a patrol plane, in this case a P2Y-2 (the report is dated 24 August 1935).

[18] Gassing was common in war games, but was not permitted when war actually broke out.

[19] However, there was interest in developing float-plane versions of standard Navy attack aircraft, which could be tender-based. Some of the pre-war tenders were designed with catapults to launch such aircraft, and they were tested (but never mass-produced or deployed).

[20] For example, the General Board aviation file (449) includes a 31 May 1938 estimate of overall patrol plane requirements prepared by Rear Admiral F. J. Horne. Horne did not include offensive operations by patrol seaplanes; he listed their roles as protection of friendly commerce, overhauling of enemy commerce near our coasts, detection of enemy raids, and strategic scouting. A war against the most likely enemy (Japan) would be preceded by a pre-war period of tension, during which the Japanese might deploy detached units into positions to launch air raids against important U.S. positions—not a bad forecast of Pearl Harbor. The only defense against such deployments would be searching seaplanes, which might be ordered to strike any enemy force they found. Numbers were defined by the likely character of a raid. The enemy carrier would probably run into position in darkness to launch from 250 miles away, two hours before dawn, so that its aircraft would appear over the target area at dawn, achieving surprise. As aircraft performance improved, the carrier could stand further off. The carrier would probably run in at 25 knots for 24 hours; she would have to be within 800 nautical miles of her objective at dawn the day before the attack. Anything that forced the carrier to launch earlier would be advantageous, since the attackers would have to arrive during darkness and so would probably be far less effective. The need to maintain patrols made it possible to estimate the number of seaplanes needed. Minimum seaplane range was 3,000 nautical miles, so that they could fly between U.S. operating bases. Because it was difficult to accelerate seaplane production, Horne wanted enough not only for initial patrols but also for operations from the closest of the Mandates, the Marshalls, which might be occupied 60 days after the outbreak of war. To maintain the desired force of 392 aircraft on M+60 would require a total of 592 seaplanes, which was more than the United States actually had in 1941.

[21] The two competitors were the Sikorsky PBS-1 and the Consolidated PB2Y-1, the latter being bought as the Coronado. Each had wing cells that could accommodate eight 1,000-pound bombs, compared to four for a PBY. The Consolidated airplane could carry another four bombs on wing racks. This load compared to a normal capacity of 4,000 pounds for the Army's standard heavy bomber, the B-17 (with another 4,000 pounds available on wing racks as overload). However, the B-17 was substantially faster (292 miles per hour at 25,000 feet for the B-17B compared to 255 miles per hour at 19,000 feet for a PB2Y-2). It also had a much higher service ceiling (36,000 versus 24,100 feet). The seaplane had much longer range: 4,275 miles maximum (no bombs) compared to 3,600 for the B-17 (no bombs). Seaplanes sacrificed considerable range if they carried heavy bomb loads: The PB2Y-2 was limited to 1,330 miles with its full 12,000 pounds of bombs. Later versions of the Coronado were more heavily loaded and sacrificed performance.

[22] Ray Wagner, *American Combat Planes of the 20th Century* (Reno: Jack Bacon, 2004) associates the new requirement with the advent of the R-3350 engine, which initially offered 2,000 horsepower (it was the engine adopted for the B-29). The competitors were the Consolidated PB3Y and the Martin PB2M, of which Consolidated asked for cancellation soon after its airplane was ordered (it was working on the B-36 at the time). In 1938, BuAer described these airplanes as a strong peacetime deterrent, with an expected endurance of 8,000 nautical miles. Horne's paper showed a range of 2,500 nautical miles at 140 knots with a bomb load of 40,000 pounds, or 7,800 nautical miles at 140 knots with 4,000 pounds of bombs. Maximum speed was given as 240 knots (276 miles per hour). If expectations were fulfilled, this airplane should be capable of carrying 20,000 pounds of bombs 3,800 nautical miles and returning. In wartime, such aircraft would be able to strike deep into enemy territory, and would present a constant threat to all enemy operations. Horne estimated that 48 could be "advantageously employed as distant overseas striking forces and as distant strategic scouting forces" (six per tender). By the time the PB2M prototype flew, its equipment was considered obsolete, and it was employed instead as a long-range transport designated JRM-1 Mars. Its significance here is as evidence of continued strong interest in long-range attack using tender-based seaplanes.

[23] This change is reflected in a revision of the "Naval Aeronautic Organization," copies of which (for various fiscal years) are in General Board 449 folder for 1937–39. A 4 September 1937 letter reorganized the Base Force patrol squadrons into five patrol wings, each a separate administrative command attached to its own aircraft tender. A further letter dated 24 September 1937 assigned the aircraft to the Scouting Force. A patrol wing consisted of two or three 12-airplane patrol squadrons (VP). At this time there were only two large tenders, *Wright* and the ex-carrier *Langley*, the other patrol wings using converted World War I minesweepers. At the least this organization required three more large tenders, of which *Curtiss* and *Albemarle* were included in pre-1939 programs. Beginning in 1938, obsolescent destroyers were converted into small seaplane tenders, in effect superseding the converted minesweepers (AVD, later reclassified as AVP). New small seaplane tenders (AVP) were ordered, as well as merchant ships for conversion into large seaplane tenders. The seaplane force expanded rapidly, to 20 squadrons in 1939 and 25 in 1941.

[24] The General Board file (420-8 of 1935–36) does not include the fighting power comparison, but it is mentioned in two of the enclosed papers (one of which, written in January 1935, thanks Admiral Kalbfus for his material) and seems to have been decisive. The opposing view was that the fleet needed at least some smaller cruisers for fleet work. The study itself, dated 10 January 1935, is in the Classified CNO/SecNav Correspondence file (RG 80) in NARA, Box 261, Folder NC (the War College was coded NC3 in the Navy filing system) together with a summary produced by the director of Fleet Training. War College Enclosure H indicated the best way to utilize a given cruiser tonnage; the director cautioned that "these academic studies represent an estimate of the average expectation for a large number of future engagements [hence] cannot be relied upon as assuring success in any one." The study covered ships displacing 8,000 to 10,000 tons, all with the same speed. The director of Fleet Training simplified the War College results somewhat. Fighting strength would be roughly proportional to some power of the ship's dimensions: Gun target would be proportional to the square and life to the cube. Number of guns and protection were taken as varying with the fourth power of the ship's dimensions (this factor did not appear in War College estimates). All of this led to an estimate that fighting power was proportional to the cube of a ship's displacement. On a fixed total displacement, the aggregate fighting power of a group of cruisers would be proportional to the square of the displacement of each of them, which gave the 10,000-ton cruiser a considerable edge over anything smaller. The War College reached that conclusion in a much more detailed way, including analysis of Orange cruisers. It pointed out that the best cruiser was roughly the type it had proposed in 1931.

[25] Letter from U.S. Fleet Cruisers, Scouting Force, 10 June 1936 in General Board 420-8 file, folder for 1935–36. A sore point at this time was that the torpedo tubes were supposed to have been removed from eight "soft" heavy cruisers in order to provide space and weight to double their anti-aircraft batteries, but that had not been done, and there was little prospect that it could be done within limits set by available money and equipment. The letter (and supporting letters) was therefore as much about possible re-installation of torpedoes in the eight cruisers as about future policy. This file does not include any rejoinder from the War College.

[26] In 1939, five of the small cruisers were still assigned to Cruisers, Battle Force, alongside eight large 6-inch cruisers. Another was flagship of Destroyers, Battle Force (and of one of its two destroyer flotillas); a second was flagship of the other flotilla. A third was flagship of Aircraft, Scouting Force, and a fourth was flagship of the Submarine Force of the U.S. Fleet. The remaining small cruiser was assigned to the Asiatic Fleet alongside a larger 8-inch gun cruiser serving as flagship (at that time *Augusta*).

[27] Studies of light cruisers (and gunboats) were requested by the General Board on 22 July 1936, shortly after the new London Naval Treaty came into force (General Board 420-8 Serial 1718-X, in File 420-8 for 1937-38). The addressees were the Chief of Naval Operations, the CinC U.S. Fleet, and the president of the Naval War College (who seems to have been the first to receive the General Board letter, stamped 22 October 1936). Because of the Vinson-Trammell Act, the letter was written as though the U.S. Navy was still bound by the tonnage limits of the old London Treaty of 1930 (as escalated, to match the British, in 1936), with no further cruiser replacements available before 1940. The cruisers of the FY38 and FY39 programs were projected under the Second Vinson Act of 1938, the 20 percent increase in allowed modern tonnage. The board, then headed by the same Rear Admiral F. B. Upham who had successfully taken the War College out from under OPNAV, asked how suitable the 10,000-ton cruisers were, and whether more should be built; the purposes and employment of other cruisers and gunboats; the types considered suitable; and the number

of each type needed. On 5 January 1937, the General Board formally asked for studies of cruisers ranging from 3,000 tons (as destroyer flotilla leaders) up, based on earlier design work. Unfortunately, the General Board file does not include the War College's reply to the General Board's request.

[28] The original letter from the Secretary of the Navy, dated 11 March 1938, simply asks that the Bureau of Construction and Repair "without delay" prepare a preliminary design for submission to the General Board. No justification is given. It is even possible that the project originated with the President; early in 1938, there was interest on his part in fast heavily armed ships. Thus, this project may actually share the same origin as the fast *Iowa*-class battleship. The Secretary's letter was copied to the various material bureaus, but *not* to the Naval War College.

[29] This paper concludes with a table of two possible designs for ships armed with nine 12-inch/50-caliber guns, dated 4 April 1938; the paper is marked "for file—Copy of Memorandum left with Board by Capt Chantry." A similar memorandum titled "Estimate of Situation: High Speed, Armored Cruisers" is the next in the file (and shares the language of the first). It is dated 19 April 1938, with the initials "A.P.F." These papers are in the General Board 420-8 file for 1937–38. This second memo provides more alternative sets of characteristics.

[30] At this time, President Franklin D. Roosevelt himself was apparently interested in building fast big-gun ships; he may have been ultimately responsible for the *Iowa* class. In other cases, files include explicit references to the President's views; the files on the super-cruiser do not. However, the Secretary's urgent request suggests presidential interest.

[31] Chantry's Class of 1936 played just such a game, Operations I (December 1935, in NWC RG 4, Box 75, Folder 11. Blue raided Orange shipping around a focal area off Singapore. The raiders included submarines. A follow-on version played in October 1936 by the Class of 1937 (in NWC RG 4, Box 80, Folder 2, and also in War Plans Division files) came closer to Chantry's ideas. As in the 1936 game, focal areas were the most profitable in which to operate. In this case, Orange made them too hot for Blue using land-based aircraft and submarines. Blue's raiding force consisted of heavy cruisers, converted cruisers, and submarines. Lessons of the game included the inability of submarines to attack commerce if they followed the usual rules of visit and search—no surprise, given World War I experience. By the time this game was played, Captain Chantry was back at the Bureau of Construction and Repair; it is not clear whether he was aware of the 1936 game. The success of Japanese submarines and aircraft in that game would have brought his rationale for Japanese construction of super-cruisers into question, particularly as it was believed at the time that the invasion of China was badly straining Japanese finances. The Blue raiding game was played again as Operations III of the Class of 1938 (November 1937), Operations III of the Class of 1939 (October 1938, and Operations III of the Class of 1940 (November 1939, in NWC RG 4, Box 91, Folder 22). No corresponding game seems to have been played by the Class of 1941.

[32] The NC (Navy Schools) file in the formerly classified papers of CNO/SecNav for 1927–39 (NARA RG 80) contains the 1935 War College paper on cruiser fighting power (as well as a 1932 paper on cruiser fighting capacity), but no later papers on warship-building policy. The 1937 papers are discussions of the charge that the game rules unrealistically favored certain fleet tactics (fighting in particular range bands). This is obviously negative evidence, and as such it can be questioned. There is, however, also no reference to War College input in General Board cruiser files (420-8) after 1935. Nor do War College files appear to include studies relevant to the super-cruiser idea.

[33] By that time, commanders afloat enthusiastically supported the big cruiser that was generally designated CA2 to distinguish it from a conventional heavy cruiser (CA1). It was not universally supported, however. In its review of the FY41 program (22 April 1940, in file 420-2 Serial 1944) the chairman of the General Board wrote the Secretary of the Navy that "the value of heavy cruisers of 26,000 tons has become increasingly questionable in view of the number of high-speed capital ships built and building by foreign powers; furthermore, from a building-time and cost standpoint, it is extremely doubtful whether their value to our Navy as compared to other types urgently needed would be sufficient to justify their construction at this time." The same memo complained that the projected heavy and light cruisers were not good enough. The big cruisers were included in some versions of the projected ten-year program, but they were clearly not

badly wanted. However, six of them were included in the big expansion program approved in July 1940 (the 70 percent expansion "Two Ocean Navy" Act). On 12 July 1940, CNO Admiral Stark circulated a memo listing ships as directed by the President, specifically including six of the big cruisers (most ships were types to be agreed upon later, and there were no specific figures for battleships and carriers). The big cruiser was ordered in preference to an oversized heavy cruiser armed with 8-inch guns. Any evaluation of choices made at the time is complicated by the more general reality that program planning proceeded in two stages. The first was an attempt to produce the needed numbers of cruisers as quickly as possible, which meant deriving any new designs from existing ones. With the London Naval Treaty dead, the redesigned ships could be somewhat larger. Thus, the wartime *Baltimore*-class heavy cruiser was derived from the last pre-war ship of that type (*Wichita*), and the standard wartime *Cleveland*-class light cruiser was an evolved *Brooklyn*. The only question was how many to build, and that may have been determined largely by yard capacity. At the same time, entirely new designs could be developed, such as the *Montana*-class battleships and the *Alaskas*. In March 1940, President Roosevelt pressed for construction of a cruiser-destroyer like the French super-destroyers, but he was brushed off. It seems interesting that although Roosevelt's naval aide asked how well destroyer-cruisers would fare against larger light cruisers, the War College was not asked for its views (using its fire-effect methodology). That was left to the Bureau of Ordnance.

[34] They were conceived as part of the Night Attack Force that was part of a fleet concept intended specifically to fight the decisive battle against the U.S. fleet. Although the War College never envisaged the Japanese fleet battle strategy (a night-attack force using torpedoes the night before a day battle), the Japanese plan to build ships for that battle was far more the sort of thing the War College would have envisioned than was Chantry's argument based on beating off U.S. surface raiders. The Japanese projected two of their super-cruisers, Hulls 795–796, as part of a program settled in the spring of 1941. Construction was postponed in November 1941, when war seemed imminent. Design work began in 1939.

[35] GB 420-8 file 1937–38, in a jacket marked "Medium Displacement Cruiser" 20 February 1939. Given the initial argument favoring large cruisers, the title is somewhat odd; however, the General Board ended up endorsing an 8,000-ton cruiser and avoiding further use of the escalation clause in the London Naval Treaty. The memo is dated 26 May 1938 with the initials A.P.E., and also dated 20 February 1939. It is marked for Admiral Reeves, who had been acting chairman for a time in 1936, but was now retired.

7. Conclusion: Games Versus Reality in the Pacific

[1] ADM 205/19, First Sea Lord papers 1939–1945. This is NID LC Report No. 18 dated 21 January 1942, "United States Naval Efficiency." It was "intended as a guide to Flag and Commanding Officers who may have to operate in conjunction with U.S. Navy forces." Document courtesy of Corbin Williamson.

[2] U.S. Naval Attaché Captain Russell Willson described British practices in a December 1937 report. He was the first foreigner allowed to attend the Tactical School. It used gaming to teach officers current standard tactics. After 1939, it was also used to develop tactics in response to requirements. However, earlier issues of the official *Progress in Tactics* mention experiments on the tactical board at the school. Willson thought the 11-week course would not seem strenuous to students at the U.S. Naval War College. "In general it may be said that the Tactical School is divorced from study of strategy, logistics, command, the broader principles of war, technical matters, and the formal estimate of the situation (called an 'appreciation' by the British). It seems that the course may be described as 'straight tactics' with particular reference to the employment of tactical units in accordance with the standard instructions.... On the game board the British forces are generally handled in accordance with standard instructions 'so the officer concerned will know what to expect when he gets out in the fleet'—but the enemy fleet are allowed a free hand 'because no one can tell what a Jap might do.'" According to Willson, students received extra points for initiative even if they were taking undue risks. In December 1937, Willson discussed Royal Navy tactical training at length with a former director of the Tactical School, who asked him whether U.S. tactical and strategic training were handled separately. Willson said that they were not, and gained the impression that the British were not entirely happy with the split between tactical and strategic training, and that they were considering establishment of a research

staff for tactics. RG 38 ONI Series (NARA), 21144 P-11-b, papers on British Tactics and Tactical Policy by Naval Attaché Captain Russell Willson, USN, report dated December 1937.

[3] The problem in the *Musashi* attack was that attacks by groups of aircraft from different carriers were not co-ordinated. The problem might be analogized to that of distributing fire properly among multiple battleships, something that certainly was understood by the inter-war Navy. It is possible that the issue never came up in the 1930s because it seemed so unlikely that more than one carrier would attack a target at a time. To some extent, however, it did arise at Midway, when three U.S. carriers were involved. In the 1944 strikes, pilots concentrated on *Musashi* because of her evident size and, apparently, because she had already been damaged. There was no strike coordinator who could distribute fire among the potential targets. The problem was unsuspected immediately after the attack because the pilots thought they had hit all the Japanese ships and had inflicted so much damage that the Japanese force had turned back. Evidence that it was going back toward Leyte Gulf apparently did not reach Admiral Halsey before he turned north. Games were not fought out in so detailed a way that such issues would have been evident; there were just not enough participants. The war games had completely failed to predict the pilots' over-optimism, and they had never raised the question of how to insure that a mass air attack properly distributed its fire. Fire distribution was a well understood issue in gunnery battle, but its air equivalent seems not to have been appreciated. The attack on the battleship *Yamato* in April 1945 was much better coordinated, probably as a result of the earlier experience.

[4] Papers describing the pre-war discussions with the British are in the files of the OPNAV War Plans Division, NARA II RG 38 Entry 355, Box 115, "Records Relating to Anglo-American-Dutch Cooperation 1938–1944." Item 1 is material on British-U.S. conversations in London, 1938–39. A draft "Estimate of the Situation" for a Blue versus Red and Orange war dated 1935 is in Box 53 (item 1).

[5] This was very much the view the British had taken during World War I. They found it difficult to provide sufficient destroyers to escort convoys (or to do other anti-submarine jobs) because without destroyers to screen it the main (Grand) fleet would be immobilized. It is often forgotten that in the absence of the British fleet, the Germans would have been free to use their surface forces to attack convoys. Surface ships were far more efficient than submarines in this task—if they could reach convoys. When two German cruisers attacked a Scandinavian convoy in November 1917, they wiped it out. A much larger German anti-convoy operation involving most of the German surface fleet, mounted in April 1918, failed only because poor German intelligence failed to discover that the convoy schedule had changed. These two episodes were well known to inter-war naval officers. They made the connection between a main fleet and trade protection clear.

[6] This was not an inevitable approach. The alternative would have been to see weapons not limited by treaty as an equalizer during the decisive battle. For example, the Japanese spent heavily on torpedoes and their tactics. A combined-arms approach to the decisive battle might have tipped the odds by using torpedo attacks to complement heavy-ship gunnery. In 1940–41, the Japanese chose instead to begin forming a torpedo-heavy night attack force that would attack the U.S. battleships on the eve of battle in a final attempt to even the odds by attrition. They seem to have envisaged the decisive battle itself as a gunnery fight. It should be added that Japanese naval aviators, who certainly did not dominate the Imperial Navy, thought that their aircraft already could and should replace all heavy-gun ships. The evidence is that at the outset of war they did not carry much weight in the service. Part of the evidence is the pre-war Japanese choice to invest enormous resources in the super-battleships *Yamato* and *Musashi* and two abortive sisters rather than in further carriers beyond the sisters *Shokaku* and *Zuikaku*. The Japanese did form the first large carrier task force, which struck Pearl Harbor. It seems likely that this concentration of naval air power was conceived primarily to attack fixed land targets rather than for fleet-on-fleet battle. Japanese tactics for fleet-on-fleet battle entailed dispersion for survival in the face of hostile carriers, but the carriers were to be concentrated to strike land targets such as Pearl Harbor.

[7] Alan D. Zimm, *Attack on Pearl Harbor: Strategy, Combat, Myths, Deceptions* (Havertown, PA: Casemate, 2011), 21–23. Zimm points out that Yamamoto's view was paradoxical. As chief of the Japanese equivalent of the U.S. Navy's BuAer, he had opposed the construction of the *Yamato*. However, in planning the Pearl Harbor attack he made battleships his principal targets, to the point that he was willing to abandon the project if it was impossible to use battleship-killing torpedoes in the relatively shallow water of Pearl Harbor. Even without those weapons, his dive bombers could have destroyed carriers present in the harbor. Zimm emphasizes the Japanese view that the attack would be extremely risky. He argues that the sense of risk was

partly the consequence of successful Japanese development of land-based torpedo bombers like the ones that sank the British *Prince of Wales* and *Repulse* in December 1941. In his view, the Japanese projected their own capabilities onto the substantial number of U.S. land-based bombers present in Hawaii in December 1941. Elsewhere in his book, he points to the likelihood that an alerted garrison at Pearl Harbor might have inflicted significant damage both on the Japanese attack force and on the Japanese carriers. This analysis is based partly on the Naval War College game rules then current. Zimm adds that the Pearl Harbor garrison had been on a practice alert until 6 December, when it was allowed to stand down.

[8] It is only fair to point out that the Japanese had an alternative to carrier aircraft: a substantial force of long-range, high-performance land-based naval bombers. In 1941, this was a unique asset. The U.S. Army Air Force had high-performance bombers capable of delivering torpedoes, but their crews did not practice such attacks and they were not of much concern to the air force leadership of the time. Only later in the war did the U.S. Navy acquire this land-based capability. At least in theory, the Japanese could shuttle their land-based bombers and covering aircraft to concentrate in threatened places in their chains of Mandate islands. In reality, it turned out that U.S. carriers could strike airfields in the Mandates before they could be reinforced in sufficient numbers. Japanese search capacity was not sufficient to support the shuttle tactic, which was vulnerable to surprise attack. In his book on Pearl Harbor, Zimm argues that the Japanese tended throughout the Pacific War not to have backup plans and thus not to evaluate their plans objectively. He attributes this lack to Japanese culture, in which failure was virtually unthinkable. The existence of a Plan B would have been admission that the basic plan might be flawed. As in other aspects of Japanese wartime behavior, however, it is also possible to attribute this behavior to something more concrete: a failure to adopt anything resembling the type of education in decision making the U.S. Naval War College provided. That education began with an estimate of enemy choices and included a test of how well that estimate had been made in the form of gaming.

[9] Halsey's choice not to detach his battleships was very controversial. When the escort carrier force called for help, Admiral Nimitz in Pearl Harbor asked where they were. Halsey later pointed out that for him the battleships were integral to his carriers, a point critics probably misunderstood because they did not understand that under some circumstances powerful Japanese surface ships might have endangered the valuable but fragile fast carriers. The situation was further complicated by Nimitz's orders to Halsey that emphasized the destruction of the surviving Japanese carriers, even if that cost considerable damage to the U.S. invasion force in Leyte Gulf. All of these issues arose in pre-war games. The orders to Halsey seem to have been forgotten in the heat of the battle. It did not help that Halsey's pilots, who had damaged the Japanese surface attack force mainly by sinking one of its battleships (*Musashi*), had been grossly over-optimistic in their evaluation of other damage they had inflicted.

[10] We now know that the aim of British World War I naval strategy, as it was understood in 1914, was to force the Germans to fight on the far side of the North Sea. The idea was that the Germans would not tolerate a blockade mounted by the British. The Royal Navy war planner drew an analogy with the Anglo-Dutch wars of the 17th century. In fact, the Germans never seem to have felt compelled to challenge the British blockade. After World War I, a senior German officer attacked the sterile German strategy. The Japanese could be expected to fight because the fleet at Singapore controlled the vital choke point (for Japanese trade) of the Malacca Straits. They threatened trade and Japanese possessions further north. However, it does not seem that the British ever thought through the possibility that the Japanese would fight closer to home, placing the British fleet in a difficult position. British assumptions as to where the decisive battle would be fought (within 24 hours of Singapore) are evident in the detailed calculations they made for required steaming endurance of new battleships and carriers.

[11] The culmination of the year-long Royal Naval College was a week-long game simulating the British Far Eastern war plan, the results of which were fed back to British war planners. No accounts of these games seem to have survived. Several versions of the war plan itself survive in British archives as versions of the War Memorandum (Eastern); the 1933 version is ADM 116/3475 and the 1938 version is ADM 116/3673. The closest equivalent to Miller's *War Plan Orange* is Andrew Field, *Royal Navy Strategy in the Far East 1919–1939: Planning for War against Japan* (London: Frank Cass, 2004). The 1933 plan envisions the dispatch of the Mediterranean and Home Fleets to Singapore. The combined fleet would arrive 28 days after dispatch was ordered. If the Mediterranean Fleet did not have to combine with the Home Fleet somewhere in the Indian Ocean, it could arrive in 22 days. This timing was considerably stretched about 1937, the reinforcement time then being

70 days. Meanwhile the defenses of Singapore were being built up; by 1936, the fortress was expected to be able to survive any practical scale of Japanese attack. The 1933 memorandum pointed further to the need for the British fleet to be able to operate from Hong Kong to deny Japan access to resources in China. An interesting part of the memorandum is analysis of the likely actions of various neutrals, including the note that both U.S. and Japanese officers expected ultimately to have to fight each other—but that the United States could not be relied on as an ally in the event of war with Japan. Part of the rationale for expecting Singapore to hold out was a judgment that the torpedo bombers based there would prevent any Japanese carriers from getting close enough to make sustained attacks. The memorandum does not include any discussion of a battle against the Japanese fleet, only the comment that the combined British fleet would clearly be dominant. The situation began to change in 1934. Once it was accepted that the British might also face a European threat, the requirement was not merely to exhaust Japan by blockade, but also to force a decisive battle to free the battle fleet to return to European waters. The British signed a naval agreement with the Germans limiting the latter to 35 percent of British strength, specifically to leave enough British naval power to face Japan. In 1935, however, the British almost found themselves fighting the stronger Italians in the Mediterranean, and they feared that in the event of war Japan might attack in the Far East. British intelligence claimed that the Japanese were maintaining an amphibious attack division that could seize Singapore. The fear of simultaneous crises east and west is evident in Admiralty papers discussing overall British strategy, in particular the nightmare of a two-ocean or two-hemisphere war. In that case, the question was whether the Japanese might choose to force a battle closer to their own home waters, hence within range of land-based aircraft. The British do not appear to have taken this possibility into account. It should have been evident in discussions of requirements levied on their own naval aircraft. The 1933 version of the War Memorandum does list possible Japanese courses of action. The strategic situation changed radically from 1937 on, as the Japanese seized control of the Chinese coast, dramatically reducing the wartime value to the British of a blocking force based at Hong Kong. The ongoing war in China also increased Japanese oil consumption and thus made that country more vulnerable to blockade. The 1937 British war-planning memorandum included several possible cases, one of which was that the fleet would be tied down by a European crisis or war. At this point, the logic was that the Japanese fleet would be drawn south by British attacks on Japanese trade and possessions such as Taiwan and the Pescadores. Near Singapore the Japanese fleet would be subject to attack not only by the British fleet, but also by RAF aircraft based at Singapore. If it could be attrited sufficiently—an echo of what the Japanese hoped to do to the U.S. fleet—the British fleet could press north. In his account of British strategy, Field points out that much of this was wishful thinking. It included the hope that no further fleet engagement would be required, and that British cruisers would be able to strangle Japanese trade as far afield as the U.S. coast. The situation changed further in 1940 when the French ceded control of air bases in their Indochinese colony (modern Vietnam) to the Japanese and thus brought large Japanese air forces within range of Singapore. By that time, too, the notion of moving the fleet to the Far East was no longer viable. Existing accounts of the "Singapore Strategy" do not indicate any examination of Japanese fleet employment alternatives other than seeking or shrinking from a fleet battle somewhere in the South China Sea. The expectation of such a battle features in pre-war British discussions of how many battleships would be needed in the Far East. The British were well aware that the Japanese might try to seize Singapore before the fleet could arrive. Detailed planning for the defense of the base prior to the arrival of the fleet seems to have been based on plans developed locally. There is some indication that the British thought that their large submarine force based at Singapore might hold the Japanese back (the U.S. Navy based numerous submarines in the Philippines for much the same reason). A local war game played to test that idea is described in an appendix to Alastair Mars, *British Submarines at War 1939–1945* (Annapolis, MD: U.S. Naval Institute Press, 1971). The submarine force had to be brought back into European waters once war broke out in Europe but not in the Far East in September 1939.

[12] The British were also concerned with the potential threat posed by long-range submarines as their fleet steamed from the Mediterranean to Singapore. For much the same reason, they fitted destroyers with high-speed mine-sweeping gear, the theory being that the same submarines might lay mines in the fleet's path. U.S. sonar installations were more primitive than those of the Royal Navy, and the U.S. Navy certainly gained considerably from interaction with the British, but the basic sets were of U.S. origin. Naval War College files include accounts of pre-war U.S. development. The British taught the U.S. Navy to install streamlined housings that made it possible to use sonar at higher speeds, and they also introduced the chemical recorder, a means of estimating the proper time to drop depth charges. Other wartime developments in sonar packaging and displays were of U.S. origin, produced by the wartime coalition of civilian academic researchers. On the

other hand, the British introduced the forward-throwing Hedgehog mortar, a vital wartime anti-submarine weapon. The navies that did not spend much on sonars were the French, Italian, and Japanese. All had had access to nascent sonar technology on an inter-allied basis in 1918. In 1939, the French were impressed by British Asdic, the implication being that they had no equivalent technology of their own. The first Japanese active sonar was approved for service in 1933, but it appears that even after that, reliance was placed mainly on passive devices. The post-war report of the U.S. Naval Technical Mission to Japan mentions considerable reliance on German assistance provided in 1942, suggesting that little had been done before that.

[13] NWC RG 4, Box 27, Folder 25. The game is labelled "Battle of Emerald Bank."

[14] The Japanese demonstrated their prowess in night combat during the Guadalcanal campaign, from Savo Island (August 1942) onward. The great demonstration of British prowess was Matapan in 1941. The pre-war U.S. Navy seems not to have appreciated a vital technical point about pre-radar night combat. Battle planning was based on what was observed just before dark, the assumption being that the enemy would not change course afterward. Thus, a night attack force had to get into position, get off its attack, and then leave before the situation changed so much that pre-battle planning was no longer relevant. U.S. observers sometimes remarked that Japanese commanders seemed to lose their nerve if their plans failed—they were inflexible. The reality was that their further information was too scanty to be a basis for continued action; better to retire and fight another day. This perception is evident in Admiral Morrison's account of the Guadalcanal battles. U.S. experience, for example in the November 1942 battles, was that the advent of radar did not solve the problem; it was necessary to do a lot more to maintain awareness of what was happening at night. U.S. lack of perception suggests that the pre-war Navy did not conduct full-scale experiments in night combat comparable to what the Japanese and the British had done.

[15] Admiral Spruance's choice to emphasize the invasion over the possibility of destroying the Japanese fleet was an excellent example of a major War College theme: formulation of the mission to be carried out. A directive generally covered several different possible priorities. The first step in formulating a tactical plan—in the "Estimate of the Situation"—was to decide what the mission really was. Surviving gaming papers emphasize the need for such formulation, and the problems many students encountered in doing so.

[16] Spruance's replacement did not reflect dissatisfaction. By this time, the Navy was using its two senior fleet commanders in tandem, one planning an operation while the other executed the previous one. The fleet and its fast carrier task force were renumbered depending on which admiral was in charge, so that it was Third or Fifth Fleet and Task Force 38 or 58 with, respectively, Halsey or Spruance commanding.

[17] It can be argued that the sheer complexity of some Japanese tactical plans was the feature of the Pacific War that could most clearly be associated with a Japanese mode of thinking. War gaming stressed the value of relatively simple plans.

[18] The Naval War College technique of estimating enemy courses of action would therefore presumably have suggested something like the kamikaze offensive had the very limited number of trained Japanese pilots been known. As it happened, by January 1944 at least, some intelligence officers were aware that Japanese pilot training had lagged badly. That, in turn, should have affected evaluation of the "Turkey Shoot" in June 1944 (the point of the January evaluation was that Japan could not afford major pilot losses). Whether anyone at the college or planning level would have taken the next step to assuming that the Japanese would adopt systematic rather than occasional suicide tactics is another question, but in the spring of 1945, a U.S. account of current kamikaze tactics made the point that they were actually more economical in terms of Japanese lives (for a given level of damage) than conventional tactics. The Pacific amphibious force's analysis of kamikaze attacks laid out the advantages they offered, particularly greater lethality per attacker, and other contemporary U.S. documents made roughly the same point. To some extent, the invasion of Saipan had demonstrated that Japanese troops and civilians preferred to die rather than surrender; at Saipan, the Japanese military forced civilians to commit suicide en masse. This too, should have been an indicator.

[19] A surviving example is "Strategical and Tactical Situation Developed in Recent Problems at the Naval War College," produced by the Department of Intelligence in November 1935, in NWC RG 4, Box 80, folder 29. It was almost certainly written by Captain van Auken.

Appendixes

[1] This discussion is based on Part B of the June 1929 Maneuver Rules.

[2] I have been unable to find sets of 1930 or 1932 rules. The 1933 edition was rewritten to group the rules more logically, so it is difficult to say whether some of the changes were introduced in 1932. Dates of pages in the 1940 rules indicate that there was no significant revision of the aircraft section. Through 1929, the rules included aircraft characteristics, but, after that, they were provided in a separate pamphlet, just as ship characteristics were separated from the game rules. Separation made it possible to limit changes in the rules without neglecting changes in aircraft.

[3] The major initial air lesson of the Sino-Japanese War that broke out in 1937 seems to have been that unescorted high-performance bombers might suffer heavy casualties at the hands of defending fighters. We now know that this experience led the Japanese to employ naval fighters as escorts for their long-range naval bombers, and ultimately to develop a very long-range high-performance fighter, the famous "Zero." It is not clear whether anyone outside Japan realized how effective Chinese fighter defense was, at least at the outset. It appears that in Spain the Germans also found that they had to escort their bombers. Again, it is not clear how obvious this was to outsiders. The Germans in particular also found themselves adjusting their fighter tactics. Such issues were not taken into account in framing rules for air-to-air combat.

[4] This page of the 1931 rules is dated May 1930, hence comes from that year's rules. A single airplane could be located approximately at 20,000 yards and accurately at 11,000; a flight of 18 at 30,000 and 15,000 yards. Ground winds would decrease ranges by 30 percent (force 4 wind).

[5] NWC RG 8, file XMAR (1935-12) dated 26 February 1935, "Manuever Rules Pertaining to Aircraft and Airplane Characteristics—Revision of." King offered revisions, which the president of the War College formally requested on 2 March. King provided a list of suggestions in a 17 May letter. He noted that the files of the director of Fleet Training showed very little definite information on which to base a rule for air-to-air combat. Reports of inter-squadron camera-gun practices gave relative numbers of hits, but not the number of hits on each attack. The average ratio corresponded closely to the current War College rule, but there was no basis for estimating the percentage of casualties on each attack. At this time, the Navy estimated the lethality of gunfire based on experiments with aircraft on the ground. A few years later, the fleet discovered to its horror that shell bursts previously ruled as lethal did not down drones. The text of the rules seemed to discourage air operations because of pilot fatigue; King wanted that softened, as otherwise it "might give the wrong impression to one unfamiliar with aircraft." He suggested reducing the effectiveness of anti-aircraft fire when the main battery was firing. The rules already assumed that secondary batteries were adversely affected by main-battery fire. Anti-aircraft batteries were generally more exposed to the blast of heavy guns. King rewrote a table showing the effect of fatigue on pilots, noting that the war allowance was 50 percent extra pilots aboard a carrier, to allow complete rest one day out of three. He added a detailed table showing time between attacks during air-to-air combat, based on speed advantage (if any) and relative altitude (service ceiling of attacker or attacked). For example, with zero speed advantage an airplane could make only a single attack. An appended note pointed out that if the attacker was no faster than the defender it could make only a single attack if the defender did not want to fight. Even that could be carried out only if the attacker first sighted the defender from a greater altitude, at least 3,000 feet greater. How many passes the attacker could make depended on the speed difference. An attacker could make a pass every three minutes if it was 50 miles per hour faster at its service ceiling. If the speed difference was 100 miles per hour, the attacker could make a pass each minute. A plane that had to remain on station, such as a spotter, could not avoid repeat attacks; King allowed ten minutes between them. King added that no such table could be absolutely accurate, since time between attacks depended on too many factors to be tabulated. However, the current rules were worse. They allowed pilots to destroy their targets instantaneously, whereas the new rules allowed for the sort of cumulative damage (in this case, a cumulative probability of fatal damage) typical of other war game rules.

[6] The most striking case of special characteristics for a foreign airplane was the British torpedo bomber–spotter–reconnaissance airplane (TSR), the Swordfish. The college rules accorded it rather better performance than it actually had.

[7] The Norden precision bombsight, developed by the U.S. Navy, used gyros to measure apparent target motion. (However, such bombsights suffered greater inaccuracy because they offered less accurate initial estimates.) With the advent of this bombsight, the U.S. Navy assumed that formations of bombers flying above the effective limit of anti-aircraft fire could hit moving warships, and such assumptions can be seen both in the rules of the late 1930s and in internal U.S. discussions of the air threat. The assumption that level bombing was newly potent also motivated interest in long-range heavy bombers, the first being the PBY Catalina. War experience showed that this idea was totally unrealistic. It is not clear to what extent pattern bombing using the Norden sight was ever tried against ships at sea. The rules did not distinguish between a U.S. Navy equipped with Norden sights and other navies or air forces without them. In 1939, both the United States and Germany had gyro-based bombsights.

[8] There were also rules concerning the use of air-launched torpedoes. A 30-knot ship could evade all torpedoes dropped more than 3,000 yards away. The assumption, which was certainly correct at the time, that torpedo bombers had to come so close to their targets explains why they made so few hits in games. During World War II, several navies, including the U.S. Navy, developed means of dropping torpedoes at greater altitude and speed, which meant greater stand-off. That greatly increased their chance of hitting, the torpedo moving at high speed until it entered the water. Pre-war rules did not reflect this possibility. In the U.S. case, the new kind of torpedo did not enter service until 1944.

[9] The table of aircraft characteristics in the 1929 rules included two-seat fighters (F8C-1s), which could dive-bomb with 500-pound bombs, meaning that they could pull out after diving with them. In the 1930s, the U.S. Navy introduced dive bombers that could deliver 1,000-pound bombs.

[10] The page giving these figures is dated December 1930. It must have been new, since the text refers back to the qualitative comments of past years. Dive bombers were not expected to lose effectiveness due to a target's maneuvers since they dropped their bombs at such low altitudes (this reasoning was not, however, explicit). During World War II, the Japanese in particular used radical maneuvers (circling at high speed) against dive bombers on the theory that the bomber actually committed to a particular path at considerably higher altitude. Horizontal bombers certainly would be affected by target maneuvers. According to the table, by maneuvering radically at high speed a large target would reduce bombing effectiveness (from 8,000 to 10,000 feet) by 20 percent. In addition, detailed tables were provided giving percentages of near-misses, which were expected to cause underwater damage.

[11] NWC RG 8, file XMAR (1936-48) dated 19 February 1936, "Aircraft Bombing—Notes for Revision of Rules on." The paper was written by Head of Operations Department Captain R. A. Dawes, and it was approved by Admiral Kalbfus, the War College president.

[12] This was a controversial change. Commanders of dive-bombing squadrons reported that there was an optimum attack altitude at which to level off after dropping the bomb, which was not the lowest safe altitude from an airplane's structural point of view, but depended instead on anti-aircraft fire. That suggested that the percentage of hits would not vary widely according to the altitude at which bombers pulled out, because the greater accuracy due to a lower pull-out would be balanced by greater inaccuracy due to anti-aircraft fire. At this time, dive bombers were subject to safety precautions: They were not to release bombs below 1,000 feet, they were not to release bombs at more than a 70 degree dive angle, and they were only to do so when the airplane was in a steady dive without slip or skid. It was argued that in wartime pilots could gain sufficient accuracy by violating the safety instructions to overcome the effect of surface fire. The decision *not* to consider wartime accuracy equal to peacetime accuracy was based on remarks by aviation officers at the War College, three of whom had recently practiced dive bombing. They found 70 degrees a practical limit because it was so difficult to maneuver at a steeper angle. They also rejected lower-level release on the ground that the concussion of the explosion would wreck the airplane before it could pull out. There was also the morale factor: It was unlikely that every pilot in an attack would take the steepest dive and release as low as possible, especially in repeated attacks. This paper also suggested a change in the anti-aircraft rules. Each gun or its equivalent would be given a percentage chance of hitting based on the average percentage of sleeves (towed targets) hit per gun in anti-aircraft practices over a period of three years, reduced like the bombing figures by 25 percent for the "fog of war." That had to be modified to allow for natural interference when multiple guns were firing at the same target (gunners might be unable to tell whether their weapons were the ones coming closest to

the target). The problem would worsen if two or more ships were firing at the same target. Arbitrary curves had to be constructed to allow for this factor. Basic performance in target practice showed a 3 percent hitting rate at 1,000 feet, falling to 2.2 percent at 9,000 feet. Only later did it turn out that these figures were grossly deceptive, partly because even hits that tore holes in a sleeve often did not stop a real airplane.

[13] It is possible that the sole wartime example of hits by horizontal bombers on fast-maneuvering ships were a few scored by formations of Japanese bombers on HMS *Repulse* in December 1941 (masthead-level bombing was a different matter). Generally, aircraft bombed from excessive altitude; in many cases, ships' commanders could maneuver away from bombs they saw falling. There also seem to have been few cases of bombers maintaining the necessary rigid formations or the necessary degree of coordination. By way of contrast, dive bombing required far less coordination, although an attack by multiple dive bombers approaching from different directions could achieve better results. Inter-war game rules do not seem to reflect these points.

[14] When, as chief of BuAer, Admiral King advised the college on revised rules for air attacks, he noted drily that students at the college could lay out air-deployed smoke screens far more precisely than any pilot could execute such plans. Smokers (aircraft laying such screens) were featured in pre–World War II fleet exercises, but they seem never to have been used during the war.

[15] That is, the bomber could set the torpedo to run at an angle rather than straight ahead to the target. This was a standard capability for surface-launched torpedoes, but it was not incorporated in air-launched torpedoes (if at all) until World War II.

[16] At this time, some discussions of the air-to-air rules included arguments that the flexible gun onboard a bomber or scout was actually superior to fixed forward-firing fighter guns, partly because the fighter pilot had to divide his attention between flying and firing his gun.

[17] This practice was abandoned by the U.S. Navy (but not by others such as the British) early in World War II, when Commander J. S. Thach proposed his "Thach weave" to counter the extremely maneuverable Japanese Zero fighter. His four-airplane tactics roughly paralleled those the Germans had developed after the Spanish Civil War. It turned out that although the second fighter could help protect the first from interference, the third concentrated far too much on keeping station, and typically it was shot down. Two pairs of fighters offered far better mutual reinforcement. The gap between exercises, in which the three-fighter formation seemed perfectly adequate, and reality, when it decidedly was not, give a sense of the limitations of the exercise-based game rules.

[18] It is difficult to square these rules with steps the U.S. Navy was actually taking. In buying new fighters, the Bureau of Aeronautics emphasized the need to escort strike aircraft to beat off enemy fighters. It would not have done so had it been nearly so optimistic about the ability of its bombers to defend themselves. On the other hand, carrier air wing composition emphasized attack over defense, either of the ship or of its strike aircraft. Typically, a carrier embarked four squadrons: one of fighters, one of dive bombers, one of scouts (dive bombers, but normally carrying 500- rather than 1,000-pound bombs), and one of torpedo bombers. When the new *Essex* class was designed in 1939–40, a major design feature was capacity for a second fighter squadron. Naval War College archives do not appear to include demands by the Bureau of Aeronautics or the fleet for a very different assessment of fighter capability.

[19] Heavy anti-aircraft guns fired explosive shells with time fuses. A fire-control system aimed the gun and automatically set the fuse in order to burst the shell near the target airplane. U.S. fire-control systems were designed to hit airplanes flying in a straight path at constant speed—horizontal bombers but definitely not dive bombers. Before the system could fire, it had to measure target course and speed by repeated observation. Again, that suited an attack on horizontal bombers, which had to line up on their targets as they set their sights. The dive bomber spent too little time diving for this process to be completed, and before it tipped over into a dive it was generally too far away to hit. Moreover, fire control required observation of shell bursts so that corrections could be applied to the firing solution that, in turn, was used to aim the guns. The dive bomber did not provide enough time for that. By way of contrast, a fast-firing automatic gun could adjust rapidly, its gunner typically watching tracers to correct aim.

Bibliography

Danlinger, Sutherland, and Charles B. Gray. *War in the Pacific: A Study of Navies, Peoples, and Battle Problems* (New York: R. M. McBride & Co, 1936).

Evans, David C., and Mark R. Peattie. *Kaigun: Strategy, Tactics, and Technology in the Imperial Japanese Navy 1887–1941* (Annapolis, MD: U.S. Naval Institute Press, 1997).

Felker, Craig E. *Testing American Sea Power: U.S. Navy Strategic Exercises 1923–1940* (College Station, TX: Texas A&M University Press, 2006).

Friedman, Norman. *U.S. Aircraft Carriers: An Illustrated Design History* (Annapolis, MD: U.S. Naval Institute Press, 1983).

———. *U.S. Cruisers: An Illustrated Design History* (Annapolis, MD: U.S. Naval Institute Press, 1984).

———. *Network-Centric Warfare: How Navies Learned to Fight Smarter Through Three World Wars* (Annapolis, MD: U.S. Naval Institute Press, 2009).

———. *British Cruisers: Two World Wars and After* (Annapolis, MD: U.S. Naval Institute Press, 2011).

———. *Fighting the Great War at Sea: Strategy, Technology, and Tactics* (Annapolis, MD: U.S. Naval Institute Press, 2014).

———. *British Battleships 1906–1946* (Annapolis, MD: U.S. Naval Institute Press, 2015).

——— and Thomas Hone. *American and British Aircraft Carrier Development* (Annapolis, MD: U.S. Naval Institute Press, 1999).

Gordon, Andrew. *Rules of the Game: Jutland and British Naval Command* (London: John Murray, 1996).

Hattendorf, John B., B. Mitchell Simpson III, and John R. Wadleigh. *Sailors and Scholars: The Centennial History of the U.S. Naval War College* (Newport, RI: Naval War College Press, 1984).

Hone, Trent, and Tom Hone. *Battle Line: The United States Navy 1919–1939* (Annapolis, MD: U.S. Naval Institute Press, 2006).

Kuehn, John T.. *Agents of Innovation: The General Board and the Design of the Fleet That Defeated the Japanese Navy* (Annapolis, MD: U.S. Naval Institute Press, 2008).

Laning, Harris. *An Admiral's Yarn*, introduction by Mark R Shulman (Newport, RI: Naval War College Press, 1999—Naval War College Historical Monograph Series, no. 14).

Lillard, John M. *Playing at War: Wargaming and U.S. Navy Preparations for World War II* (Dullas, VA: Potomac Books, 2016).

Maurer, John, and Christopher M. Bell. *At the Crossroads Between War and Peace: The London Naval Conference of 1930* (Annapolis, MD: U.S. Naval Institute Press, 2014).

Melhorn, Charles M. *Two-Block Fox: The Rise of the Aircraft Carrier 1911–1929* (Annapolis, MD: U.S. Naval Institute Press, 1974).

Miller, Edward S. *War Plan Orange: The U.S. Strategy to Defeat Japan 1897–1945* (Annapolis, MD: U.S. Naval Institute Press, 1991).

Moretz, Joseph. *Thinking Wisely, Planning Boldly: The Higher Education and Training of Royal Navy Officers 1919–1939* (Solihull, UK: Helion, 2014).

Morrison, Elting E. *Admiral Sims and the Modern American Navy* (Boston: Houghton Mifflin, 1942).

Morton, Louis. *Strategy and Command: The First Two Years* (Washington, DC: Government Printing Office, 1961—U.S. Army in World War II: The War in the Pacific series).

Nofi, Albert A. *To Train the Fleet for War: The U.S. Navy Fleet Problems, 1923–1940* (Newport, RI: Naval War College Press, 2010—Naval War College Historical Monograph Series, no. 18)

Peattie, Mark R. *Sunburst: The Rise of Japanese Naval Air Power 1909–1941* (Annapolis, MD: U.S. Naval Institute Press, 2001).

Roskill, Stephen. *Naval Policy Between the Wars, Vol. I: The Period of Anglo-American Antagonism, 1919–1929* (London: Collins, 1968).

———. *Naval Policy Between the Wars, Vol. II: The Period of Reluctant Rearmament* (London: Collins, 1976).

U.S. Navy Office of Naval Intelligence Historical Section. *American Naval Planning Section, London* (Washington, DC: Government Printing Office, 1923).

U.S. Naval War College. *Sound Military Decision* (Annapolis, MD: U.S. Naval Institute Press, 1994— republication of 1942 edition).

Vlahos, Michael. *The Blue Sword: The Naval War College and the American Mission 1919–1941* (Newport, RI: U.S. Naval War College Press, 1980—Naval War College Historical Monograph Series, no. 4)

Wagner, Ray. *American Combat Planes of the 20th Century* (Reno, NV: Jack Bacon, 2004).

Wheeler, Gerald E., *Admiral William Veazie Pratt, U.S. Navy: A Sailor's Life* (Washington, DC: Naval History Division, 1974).

Zimm, Alan D. *Attack on Pearl Harbor: Strategy, Combat, Myths, Deceptions* (Havertown, PA: Casemate, 2011).

Index

Aircraft carriers, 4, 6–7, 13–16, 18, 20–21, 23–28, 62, 66, 68, 73–103, 106, 110, 113–16, 126, 128, 132–35, 137–41, 146ff, 149ff, 150–53, 155, 161–63
    Origins, 15, 73, 170
    British, 1, 15, 22ff, 68, 74, 89–90, 113, 147, 170–71, 174ff
    Japanese, 1–3, 13–14, 47ff, 58, 73, 113, 147, 167
    "Battle-line carriers," 98–101
    Size of air wing, 2, 78, 91, 99
    Armored versus unarmored flight decks, 2–3, 79ff, 101–103
    Ideal size, 12, 24, 73, 83–86, 89–91, 97–98, 147
    In World War II, 1–2, 9ff, 10ff, 15, 26–27, 40ff, 49ff, 50ff, 168, 170–71, 175–77
Airships, 19, 24ff, 133, 140
Asiatic Fleet (U.S.), 18, 125
Atlantic Fleet (U.S.), 18, 23

Battlecruisers, 36, 59, 61, 137, 171, 197
    Cancellation and conversion of *Lexington* and *Saratoga*, 12, 23, 73, 91, 110
    U.S. consideration of in the 1930s, 148–51
    In the Japanese navy, 44ff, 122, 159
Battleships, 14, 17–18, 22ff, 25–28, 31, 36, 41, 50ff, 52, 57–61, 64, 66–69, 76ff, 79–81, 84, 87ff, 88, 96, 98, 104, 107, 110–11, 121ff, 124–27, 131–37, 145–52, 156–59, 161–67, 170–73, 178
    U.S. Navy orientation toward, 1, 3–4, 9ff, 11, 15, 21ff, 27, 101, 104–105
    Utility vis-à-vis aircraft carriers, 14–15, 98, 100–101, 105–106, 116
Battles
    Cape Matapan (1941), 62
    Coral Sea (1942), 3, 134
    D-Day/Normandy (1944), 1, 17

Guadalcanal campaign and associated battles (1942–43), 57, 62, 163, 176, 179, 199
Jutland (1916), 45, 47–49, 134
Leyte Gulf (1944), 92ff, 129, 163, 176–77
Midway (1942), 2–3, 11, 15, 16ff, 27, 49ff, 57, 77, 84, 89ff, 96, 134, 162ff, 176, 193, 199
Pearl Harbor (1941), 15, 26–27, 47ff, 58, 164ff, 166–67, 170–71, 173, 176
Philippine Sea (1944), 11, 15, 168, 176
Tsushima (1905), 163–64

Bureau of Aeronautics (BuAer), 14, 19–20, 75, 78, 85ff, 100–101, 105–106, 129, 141, 151–54
Bureau of Construction and Repair (BuC&R), 19, 83, 85, 90, 93, 102, 115, 119, 140, 155–57
Bureau of Navigation (BuNav), 33, 142
Bureau of Ordnance (BuOrd), 19, 57, 59, 156

Cruisers, 8, 18, 25–26, 29, 45–46, 49–50, 60–62, 64, 66, 79, 101, 107, 109–39, 145–47, 155, 158–59, 168, 176
 Cruisers, British, 45–46, 66, 73, 78, 89, 110, 112, 120, 125, 147
 Cruisers, "flight-deck," 15–16, 62, 92, 97, 99–101, 107, 113–18, 123–26, 129–33, 135, 139–41, 145
 Cruisers, heavy ("Treaty cruisers"), 57, 84, 107, 110, 112–13, 122–27, 146, 159
 Cruisers, Japanese, 122–23, 125, 158
 Cruisers, light/scout, 23, 28, 80, 107, 110, 116–17

Destroyers, 10, 25–29, 35, 46, 48–50, 56, 62, 66, 78–79, 87–88, 107, 109–13, 115, 117–18, 120, 122–28, 131–32, 137-138, 145, 150, 155–56, 161, 166ff, 173–74, 177
DumanquilasBay, Philippines, 68, 143

Fleet Problems, 8–10, 13ff, 28, 38ff, 51, 53, 55, 62, 71, 97, 100, 116, 183
Fletcher, Frank Jack, ADM, USN, 15, 73
Flying boats, 70, 90ff, 104, 154
 German Dornier Do-X, 24, 70, 152

General Board of the U.S. Navy, 20, 25, 28, 36–37, 41, 58, 65, 97, 100–101, 103, 105–106, 110, 112, 115–21, 124, 127–28, 140–41, 143, 147–48, 150–51, 153
 Relationship to the Naval War College, 31–35, 60, 65
 Responsibility for warship characteristics, 9, 15, 20, 29, 84–85, 91, 98, 105, 140, 154–55, 159, 168
Gibraltar, 66, 132, 135–36, 165
Guam, 24, 67, 79

Halsey, William F., FADM, USN, 171, 176–77

Imperial Japanese Navy, 1–5, 8–9, 20–23, 44ff, 46, 47ff, 49ff, 61–63, 84, 89ff, 96, 102, 113, 121ff, 126, 141, 151–54, 158
 Exercises, 58, 62, 165
 Interwar development, 12–17, 23–26, 73–74, 91, 93, 118, 147–48
 Planning for World War II, 165–67
 In World War II, 84, 160–80

Jellicoe, John, Admiral of the Fleet, RN, 45, 47–48, 52

Kalbfus, Edward C. ADM, USN, 70, 143, 144ff, 148, 150
Kamikaze attacks, 2, 79ff, 161, 177–78.
King, Ernest J., FADM, USN, 75, 100, 141, 185

Laning, Harris, ADM, USN, 38ff, 40ff, 42, 55, 58, 63, 69, 83, 97–99, 115–18, 125, 140, 142, 146ff, 173–75
London Conference (1929–30), 61, 64, 93, 111, 112, 113, 117ff, 147
London Naval Treaty (1930), 25, 26ff, 45, 60, 64, 97, 98, 103, 115, 116, 125, 140, 145, 155, 173
London Naval Treaty (1936), 25, 104, 121ff, 143, 146ff, 147, 151, 156, 157, 159

MacFall, Roscoe, CPT, USN, 2, 40ff, 174, 175
Mahan, Alfred Thayer, RADM, USN, 17
Malampaya Sound, Philippines, 68,
Manila, Philippines, 21–22, 24, 66–70, 104, 123, 154
    and "through ticket" strategy, 7, 22, 67–72, 104, 123, 154
McCandless, Bruce, RADM, USN, 57
Moffett, William A., RADM, USN, 78, 101, 113–14, 129, 140–41, 155, 168

Nimitz, Chester W., FADM, USN, 2, 6, 160–61, 165, 172, 175–77

Office of the Chief of Naval Operations (OPNAV), 2, 4, 19, 25, 28, 33, 36–37, 41, 58, 61, 63, 65, 107, 112, 142, 155
OPNAV War Plans Division, 4, 6, 21, 41, 63–65, 67, 69–70, 91, 101, 104, 106, 131, 141, 161, 163, 187

Pacific Fleet (U.S.), 18, 135
Pacific War, see *World War II*
Pearl Harbor, HI (also see Battle of Pearl Harbor), 18, 38ff, 69–70, 104, 135, 154
Philippines, 18, 24, 99, 122, 143, 146, 172
    Vulnerability to Japanese attack, 20–22, 25, 53, 67, 71, 167, 173
    Recapture of in USN war plans, 21, 66–67, 70, 72, 79, 123, 151, 154, 164ff
Pratt, William V., ADM, USN, 16ff, 41–42, 111–12, 118–19, 152, 168
    CINCUS, 111
    Chief of Naval Operations, 4, 41, 98, 100, 104,112, 114–16, 118, 127, 140, 142
    War College President, 4, 41, 42, 85–93, 97, 100, 111, 141
    Arms-control negotiator, 61, 93, 97, 111, 115, 117ff

Reeves, Joseph M., ADM, USN, 2, 14ff, 16ff, 38ff, 60, 83, 102, 141, 187
    Innovations while captain of *Langley*, 93–97
Robison, Samuel S, ADM, USN, 175
Royal Navy, British, 7, 15–16, 29, 35–37, 39, 45–49, 51, 57–58, 63, 66, 68, 74, 79ff, 95–96, 109–10, 147, 149ff, 158, 161–64, 170, 175
    Exercises, 35, 48,
    Interwar plans, 171–73

    In World War I, 47, 52, 78, 160

San Diego, CA, 12ff, 18, 28ff, 95, 135, 167
San Pedro (Los Angeles), CA, 18, 111ff, 173
Sims, William S., ADM, USN, 4, 9ff, 22ff, 28–29, 33, 35–39, 41, 44ff, 46–49, 58–61
    Commander, U.S. Naval Forces Operating in Europe, 29, 35–37, 111
    War College president, 32ff, 35, 37–39, 54–55, 63, 75
Spruance, Raymond F., ADM, USN, 2, 7, 10, 15, 73, 162ff, 176–77
Standley, William H., ADM, USN, 142, 147, 151
Submarines, 5, 7, 10, 25, 49, 62, 75, 80, 88, 91, 96, 100–101, 106, 107, 110, 115, 126, 131, 133, 145, 150, 161, 164, 167
    As scouts, 19, 24ff, 28ff, 29
    British, 29, 134, 137–38
    Japanese, 3, 29, 69–70, 81, 89ff, 99, 123, 169–71, 173–75
    Unrestricted submarine warfare, 23, 35, 168–69

"Treaty system," interwar, 23–27, 68, 70, 101, 105, 116, 121ff, 141, 143, 145–48, 150, 152, 158, 165, 167
Tawi Tawi, Philippines, 68

United States Fleet (interwar U.S. Navy formation), 8, 18, 111, 125, 148, 175

van Auken, Wilbur, CPT, USN, 63–64, 66, 69, 130–31, 143, 163, 178

Vessels and classes
    HMS *Argus*, 22ff, 93
    HMS *Courageous*, 74, 89
    HMS *Dreadnought*, 31–33
    HMS *Glorious*, 74, 89, 171
    HMS *Hood*, 46ff, 57
    HMS *Malaya*, 22ff
    HMS *Prince of Wales*, 27
    HMS *Queen Elizabeth*, 61
    HMS *Sussex*, 120
    HMS *Vindictive*, 89.
    IJN *Akagi*, 73
    IJN *Hyuga*, 44ff
    IJN *Ise*, 44ff
    IJN *Kaga*, 73
    IJN *Kirishima*, 44ff
    IJN *Kongo*, 44ff, 148, 159
    IJN *Mogami*, 123,
    IJN *Musashi*, 14–15, 27, 163

IJN *Nagato*, 44ff
IJN *Ryujo*, 113.
IJN *Yamato*, 14–15, 157
KMS *Bismarck*, 27, 46ff
KMS *Deutschland/* "pocket battleships," 158
USS *Arizona* (BB-39), 57
USS *Barracuda* (SS-163), 28ff
USS *Bass* (SF-5), 28
USS *Brooklyn* (CL-40), 117ff, 118, 127, 150, 155–56, 159
USS *Canberra* (CA-70), 166ff
USS *Claxton* (DD-571), 166ff
USS *Cleveland* (CL-55), 156
USS *Dixie* (AD-14), 174ff
USS *Enterprise* (CV-6), 76ff, 87ff, 89ff, 98, 103, 140, 150
USS *Essex* (CV-9), 50, 77ff, 82ff, 98, 102–104
USS *Harmon* (DE-678), 174ff
USS *Holland* (AS-3), 28ff
USS *Hornet* (CV-8), 25, 49ff, 104–105, 107, 146ff
USS *Iowa* (BB-61), 105, 151, 159
USS *Killen* (DD-593), 166ff
USS *Langley* (CV-1), 14ff, 73, 75, 78, 83, 93–96, 137
USS *Lexington* (CV-2), 12–13, 21ff, 24, 73–75, 83–84, 86, 92, 97, 100, 102–103, 134, 188
USS *Narwhal* (SC-1), 28ff
USS *Nautilus* (SS-168), 28ff
USS *Northampton* (CA-26), 119
USS *North Carolina* (BB-55), 166
USS *North Dakota* (BB-29), 33
USS *Omaha* (CL-4), 116, 120, 122, 141, 155, 156
USS *Pensacola* (CA-24), 120
USS *Randolph* (CV-15), 177
USS *Ranger* (CV-4), 74ff, 76ff, 91–92, 97–98, 101–102, 113, 115
USS *San Francisco* (CA-38), 57, 120
USS *Saratoga* (CV-3), 12–13, 24, 38ff, 73–75, 83–84, 92, 97, 116, 134, 188
USS *Washington* (BB-56), 166
USS *Wasp* (CV-7), 101–102, 150
USS *Yorktown* (CV-5), 76ff, 77ff, 97–107, 140, 146ff, 150, 192

War Games
    "Battle of "Siargao" (Class of 1923 Tactical III), 79–84.
    Class of 1932–34 cruiser games, 129–39

    Class of 1932, Operations II, 129–31
    Class of 1933, Operations II, 64,
    Class of 1933, Operations IV, 3–4, 7, 22, 29, 68–72, 142–43
    Class of 1934, Operations/Tactics IV, 135–39
    Evidence for high aircraft/pilot attrition, 2, 52ff, 81, 92–93, 105, 152, 176–77
    Influence on USN war plans, 2, 4, 6–7, 9, 21–22, 33–35, 41–42, 55, 61, 63–72, 129–39, 142–43, 148, 161, 172–73, 178, 180
    Relationship to fleet exercises, 2, 4, 6, 8–10, 17, 20, 27–28, 30, 55, 70, 160, 178

Washington Conference (1921–22), 13ff, 111, 113
Washington Treaty (1922), 25, 41, 60, 66, 74ff, 78, 84, 90, 109–111, 114–15
World War I, 1, 7, 11–12, 15–16, 17, 29, 34, 49, 52, 111
World War II, 1–7, 11–12, 23, 26, 73, 84, 160–80

Yamamoto Isoroku, Admiral, IJN, 166–67

About the Author

NORMAN FRIEDMAN has published more than 40 books, most of them concerned with the intersection of defense policy, strategy and tactics, and technology. They include design histories of most U.S. warships and many of their Royal Navy equivalents, and *Network-Centric Warfare*, a history of naval command-and-control and information warfare. Dr. Friedman's Cold War history, *The Fifty-Year War*, won the Westminster Prize awarded by the Royal United Service Institute as the best military history book of its year; his *Fighting the Great War at Sea* won the Lyman Award from the North American Society of Oceanic Historians; and his *Seapower as Strategy* won the Samuel Eliot Morrison award given by the Naval Order of the United States. Friedman's most recent book is *Fighters over the Fleet*, a history of carrier-based fighters and their role in fleet air defense. Works written under government contract include a history of U.S. Navy air defense missiles and a history of the joint program to develop the MRAP (mine-resistant, ambush-protected) vehicle. Dr. Friedman spent over a decade at a prominent defense-oriented think tank, another decade in the office of the Secretary of the Navy, and two years in Marine Corps Headquarters. He writes a monthly column on "World Naval Developments" for the U.S. Naval Institute *Proceedings*, and has published numerous articles in other defense journals. Dr. Friedman received his Ph.D. in theoretical physics from Columbia University.

CPSIA information can be obtained
at www.ICGtesting.com
Printed in the USA
LVHW100809191119
637822LV00016B/808/P